The Plasma State

The Plasma State

Juda Leon Shohet
University of Wisconsin

ACADEMIC PRESS New York and London

ACADEMIC PRESS, INC.
111 Fifth Avenue, New York, New York 10003

United Kingdom Edition published by
ACADEMIC PRESS, INC. (LONDON) LTD.
Berkeley Square House, London W1X 6BA

LIBRARY OF CONGRESS CATALOG CARD NUMBER: 70-162938

PRINTED IN THE UNITED STATES OF AMERICA

Contents

Preface ix

1 ***The Plasma State*** 1

Suggested Reading 12

2 ***Collisions and Collisional Processes***

A. *Elementary Concepts* 14
B. *Conservation of Energy, Momentum, and Angular Momentum* 17
C. *The Collision Cross Section* 21
D. *Rutherford Scattering* 23
E. *Diffusion and Mobility* 27
Suggested Reading 36
Problems 36

3 ***The Motion of Isolated Charged Particles***

A. *The Motion of a Charged Particle in a Magnetic Field* 39
B. *Crossed Electric and Magnetic Fields* 41
C. *Gravitational Fields* 43
D. *Magnetic Field Gradients—Magnetic Mirrors and Cusps* 44
E. *The Effects of AC Electric Fields* 52
Suggested Reading 65
Problems 66

4 *The Beginnings of Collective Phenomena—Plasma Statistical Mechanics*

A. *Fundamental Definitions* 69
B. *Conservation Laws for Systems* 75
C. *Density and Distribution Functions and Averages* 77
D. *Liouville's Theorem* 80
E. *The Microcanonical Ensemble* 83
F. *Development of the Distribution Laws* 84
G. *Velocity, Speed, and Energy Distribution Functions* 93
H. *Application of Ensemble Theory to a Plasma* 96
Suggested Reading 98
Problems 99

5 *Further Aspects of Collective Phenomena: Statistics of Collisions and Fluid Behavior*

A. *Adiabatic Invariants and Constants of the Motion* 101
B. *Hamilton–Jacobi Theory* 111
C. *Collisions in the Boltzmann Equation: Fokker–Planck Methods* 115
D. *Diffusion and Mobility from the Boltzmann Equation* 119
E. *The Moments of the Boltzmann Equation* 124
Suggested Reading 130
Problems 130

6 *Simple Applications of the Fluid and Statistical Models of a Plasma*

A. *Hydromagnetics (Magnetohydrodynamics)* 134
B. *Hydromagnetic Equilibrium* 138
C. *Plasma Diamagnetism* 141
D. *Magnetic Confinement* 143
E. *Hydromagnetic Wave Motion and Diffusion* 146
F. *Plasma Parameters* 149
G. *Wave Propagation in Plasmas—Introductory Remarks* 157
Suggested Reading 163
Problems 164

7 *Waves in Cold Plasmas*

A. *Waves and the Wave Equation* 167
B. *Eigenvalues of the RLC Circuit (Ordinary Differential Equations)* 169

C. *Eigenvalues of the Wave Equation (Partial Differential Equations)* 170
D. *Eigenvalues of Waves in Cold Plasmas—The Dispersion Relation* 173
E. *Representations of the Eigenvalues* 179
F. *Nonzero Temperature Effects* 193
Suggested Reading 199
Problems 200

8 ***Waves (Stable and Unstable) in Hot Plasmas***

A. *Transform Theory* 203
B. *Applications of Transform Theory to the Wave Equation and Boltzmann Equation* 208
C. *The Electrostatic and Long-Wavelength Approximation* 211
D. *The Dispersion Relation for Longitudinal Electrostatic Waves, I* 213
E. *Complex Contour Integration and Analytic Continuation* 217
F. *The Dispersion Relation for Longitudinal Electrostatic Waves, II* 222
G. *The Nature of the Eigenmodes—Landau Damping* 227
H. *Waves and Instabilities—Criteria* 230
I. *Branch Lines* 234
J. *Position-Space Damping* 236
K. *Confinement and Equilibrium of a Plasma—Hydrodynamic and Microinstabilities* 242
Suggested Reading 244
Problems 245

9 ***Plasma Kinetic Theory—Nonequilibrium Statistical Mechanics***

A. *Collisional Velocity Averages* 249
B. *Validity of Collisional Models* 263
C. *Development of the Bogoliubov, Born, Green, Kirkwood, Yvon (BBGKY) Hierarchy* 265
D. *Cluster Expansion* 274
Suggested Reading 284
Problems 285

10 ***Radiation Processes and Correlation Functions***

A. *Generalized Emission and Absorption Processes* 286
B. *Radiation Fields from a Charged Particle* 291
C. *The Radiation Spectra* 297
D. *Bremsstrahlung* 300
E. *Cyclotron Emission* 306
F. *Fluctuations and Correlations in Plasmas* 311
Suggested Reading 318
Problems 319

Appendix

A. *Frequently Used Physical Constants* 321
B. *Useful Vector Indentities* 322
C. *Units and Dimensions* 323

INDEX 324

Preface

This text should provide the student with a sound grasp of the fundamental principles of the plasma state. In a book of this length not every important topic can be covered; for the most part, the selection of topics and the emphasis given to each should allow for a good deal of flexibility in the ways a lecturer can approach this course.

I have been teaching this material to seniors and first year graduate students of physics and engineering over the past eight years. Generally they have had for background a knowledge of basic electromagnetic theory and mechanics. Those who have also taken a course in atomic physics have found it useful. Because the training of students beginning the study of plasmas is quite diverse, some necessary mathematical and physical concepts are developed in the text. These include Fourier and Laplace transforms, statistical mechanics, dispersion relations, complex contour integration, radiation, and correlation functions.

In organizing the book I have treated some topics more than once to permit the student to recall previous work and to show applications of material presented earlier to more advanced subjects. Necessary formulas and equations are presented in detail. Except where a derivation is reasonably attainable by most students I have eschewed the use of "it can be shown that" in favor of carefully presenting the steps that lead from one equation to another. Yet I have tried to limit the mathematics, preferring references to experiments where possible. However, there is not much discussion of specific experiments and experimental devices since I feel this kind of analysis is best presented to the student after he has been introduced to the subject.

References occur at the ends of chapters to ensure proper credit and to aid understanding. The most important of these are briefly described. The chapter-end problems are very important to attaining mastery of the material.

Most have been tested in homework assignments and examinations. Some of them are designed to be done numerically if a small computer is available.

Rationalized mks units are used throughout with two exceptions: Some magnetic field values are expressed in gauss and some particle densities in number per cubic centimeter.

I wish to express my thanks to those whose discussions with me made this a much better book than would otherwise be the case. I would especially like to thank J. A. Tataronis and T. Mantei for their help and assistance. And to my students, many of whom devoted much time and effort to make this material understandable to themselves and thus, I hope, helped me to make it so to others, this book is gratefully dedicated.

1 The Plasma State

This book is concerned with a material of which more than 99 percent of the matter in the universe is composed, yet that is not readily seen on earth. This elusive material, which is difficult to produce in the laboratory, often completely disappears in times much less than the "twinkling of an eye."

Yet, this subject has been known to man from his earliest beginnings. When our long distant ancestors observed the violent lightning displays, the aurora borealis, the stars, and fire, in all cases they were looking at this material: *plasma.*

What is different about a plasma from the ordinary objects that we see around us? In many cases, nothing at all, for even solid materials like copper, silver, and gold may often be considered plasmas.

Basically, most persons know of the three "common" states of matter on earth: solid, liquid, and gas. A substance may exist in one or more of these states, depending upon certain values of quantities like temperature and pressure.

In general, if we start with a solid substance, we may change it to a liquid by adding energy to it, perhaps by heating with a flame. The substance melts when the "heat of fusion" of the substance has been added at the melting temperature. If the temperature is raised enough, the liquid evaporates when the "heat of vaporization" has been added and a gas is formed. Obviously, then, it is possible to have, for example, solid iron, liquid iron, and gaseous iron.

Let us carry the process still further. If more energy is added to the gas, a point is reached when the gas *ionizes*; in other words, when the "heat of ionization" is added. We now have created a "soup" of ions and electrons, which we call a plasma. The ionizing process divides the molecules and atoms of the gas into separate electrons and ions. Figure 1.1 shows this pictorially.

Note that it is possible to reverse the process and go from the plasma state to the gaseous state, and so on, by removing energy, as shown in the figure. The dotted arrows show possible transitions, one of which, for example, creates a solid state plasma, without the necessity of having the substance exist in the liquid and gaseous phases at all.

These methods of producing plasma occur in nature. For example, a flame or a fire is usually hot enough so that a measurable amount of ionization occurs. Internal stellar material is almost all plasma, because the temperatures there are of such great magnitude. The aurora borealis and the ionosphere,

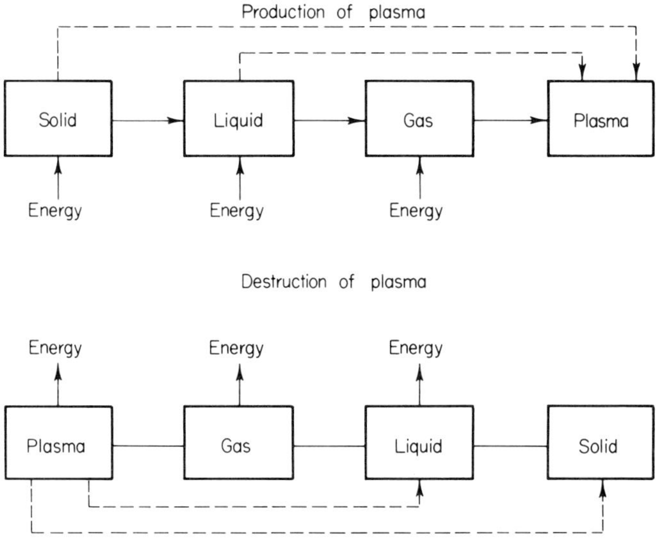

Fig. 1.1. Production of the plasma state.

most fascinating phenomena, are both plasma. Lightning and other electrical discharges are plasma as well.

Man-made plasmas appear commonly in the "neon" sign, mercury and sodium vapor lamps, fluorescent lamps (although the light that we see is from a phosphor on the inside surface of the tube, and not from the plasma), "glow" tubes, mercury arc rectifiers, arc welders, plasma torch paint sprayers, and even explosions.

Much research is performed in the laboratory under reduced pressure on plasmas made from materials that are gases at room temperature and atmospheric pressure, such as hydrogen, helium, nitrogen, and argon. The very colorful and striking displays of such "glowing gas" have been observed since about the turn of the century.

A simple definition of a plasma might be: A large collection of approximately equal numbers of positively and negatively charged particles.

Note that a plasma tends to be *electrically neutral.* That is, it carries no net electrical charge.

When a plasma is produced, several actions are usually taking place simultaneously and continuously. It is important to consider them and to have some idea as to what effects they produce on the behavior of the plasma.

First, to effect *ionization* of a neutral atom or molecule, enough energy must be added to it to allow one or more of the bound electrons to escape from the nucleus. The energy may be added either by a photon of radiation or by means of a collisional process between two particles. If the energy to produce ionization is very low, as is the case, for example, for the element *cesium*, the vapor of the element may be allowed to come in contact with an electrically heated "hot plate." The energy supplied by this plate is sufficient to produce nearly 100 percent of this *thermal* ionization under the proper conditions. Ionization of neutral atoms by collisional interaction is strikingly observed in a Geiger counter when alpha particles that enter the Geiger tube cause a brief burst of ionization in this partially evacuated tube. The ionosphere is an example of a plasma that is produced by radiation ionization. Noticeable changes in the ionosphere occur from day to night as the sun's radiation is removed.

The collisional ionization process also is very common in laboratory plasmas. Since the ions and electrons making up the plasma are electrically charged (although the plasma is usually not), they may be accelerated by means of externally applied combinations of dc or ac electric, magnetic, and/or electromagnetic fields. This acceleration can be sufficient to increase the speed of the plasma particles until they cause ionization of neutral atoms by collision. Note that there are always a few charged particles present even in "un-ionized" gas, so that this process may be used to "ignite" a plasma.

At the same time, the process opposite to ionization is occurring, that is, *recombination.* In this effect, electrons and ions come back together to form neutral atoms. This situation, if it were to exist by itself, would eventually result in the complete disappearance of the plasma state. However, if ionization is also occurring at the same time, the plasma can continue to exist. Radiation is often emitted as a result of the recombination process.

In most earthbound cases, a plasma will not be 100 percent ionized, and there will always be some neutral atoms present in the "soup." These particles collide with the electrons and ions in the plasma, often causing them to lose energy. On the other hand, they may, by a collisional process called *charge exchange*, force the ion to take an electron away from the neutral, which, in effect, results in the two particles having "changed places." This may also result in a loss of energy from the plasma itself, since if a "hot" ion and a "cold" neutral exchange charges, a "cold" ion and a "hot" neutral result. Other particle interaction processes that are often continuously occurring in a plasma will be mentioned subsequently.

Typical laboratory plasmas, produced under vacuum, appear at pressures of from 10^{-6} to 10 Torr and have densities of from 10^7 to 10^{18} particles/cm^3 when conditions to produce ionization are achieved. On the other hand, it is possible to produce plasma at pressures equal to or greater than 1 atm, or well below 10^{-6} Torr, and in solids and liquids as well. Average particle energies vary from less than 1 eV to more than 1 MeV for electrons (1 eV = 11,605°K). Figure 1.2 shows typical values of temperature and density for various kinds of plasmas.

At the present time, one of the most important branches of plasma research is in the area of *controlled thermonuclear fusion.* Essentially, the goal of this

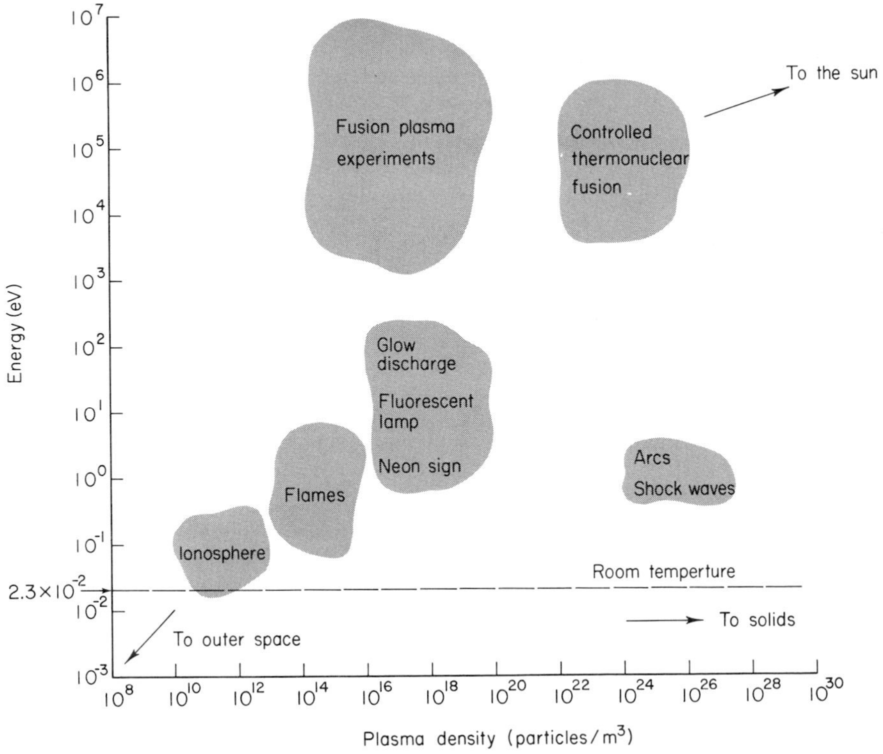

Fig. 1.2. Energy–density diagram of the plasma state.

program is to make a miniature star on earth, generating energy by means of the same reaction that takes place in the sun. The fundamental process is that of the nuclear fusion of heavy isotopes of hydrogen into helium, with the subsequent release of energy. In order to achieve the fusion process, the reactants must be heated (to some millions of degrees) and then kept together in sufficient concentration for a period of time to allow the reaction to become exothermic. The high energies required almost invariably result in the reactants being ionized, hence the plasma aspect in controlled thermonuclear fusion.

Since the plasma must be kept both hot and in a concentrated form, the first problem is the choice of a suitable means of production and storage (confinement) of the plasma. The high energies can more easily be achieved under low pressures (between 10^{-8} and 10^{-3} Torr), so a vacuum system is usually required. Then, the extremely high energies of the particles must not be allowed to dissipate by, for example, allowing the plasma to come in contact with the walls of the vacuum chamber, so some form of confining scheme to keep the plasma away from the walls must be used.

Parenthetically, it should be noted that nature has achieved this result already in the sun and stars. For this "controlled fusion reactor" the vacuum system is outer space itself, and the confinement scheme is the self-gravity of the stars. On earth, the mass of fusion reactors will obviously always be far too small to make use of this method.

The approach for confinement up to this time has been to capitalize on the electrical properties of plasma particles. Once ionized, the plasma particles *are* affected by electric and magnetic fields. Should a scheme be made which then could confine a thermonuclear plasma so that it does not interact with the walls and could keep it concentrated for a long enough time, the problem will have been essentially solved. This method has used externally applied electric, magnetic, and/or electromagnetic fields to attempt to produce adequate confinement. Often, the same fields used for ionization can be used for confinement.

Controlled fusion can be achieved by having a large number of hot plasma particles together for a short period of time or a smaller number of particles together for a longer period of time. That is, the product of plasma density times confinement time is roughly a constant. As a consequence, many fusion experiments that operate in the high density–short confinement time regime can utilize a pulsed mode of operation, which often allows the apparatus to be less massive.

The discussion above has brought out the point that in order to produce confinement, we must know how the charged particles making up the plasma move in electric, magnetic, and gravitational fields. (The latter are, of course, inescapable, not only on earth, but throughout the universe.)

The basic equation of motion that charged particles obey in an electric, magnetic, and gravitational field is

$$\mathbf{F} = m\frac{d\mathbf{v}}{dt} = q\mathbf{E} + q(\mathbf{v} \times \mathbf{B}) + m\mathbf{g} \tag{1.1}$$

This is Newton's Second Law for a particle of mass m and charge q moving with velocity $\mathbf{v}$ in electric field $\mathbf{E}$, magnetic field $\mathbf{B}$, and gravitational field $\mathbf{g}$. This equation, deceptively simple, should give us the paths of all particles, if we knew the values of $\mathbf{E}$, $\mathbf{B}$, and $\mathbf{g}$ at every instant of time and at all points in space. The effects of relativity and radiation should also be included, if necessary.

The net electric and magnetic fields may be obtained from a solution to Maxwell's equations, which are

$$\nabla_r \times \mathbf{E} = -\frac{\partial \mathbf{B}}{\partial t} \tag{1.2}$$

$$\nabla_r \times \mathbf{H} = \mathbf{J}_E + \frac{\partial \mathbf{D}}{\partial t} \tag{1.3}$$

$$\nabla_r \cdot \mathbf{D} = \rho_E \tag{1.4}$$

$$\nabla_r \cdot \mathbf{B} = 0 \tag{1.5}$$

where $\mathbf{H}$ is the magnetic flux density in amperes per meter, $\mathbf{D}$ is the electric flux density in coulombs per square meter, $\mathbf{J}_E$ is the electric current density in amperes per square meter, and ρ_E is the electric charge density in coulombs per cubic meter. We may also use the constitutive relations

$$\mathbf{D} = \varepsilon\mathbf{E} \tag{1.6}$$

and

$$\mathbf{B} = \mu\mathbf{H} \tag{1.7}$$

where ε and μ are the permittivity and permeability of the medium, along with the equations for conservation of charge and Ohm's Law, which are

$$\nabla_r \cdot \mathbf{J}_E + \frac{\partial \rho_E}{\partial t} = 0 \tag{1.8}$$

and

$$\mathbf{J}_E = \sigma(\mathbf{E} + \mathbf{v} \times \mathbf{B}) \tag{1.9}$$

respectively; σ is the conductivity of the medium. Equations (1.2)–(1.5), (1.8), and (1.9) may be expressed in integral form as well. ∇_r is the gradient operator in position space.

For various purposes we may need to define a vector potential **A** such that

$$\mathbf{B} = \nabla_r \times \mathbf{A} \tag{1.10}$$

and a scalar potential ϕ such that

$$\mathbf{E} = -\nabla_r \phi - \frac{\partial \mathbf{A}}{\partial t} \tag{1.11}$$

Since $\nabla_r \cdot \mathbf{A}$ has not been specifically defined, the choice is arbitrary. Often we may use the *Lorentz gauge* definition for the divergence of **A**, that is,

$$\nabla_r \cdot \mathbf{A} = -\frac{1}{c^2}\frac{\partial^2 \phi}{\partial t^2} \tag{1.12}$$

where c is the velocity of light.

Another gauge that is sometimes used is the *transverse gauge*, which is expressed as

$$\nabla_r \cdot \mathbf{A} \equiv 0 \tag{1.13}$$

Proper choice of gauge often results in simplification of the equations describing a problem.

From this statement of the problem, the complications begin to arrive. First, we must have more than one particle (usually 10^8–10^{16} particles/cm^3) in a plasma. This means that each particle, if it is charged, sets up its own electrostatic, magnetic, and radiation (if it is accelerated) fields, which react on all the other particles. In principle, we could compute the net effects of each particle on every other particle, as well as include the effects of externally imposed **E**, **B**, and **g**, and still follow the trajectories by solving the equations of motion simultaneously. However, we would need a computer that could follow the billions and billions of particles making up a plasma all at once. Such computers are not available, but substantial progress in the realm of *plasma simulation* has already been achieved on computers. This simulation involves the simultaneous solution of the equations of motion of a "superparticle," which is often defined as a rigid "cloud of particles" moving through other such "clouds." Up to 100,000 or 500,000 of such interacting superparticles *can* be moved simultaneously on a high-speed digital computer. However, in spite of this progress, there are many cases where such methods still will not be sufficient to adequately describe a given problem.

Without using these tracking methods, a statistical approach is often required. We divide the plasma into energy levels, and calculate the number of particles in each level. If the levels are close together, we may utilize a continuous function for the particle energy distribution. The means by which this function changes with time and in space are often used to determine plasma behavior.

Finally, a conducting fluid approximation may be made, where the plasma is assumed to be a fluid that has an electrical conductivity. The momentum, continuity, energy, and other equations of this type may then be used as in fluid mechanics.

We summarize the methods of study by using the following diagram.

$$\text{Single-particle motion} \xrightarrow{\text{becomes}} \text{Statistical treatment} \xrightarrow{\text{becomes}} \text{Fluid model}$$

as

$$0 \longrightarrow \text{Number of particles} \longrightarrow \cdots \infty$$

When the frequency of interactions between particles is not very great, the single-particle approach may be used with success. Often, only the equation of motion for one particle need be solved to give meaningful results. As the density increases and/or the interactions become more numerous, the statistical approach must be used, culminating, in the highest density limit, with the fluid mechanical approach. Each method slowly blends into the other as the plasma density changes.

In addition, another problem may appear. All the particles may not be ionized. That is, we already know that some neutral atoms are often present in the plasma. They are not as directly affected by the electric and magnetic fields as the charged particles. They do, however, make collisions with the charged particles and affect their motion as a result.

As expected from the previous discussion, there are several different kinds of collisional processes that occur among the ions, electrons, and neutrals in the plasma: momentum transfer; recombination; ionization; attachment; charge exchange; excitation; sputtering.

Momentum transfer, as the name implies, results in a transfer of momentum between colliding particles. *Recombination*, mentioned previously, occurs when an electron and an ion recombine to form a neutral atom. *Ionization*, again mentioned previously, produces a newly formed ion and electron pair after a neutral atom has been involved in a collision, or has absorbed a photon. *Attachment* produces a negative ion after a collision between an electron and a neutral, and *charge exchange*, previously described, produces a transfer of charge between an ion and a neutral. *Excitation* or *deexcitation* of a neutral atom or ion may also occur when a collision results in a jump in energy level of bound electrons in the atom or ion. *Sputtering* results when energetic particles strike the walls of the plasma vessel or electrodes, if any, and heavy "impurity" atoms are knocked off into the plasma.

Most or all of these collisional processes, if they involve charged particles, produce or absorb radiation while the collision is taking place, if the charged particle is accelerated or decelerated. We call an interaction process a collision if it occurs "quickly" compared to the time scale of the problem.

We further note that collisions may be considered *elastic* or *inelastic.* If we may assume that, during the collision, the *internal* energies of the colliding particles are left *unchanged*, the collision is said to be elastic. Otherwise, the collision is inelastic. All collisional processes are, to some degree, inelastic, but in many cases the elastic treatment, which is far simpler to treat, will suffice.

In spite of this, however, the proper treatment of collisional interactions is still one of the most difficult problems in plasmas.

Finally, after all might appear in order, just when the conditions for fusion may be correct, the plasma may suddenly leak out of its confinement region too rapidly or fly apart, that is, become unstable. Each of the three approaches to interpreting the behavior of plasmas predicts various instabilities, and at the present time, a large group of instabilities have already been discovered theoretically and observed experimentally.

Another important question to be asked is how plasmas are produced. There are almost as many ways of producing plasmas as there are instabilities.

To make plasma, what is basically required is that the energies of the neutrals or of the few charged particles that are always present be made high enough to cause a large-scale collisional ionization breakdown of the material. The energies required may be quite low; room temperature for solid metals to perhaps only a few thousand degrees, as for some of the vapors of heavy metals, to millions of degrees for ceramics and good dielectrics. Energies in plasma work are often expressed in electron volts (eV) instead of degrees Kelvin (°K). Previously we stated that 1 eV is equivalent to a temperature of 11,605°K, and is the energy that one electron obtains by being accelerated through a potential of 1 V.

Some of the current methods for production and heating of a plasma may be grouped as shown in Table 1.1.

We shall now briefly explain how some of these methods operate. The first, thermal production of a plasma, simply involves heating, perhaps by combustion, the material, which is usually a gas, until ionization occurs. Flames of various substances are partially ionized and the degree of ionization is often enhanced by the addition of various alkali metals to the flame mixture. This type of plasma may exist at atmospheric pressure. Contact ionization, as discussed previously, is observed in a vacuum where a metal is allowed to evaporate and strike a heated plate. Such plasmas are usually very highly ionized and quite stable, but the particles are usually of low energies.

The dc discharge usually involves two conducting electrodes, set apart in vacuum. When a sufficiently high dc potential is applied between the electrodes, the gas breaks down and a *glow discharge* plasma is formed. If the pressure is raised high enough (perhaps atmospheric or greater) and/or the energy of formation is very high, the breakdown becomes quite violent and the discharge then becomes an *arc* or possibly a *shock wave plasma.* Electric

Table 1.1

Thermal production	dc Discharge
flames	cold electrodes
contact heating	hot or emitting electrodes
Beam injection	ac and rf Discharges
plasma beam	electrodes (cold)
electron beam	hot (emitting) electrodes
ion beam	electrodeless
energetic neutral beam	ion or electron cyclotron resonance[a]
Photoionization	Pulsed discharge
	capacitor discharge] → shock waves
	inductive discharge] → shock waves
	(Any of the other methods can usually be pulsed.)

[a] A dc magnetic field is required for this method (most of the methods above can be utilized in a magnetic field).

welders, arc lamps, and lightning are examples of arcs. The electrodes may be fed from a dc power supply or excited by a pulsed energy storage device, such as a capacitor bank, inductance coil, or delay line. In addition, if the negatively charged electrode is heated, it may inject electrons by thermal emission, which increases the amount of ionization.

An ac or rf discharge can be made by applying a high ac or rf potential to the same electrode configuration. However, the use of electrodes may not be required if some kind of mechanism, such as an inductive loop or microwave cavity, is used to match the output of the ac or rf source to the plasma. Other mechanisms for coupling energy into the plasma are sometimes used, such as waveguides or various other microwave structures. Turbulent heating of a plasma by stochastic fields and various plasma waves can add energy effectively. Often it may be necessary to pre-ionize plasmas that are made in these ways by some other means so that efficient energy transfer and efficient coupling can occur. The phenomenon of photoionization is an extension of the rf discharge to optical frequencies, where sometimes lasers are used to produce ionization. At these wavelengths, the photons themselves create the ionization.

If a dc magnetic field is placed in the plasma region, one of the first effects to be observed is a general improvement in ionization, density, and temperature. This is a good example of the confining effects of a magnetic field. We may also find that energy can be added at the electron or ion cyclotron frequencies. This condition provides an extremely efficient method for coupling energy into a plasma and is often used to produce very high energy plasmas.

The last category of plasma production to be discussed is that of injection. Many systems for production of highly intense directed beams of plasma, electrons, ions, or neutral atoms are used to produce plasma. The basic method is to inject one of these types of beams into a specially prepared region of plasma confinement where ionization of the neutrals present can occur.

Although the controlled fusion problem has been stressed in this introduction, it is by no means the only branch of plasma research.

There are some very interesting astronomical phenomena involving plasmas. The aurora and other extremely tenuous plasmas (less than 10^8 electrons/cm^3) are under study, both for their effects on communications as well as for the knowledge of extraterrestrial plasmas that we can gain from them. For example, the ionosphere seems to be a plasma that is *both stable and confined* by one of the mechanisms used for this purpose in the laboratory: a magnetic field, which in this case is that of the earth. In addition, when radio waves, light, or cosmic rays reach us from other sections of the universe, they have undoubtedly crossed vast reaches of plasma. What happens to radiation as it crosses plasma? Are there possibilities that the frequency, wavelength, or intensity are changed radically? Studies of laboratory plasmas often help to solve these problems and theories are developed to explain this behavior.

Another intriguing aspect of plasma investigation is that of rocket propulsion. The typical blast-offs of earth-launched rockets are produced by chemical explosion. These are capable of producing the extremely high values of thrust required to leave the earth's gravity. For producing small amounts of thrust for long periods of time, however, they are not as efficient as a *plasma rocket engine.*

By using electric and/or magnetic fields to accelerate plasma, a very highly efficient rocket (measured in terms of thrust per unit mass expended) can be made. Such rockets provide the means to navigate to distant stars or planets after escaping from the earth's gravity. These rockets generate only a small amount of thrust, but do so for a very long time.

Plasmas may also be used as circuit elements. For example, electromagnetic radiation impinging on a plasma may produce nonlinear effects resulting in growth of the amplitude of the radiation or in excitation of harmonics and subharmonics. At times, however, plasmas are unwanted. The examples of arcs and the need for their supression in switching circuits shows situations of this type.

Finally, the realm of solid state plasmas is only beginning to be touched. Many of the same experiments and theories in gaseous plasmas may be set up or verified using solid state plasmas at a fraction of the cost of the gaseous state experiments. The extremely high plasma densities that are normally found in solid plasmas (more than 10^{19} particles/cm^3) permit modifications and extensions of theories developed for lower density plasmas. The most likely use for solid state plasmas is in circuitry. The plasma effects produced in compact materials that are in the solid state are a natural complement to existing electronic circuitry.

To begin the study of plasmas is to enter into new, diverse, and complicated theories and concepts. The endless fascination of bringing the pieces of the puzzle together has captivated the imagination of this writer and will also, hopefully, capture you, the reader, as well.

Suggested Reading

Bishop, A. S., "Project Sherwood: The U.S. Program in Controlled Fusion," Addison-Wesley, Reading, Massachusetts, 1958. This is an account of fusion experiments and programs.

Boyd, T. J., and Sanderson, J., "Plasma Dynamics," Barnes and Noble, New York, 1970.

Cobine, J. D., "Gaseous Electronics," Dover, New York, 1958. A large collection of empirical data on glow and arc discharges are presented.

Glasstone, S., and Loveberg, R. H., "Controlled Thermonuclear Reactions," Van Nostrand, Princeton, New Jersey, 1960. This work provides good general background.

Holt, E. H., and Haskell, R. E., "Plasma Dynamics," Macmillan, New York, 1965. This good treatment uses tensor notation throughout.

Rose, D. J., and Clark, M., "Plasmas and Controlled Fusion," The M.I.T. Press, Cambridge, Massachusetts, 1961. This book is a complete treatment with applications to experiments.

Spitzer, L., Jr., "Physics of Fully Ionized Gases," 2nd ed., Wiley (Interscience), New York, 1962. An accurate, *concise* treatment of the subject is presented.

Tananbaum, B. S., "Plasma Physics," McGraw-Hill, New York, 1967.

Uman, M., "Introduction to Plasma Physics," McGraw-Hill, New York, 1964. This work gives concise general coverage.

2 Collisions and Collisional Processes

The brief introductory remarks to the study of plasmas made in Chapter 1 show that one of the most fundamental needs for understanding plasmas is the knowledge of the nature of the interactions between particles that make up a plasma.

The simplest reaction between charged particles is the Coulomb force. This can be handled somewhat simply if there are only two particles and the particles do not come too close to each other (a two-body problem).

Since there are many billions of plasma particles, in principle, each and every plasma particle reacts with each and every other one. Should there be an external electric or magnetic field, this must be included as well.

If we had a large, extremely fast computer, we could handle the plasma problem by following the motion of each individual plasma particle. We would have to include the following effects:

(a) Coulomb forces from each other charged particle;
(b) external electric and magnetic fields;
(c) the effects of relativity;
(d) radiation due to the acceleration of each particle;
(e) gravitational forces;
(f) collisions with neutral atoms;
(g) ionization–recombination behavior.

We would then be able at a given instant of time, to calculate the net force on each particle. By a finite difference procedure, we could move all the plasma particles along in time and follow, as on a television screen, the motion of all the plasma particles.

Such a computer is, of course, not going to appear for a very long time, if at all. However, it is possible to use this method of *plasma simulation* as previously mentioned in Chapter 1, quite effectively if a "clump" or a superparticle consisting of a large number of plasma particles is used instead. Such calculations usually include only some of the effects due to (a), (b), and (c). The results of these "computer experiments" are still often quite useful and are of great help in understanding plasma phenomena.

However, these methods are very often lossless. That is, no damping of energies by collisions can occur. Then, the remaining effects of collisions are usually exaggerated, much more than in reality.

In order to consider collision phenomena amenable to analytic solution, there must be a method for carefully considering collisions. The results of a collision ("short"-time interaction) may be a net change in energy, velocity, momentum, or position of the particles making the collision.

The next question is, how many particles are reacting in a collision? In principle, all particles with a charge are *always* colliding with each other (N-body interaction). However, we will assume initially that the interaction can be separated into *two-body interactions*; that is, only two particles need to be considered for each interaction.

This chapter begins with a set of elementary concepts that are of great value in understanding collision phenomena. After covering the behavior of a single set of two interacting particles, the behavior of a large collection of particles is considered.

A. ELEMENTARY CONCEPTS

We begin investigating collisions by considering particles that have some mechanism of interaction with one another. We call this a collision if it takes place over a short time. Collisions may be long range or short range, depending on the nature of the interaction. Often, long-range interactions, which occur when the particles are far apart, may be significant in plasma behavior.

In any event, we can define some quantities that will help us to work with these collision processes. The first quantity is the *collision cross section*.

We consider a collection of particles characterized by a density N per cubic meter. Let us fire a beam of test particles at this group, called the target. Suppose that the beam of particles has a density of n per cubic meter. Figure 2.1 shows the nature of the system under discussion.

We now determine the number dn of particles that have "collided" while traversing a distance dx. The density of colliding particles dn ought to be proportional to the beam density n, the density of target particles N, and the distance traveled by the beam dx. The relation for dn is then

$$dn = \sigma n N \, dx \tag{2.1}$$

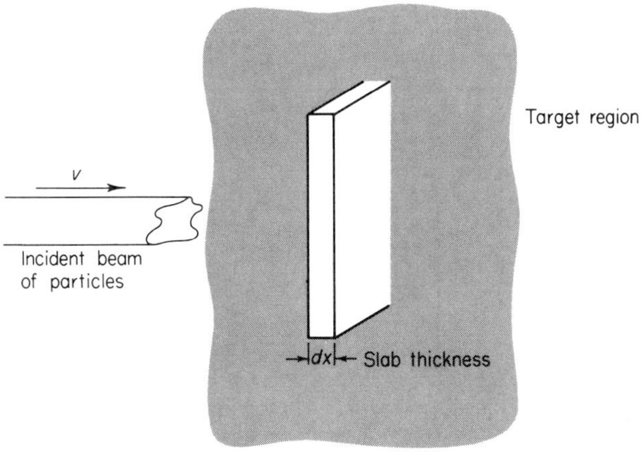

Fig. 2.1. Collisions between a beam of particles and a target.

σ, the proportionality constant, is a quantity that we will call the *collision cross section.* It may be a function of various quantities, especially the energies of the interacting particles and the forces between the particles, which change with their separation. It should be mentioned that Eq. (2.1) is a very simple expression and must be verified by experimental measurements to show the proper value for σ.

If we consider that dn particles have made a collision when the beam traverses a slab of thickness dx, then we can compute the number of particles "lost" from the beam after it has traveled a distance x, by integrating Eq. (2.1):

$$dn_{\text{lost}} = -\sigma n N \, dx$$

Therefore

$$n_{\text{left}} = n_0 \exp(-N\sigma x) \tag{2.2}$$

where n_{left} are those particles remaining in the beam that have not collided and n_0 is the beam density at $x = 0$. We assume that $N \gg n$, so that N is not perturbed, and that the most likely collision will be between a beam particle and a target particle.

We can now define a second quantity l, the *mean free path*, by considering the average distance that a particle travels under these conditions. We compute l as shown in Eq. (2.3)

$$\frac{\int_0^\infty x \exp(-N\sigma x) \, dx}{\int_0^\infty \exp(-N\sigma x) \, dx} = \frac{1}{N\sigma} = l \tag{2.3}$$

by the standard methods of averaging.

The *collision frequency* (number of collisions per second that a particle experiences) can now be defined as

$$\nu = \frac{v}{l} = vN\sigma \tag{2.4}$$

since a particle moving with velocity v travels a distance l in time $1/\nu$.

We can now consider the problem of determining for how long a time a particle may exist without making a collision.

Note that the exponent of Eq. (2.2) is $-N\sigma x$. Therefore, if $x = vt$ and $N\sigma = 1/l$, then $-N\sigma = -vt/l = -\nu t$, where ν is the collision frequency. We also call the reciprocal of ν, τ, the mean collision time. Thus, a particle has a chance $\exp(-\nu t)$ of surviving t seconds without a collision. Equation (2.2) may then be written as

$$n_{\text{left}} = n_0 \exp(-N\sigma x) = n_0 \exp\left(-\frac{x}{l}\right) = n_0 \exp(-\nu t) \tag{2.5}$$

The dimensions of the cross section σ are area. Often the *barn*, 10^{-24} cm^2, is used as a unit for σ. Other units might be square angstroms (Å^2) or square microns (μ^2).

Collision frequencies for several processes, such as electron–ion or electron–neutral interactions, may be defined separately. A total net collision frequency may be defined as the sum of the individual interaction rates; that is,

$$\nu_{\text{total}} = \nu_{\text{ee}} + \nu_{\text{ei}} + \cdots = \sum_k \nu_{jk} \tag{2.6}$$

for the given jth species of particle. This implies that all the interactions are occurring simultaneously and independently of each other. This can be the case in practice for simple interactions in a plasma. Under this assumption, mean free paths would then add reciprocally as shown in Eq. (2.7).

$$\frac{1}{l_{\text{total}}} = \frac{1}{l_{\text{ee}}} + \frac{1}{l_{\text{ei}}} + \cdots = \sum_k \frac{1}{l_{jk}} \tag{2.7}$$

Cross sections can be measured experimentally for many types of interactions. Unfortunately, they often do not lend themselves to representation by an analytic function of their parametric variables, and so empirical formulas have to be used. Typically, collision cross sections are presented on a graph as a function of the energies of the colliding particles. Usually, the cross section for most collisions is low at low energies, reaches a maximum, and then drops off again at high energies. Many varieties of cross sections have been obtained experimentally and tabulated.

B. CONSERVATION OF ENERGY, MOMENTUM, AND ANGULAR MOMENTUM

We now will obtain a more quantitative representation of the collision cross section and collision phenomena. We first determine the end behavior of two colliding particles.

Consider the elastic encounter of two particles of masses m_1 and m_2. We shall assume that the force between them lies along a line joining their centers. The encounter may actually be continuous, and act over extremely long ranges, depending upon the magnitude of the interaction. However, we shall consider that an asymptotic set of positions and velocities at $t = -\infty$ and $t = +\infty$ of the particles is possible.

Figure 2.2 shows the geometry of this encounter. We shall consider that the velocities before the encounter are $\mathbf{v}_1$ and $\mathbf{v}_2$ and that the velocities after the encounter are $\mathbf{v}_1'$ and $\mathbf{v}_2'$, respectively. $\mathbf{G}$ is the velocity of the center of mass. The exact details of the encounter do not matter in this development. Only the end points of the collision need be known.

Let m_1 and m_2 be the masses of the two particles. Then, we shall define

$$m_0 = m_1 + m_2, \qquad M_1 = \frac{m_1}{m_0}, \qquad \text{and} \qquad M_2 = \frac{m_2}{m_0} \tag{2.8}$$

Note that

$$M_1 + M_2 = 1. \tag{2.9}$$

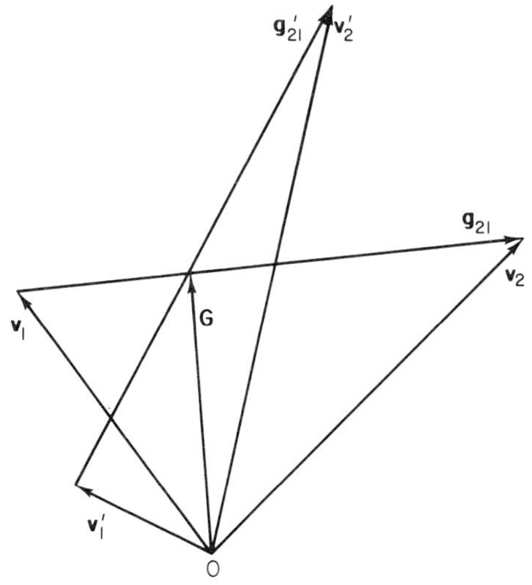

Fig. 2.2. Velocity coordinates of two colliding particles.

The system has a center of mass. Its momentum is conserved in this interaction. This is stated as Eq. (2.10).

$$m_0 \mathbf{G} = m_1 \mathbf{v}_1 + m_2 \mathbf{v}_2 = m_1 \mathbf{v}_1' + m_2 \mathbf{v}_2' \tag{2.10}$$

Let $\mathbf{g}_{21}$, $\mathbf{g}_{21}'$ and $\mathbf{g}_{12}$, $\mathbf{g}_{12}'$ denote the initial and final velocities of the second particle relative to the first and the first relative to the second, respectively.

Therefore

$$\mathbf{g}_{21} = \mathbf{v}_2 - \mathbf{v}_1 = -\mathbf{g}_{12} \tag{2.11}$$

$$\mathbf{g}_{21}' = \mathbf{v}_2' - \mathbf{v}_1' = -\mathbf{g}_{12}' \tag{2.12}$$

As can be ascertained from Fig. 2.2,

$$|\mathbf{g}_{21}| = |\mathbf{g}_{12}|$$

We can call this g. Also,

$$|\mathbf{g}_{21}'| = |\mathbf{g}_{12}'|$$

We can call this g'. We can then express $\mathbf{v}_1, \mathbf{v}_2, \mathbf{v}_1'$, and $\mathbf{v}_2'$ in terms of $\mathbf{G}$, $\mathbf{g}_{21}$, and $\mathbf{g}_{21}'$ as shown.

$$\mathbf{v}_1 = \mathbf{G} + M_2 \mathbf{g}_{12} \qquad \mathbf{v}_2 = \mathbf{G} + M_1 \mathbf{g}_{21} \tag{2.13}$$

$$\mathbf{v}_1' = \mathbf{G} + M_2 \mathbf{g}_{12}' \qquad \mathbf{v}_2' = \mathbf{G} + M_1 \mathbf{g}_{21}' \tag{2.14}$$

If we know $\mathbf{G}$ and $\mathbf{g}_{21}$ or $\mathbf{G}$ and $\mathbf{g}_{21}'$, we are able to find $\mathbf{v}_1, \mathbf{v}_2, \mathbf{v}_1'$, and $\mathbf{v}_2'$.

The potential energy before and after the encounter is also assumed to remain constant throughout the encounter. Therefore, the equation of conservation of energy should give

$$\tfrac{1}{2}(m_1 v_1{}^2 + m_2 v_2{}^2) = \tfrac{1}{2}(m_1 v_1'^2 + m_2 v_2'^2) \tag{2.15}$$

By substituting Eqs. (2.13) and (2.14) into (2.15) we can find that

$$\tfrac{1}{2}(m_1 v_1{}^2 + m_2 v_2{}^2) = \tfrac{1}{2} m_0 (G^2 + M_1 M_2 g^2) \tag{2.16}$$

and

$$\tfrac{1}{2}(m_1 v_1'^2 + m_2 v_2'^2) = \tfrac{1}{2} m_0 (G^2 + M_1 M_2 g'^2) \tag{2.17}$$

Therefore, g' must be equal to g due to this assumption of conservation of energy.

From now on in this development velocities will be expressed in terms of the center of mass velocities and the relative velocities. We can transform, therefore, to a frame of reference that moves along with one particle. The speed of the second particle is then g both before and after the collision. This last result implies that only the direction of the relative velocity is changed, not the magnitude of it. This is, of course, not true in an actual inelastic collision. The examples of radiation by collisional processes, changes in vibrational states, ionization, excitation, and so on are too numerous to disregard. However, for the purpose of simplicity, and noting that many of the collision processes involved in plasmas may be *approximated* by this type of elastic encounter, we proceed further with the development. The type of collisional process that we now investigate is called *Rutherford scattering*. It is elastic scattering from a $1/r^2$ Coulomb force.

It should now be noted that conservation of momentum and energy alone are not enough to completely determine the solution to the problem. Two additional variables are required. To find them, examine Fig. 2.3.

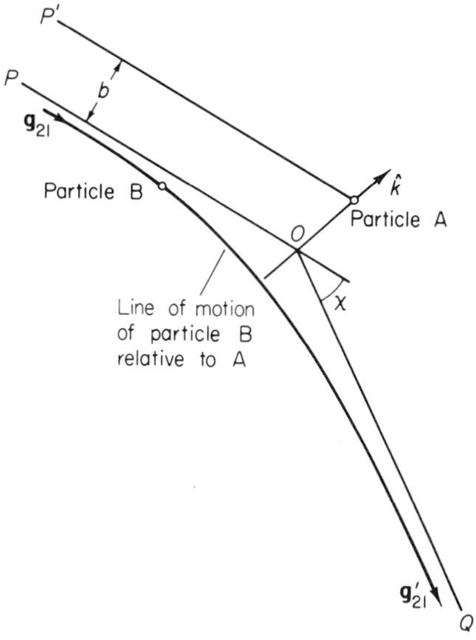

Fig. 2.3. Path of a colliding particle B relative to another colliding particle A.

Suppose that molecules A and B are interacting. A may be fixed, or a set of axes with the origin at point A may be moving along with it. The interaction between A and B is confined to the line joining their centers. The motion of the particles will then be in some plane passing through point A. This can be seen by noting that the *angular momentum* of the interacting particles is also conserved over the collision. (Angular momentum is the moment of the momentum.) In this geometry, particle B has all the momentum. Its moment of momentum about A must be constant before, during, and after the collision. The direction of the angular momentum is out of the plane of the paper at the beginning of the collision and must therefore be pointing out of the paper throughout the interaction. The result is that the motion must take place in a plane through point A. Note that an infinite number of planes of interaction are possible, but they must all pass through A. Let line

(1) be the path described by the particle B. The asymptotes of this path, $\overline{OP}$ and $\overline{OQ}$ are in the directions of the initial and final relative velocities.

The line $\overline{P'A}$ is parallel to $\overline{OP}$ and is therefore in the direction of $\mathbf{g}_{21}$. Its direction is fixed by $\mathbf{v}_2$ and $\mathbf{v}_1$. However, the orientation of the plane defined by lines $\overline{OP}$ and $\overline{OQ}$ about the line $\overline{P'A}$ is *not* fixed and is therefore *an additional* variable. Let us specify it by denoting the angle that the plane makes with a fixed plane by an angle ε. Figure 2.4 shows such an arrangement.

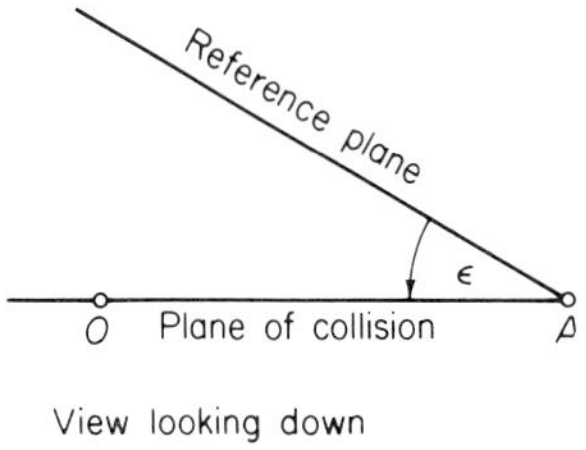

Fig. 2.4. Diagram showing the angle that the plane of collision makes with a reference plane.

When, after a collision, $\mathbf{g}_{21}$ becomes $\mathbf{g}'_{21}$, it has been deflected through an angle χ, which is called the scattering or deflection angle. It depends on the magnitude of the initial relative velocity *and* on the distance b, the minimum distance between the particles, measured between $\overline{P'A}$ and $\overline{OP}$, and, of course, the interaction mechanism. This quantity b is called the *impact parameter* and is another required variable.

There is then a functional relation between χ, b, g, and ε. We can usually determine χ as a function of b, g, and ε.

The *apse line* is line $\overline{OA}$ and the path of the particle B is symmetric about the apse line. If we define a unit vector $\hat{k}$ as shown in Fig. 2.3, and if $\mathbf{g}_{21}$ and $\hat{k}$ are known, we can find $\mathbf{g}'_{21}$ by this geometrical construction. Note that the components of $\mathbf{g}_{21}$ and $\mathbf{g}'_{21}$ in the direction perpendicular to $\hat{k}$ are the same in magnitude *and* direction, while the components in the $\hat{k}$ direction have opposite sign.

We now must solve for the deflection angle in terms of the nature of the force between particles and the particle speed and impact parameter. In some cases the interaction is dependent on angle ε.

We now draw Fig. 2.5, showing the relationship between appropriate quantities. This is Fig. 2.3 except that we now include t and ε as variables. The lettering ε, P, P', b, O, A corresponds to previous figures. In the figure, $\mathbf{g}_{21}$ goes through the solid cylinder. In a time dt, any molecule that is inside this cylinder will be considered to suffer a collision. From Fig. 2.3, we can determine that the unit vector $\hat{k}$ makes an angle of $\frac{1}{2}(\pi - \chi)$ with $\mathbf{g}_{21}$. The plane described by the vectors $\hat{k}$ and $\mathbf{g}_{21}$ makes the angle ε with a fixed plane through $\mathbf{g}_{21}$, so that $\frac{1}{2}(\pi - \chi)$ and ε specify the direction of $\hat{k}$.

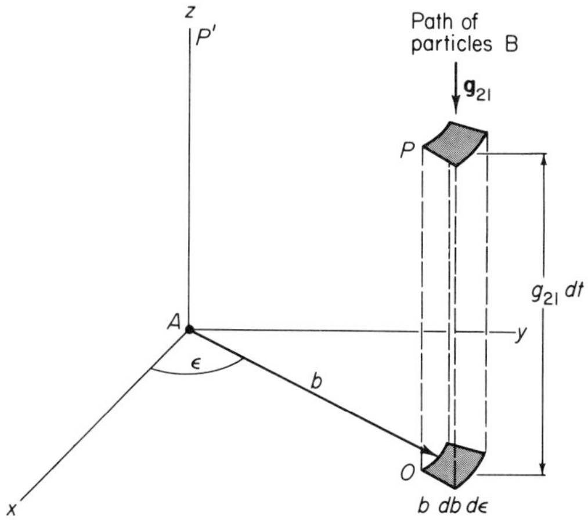

Fig. 2.5. Collisions of a group of particles with particle *A*.

C. THE COLLISION CROSS SECTION

We now consider a solid angle coordinate, which will be used for locating the particles deflected by particle A as shown in Fig. 2.6. The incremental solid angle in which we are interested is that angle into which a particle whose impact parameter lies between b and $b + db$ will be scattered if it is deflected through an angle χ. The magnitude of this solid angle is

$$d\Omega = \sin\chi \; d\chi \; d\varepsilon \tag{2.18}$$

The incident conditions necessary to obtain a given particle flux through the solid angle are specified by the following relation.

$$|\mathbf{g}_{21}|\, b \; db \; d\varepsilon = |\mathbf{g}_{21}'| \sin\chi \; d\chi \; d\varepsilon \, \frac{d\sigma}{d\Omega} \tag{2.19}$$

Equation (2.19) states that the number of particles per second incident on the area $b\;db\;d\varepsilon$ in the (x, y) plane of Fig. 2.5 is equal to the number per second deflected into the solid angle $d\Omega$. We call $d\sigma/d\Omega$ the *differential* collision cross section; it gives us the probability that the particle will be deflected through the angle χ, given the appropriate impact parameter, relative velocity, and angle ε.

Note that in Fig. 2.6 and Eq. (2.19) the incident particle location is expressed in *cylindrical* coordinates, whereas the scattered particle location is expressed in *spherical* coordinates. There is no loss of generality in doing this.

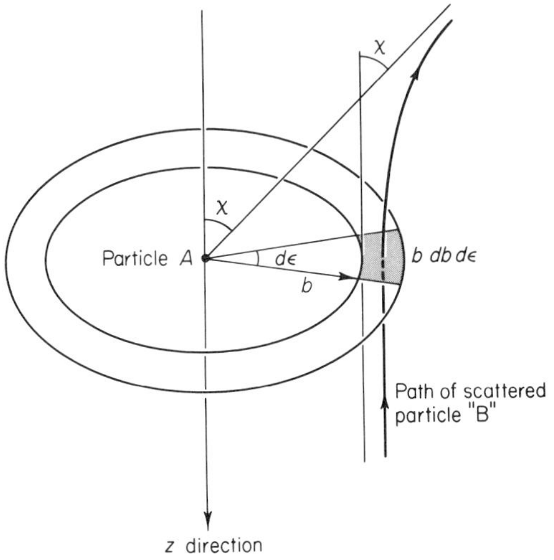

Fig. 2.6. Solid angle coordinates.

In general, if the interaction is independent of ε, we can solve (2.19) for the differential scattering cross section, namely,

$$\frac{d\sigma}{d\Omega} = \frac{b}{\sin \chi} \frac{db}{d\chi} \tag{2.20}$$

The maximum scattering angle χ is π radians, which corresponds to a 180° reversal of the particle motion. The total cross section ought to be the integral of Eq. (2.20) over all possible solid angles, described by Eq. (2.18). Note that at $\chi = 0$ the differential cross section becomes infinite, which tells us that the probability of a very small angle deflection is extremely high. This does present a difficulty in the integration.

This problem can be resolved by noting that "zero-angle" collisions can occur only for those impact parameters which are large enough so that the interaction potential is extremely small. This means that very little momentum is transferred in this type of collision and we can weight (2.20) by an appropriate function that removes the singularity at $\chi = 0$. This enables us to utilize an effective collision cross section for *momentum transfer*. The choice of this function is somewhat arbitrary, although later on we shall find some justification for a particular type of weighting function.

Such a weighting function might be

$$(1 - \cos \chi)$$

so that we can now obtain a new differential scattering cross section:

$$\frac{d\sigma_m}{d\Omega} = \frac{d\sigma}{d\Omega}(1 - \cos\chi) = \frac{b}{\sin\chi}\frac{db}{d\chi}(1 - \cos\chi)$$

σ_m is the *momentum transfer* cross section. We may also define a momentum transfer collision frequency ν_m in terms of this new cross section.

The total cross section is then found by integrating $d\sigma_m$ over all deflection angles possible.

$$\sigma_m \text{ total} = \int_0^{2\pi}\int_{\chi=0}^{\pi} \frac{b}{\sin\chi}\frac{db}{d\chi}(1 - \cos\chi)\sin\chi\, d\chi\, d\varepsilon \qquad (2.21)$$

It should be again emphasized that the choice of this weighting function is completely unrestricted and there are some other functions or methods that are often used to remove the singularity for $\chi = 0$.

As we may realize, many other types of encountering processes are possible in addition to nearly elastic scattering. Radiative transitions, in which a photon of radiation is emitted or absorbed, are often prevalent, even when the particles remain free both before and after the interaction. Such a process is called *bremsstrahlung* (braking radiation). More will be said about this later. Plasma particles may recombine with each other, may ionize partially ionized or neutral atoms, may excite internal energy states in other particles, and so on. Each process has its own cross section and the relative importance of each process may be obtained by investigating the cross section for the particular set of experimental conditions utilized.

D. RUTHERFORD SCATTERING

The Rutherford model of Coulomb scattering of charged particles will now be fully developed. We neglect the shielding effects of adjacent electrons and/or atoms and consider the scattering problem to be a two-body one. We shall also assume the velocities of the particles to be nonrelativistic.

The equations of motion of the two particles are, assuming that the only force on each particle is the Coulomb force from the other particle,

$$m_1\ddot{\mathbf{r}}_1 = \frac{q_1 q_2(\mathbf{r}_1 - \mathbf{r}_2)}{4\pi\varepsilon_0|r_1 - r_2|^3} \qquad m_2\ddot{\mathbf{r}}_2 = -\frac{q_1 q_2(\mathbf{r}_1 - \mathbf{r}_2)}{4\pi\varepsilon_0|r_1 - r_2|^3}$$

m_1, m_2, q_1, q_2 are the masses and charges of the two particles. Their paths are shown in Fig. 2.7, if the charges are of the same sign. ε_0 is the permittivity of free space.

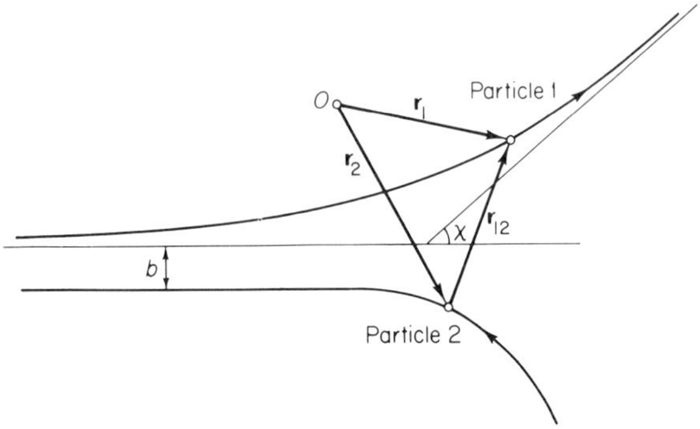

Fig. 2.7. Trajectories of two colliding particles.

We can combine these motions as was done previously, to get

$$m_1\dot{\mathbf{v}}_1 + m_2\dot{\mathbf{v}}_2 = (m_1 + m_2)\dot{\mathbf{G}} = 0$$

Now since

$$\dot{\mathbf{v}}_1 - \dot{\mathbf{v}}_2 = \dot{\mathbf{g}}_{12},$$

we have

$$\dot{\mathbf{g}}_{12} = \frac{m_1 + m_2}{m_1 m_2} \frac{q_1 q_2}{4\pi\varepsilon_0 (r_{12})^3} \mathbf{r}_{12} \tag{2.22}$$

Equations (2.22) state that $\mathbf{G}$ = constant and $\mathbf{r}_{12}$ varies as if a single particle of mass $m_1 m_2/(m_1 + m_2)$ were moving around a *fixed* force center. We call

$$m_r = \frac{m_1 m_2}{m_1 + m_2}$$

the reduced mass.

Let us ignore the center of mass motion. If we do this, and transform to polar coordinates r and θ, then

$$\mathbf{r}_{12} = r\hat{a}_r \tag{2.23}$$

$$\dot{\mathbf{r}}_{12} = \dot{r}\hat{a}_r + r\dot{\hat{a}}_r = \dot{r}\hat{a}_r + r\dot{\theta}\hat{a}_\theta \tag{2.24}$$

$$\ddot{\mathbf{r}}_{12} = \ddot{r}\hat{a}_r + 2\dot{r}\dot{\theta}\hat{a}_\theta + r\ddot{\theta}\hat{a}_\theta - r\dot{\theta}^2\hat{a}_r \tag{2.25}$$

where $\hat{a}_r$ and $\hat{a}_\theta$ are the unit vectors in the r and θ directions, respectively. Note that

$$\dot{\hat{a}}_r = \dot{\theta}\hat{a}_\theta \qquad \dot{\hat{a}}_\theta = -\dot{\theta}\hat{a}_r$$

Equation (2.22) can be written in component form as follows.

$\hat{a}_r$ *Component*

$$\ddot{r} - r\dot{\theta}^2 = \frac{K}{r^2} \qquad \text{where} \quad K = \frac{q_1 q_2}{4\pi\varepsilon_0 m_r} \tag{2.26}$$

$\hat{a}_\theta$ *Component*

$$r\ddot{\theta} + 2\dot{r}\dot{\theta} = 0 \tag{2.27}$$

If (2.27) is multiplied by r and integrated we get

$$r^2\dot{\theta} = \text{constant} = -gb = \text{angular momentum} \tag{2.28}$$

which is always conserved. We now determine the maximum deflection angle χ as a function of r. First, we let $u = 1/r$. Then

$$r = \frac{1}{u} \tag{2.29}$$

and

$$\dot{r} = -\frac{\dot{u}}{u^2} = -\dot{\theta}\,\frac{du/d\theta}{u^2} = gb\,\frac{du}{d\theta} \tag{2.30}$$

We can differentiate Eq. (2.30) with respect to time, giving

$$\ddot{r} = gb\dot{\theta}\,\frac{d^2u}{d\theta^2} = -g^2b^2u^2\,\frac{d^2u}{d\theta^2} \tag{2.31}$$

By substituting (2.31), (2.30), and (2.28) into (2.26), we get

$$-g^2b^2u^2\,\frac{d^2u}{d\theta^2} - g^2b^2u^3 = Ku^2$$

or

$$\frac{d^2u}{d\theta^2} + u = -\frac{K}{g^2b^2} \tag{2.32}$$

Integrating this gives

$$u = \frac{1}{r} = A\cos(\theta + \delta) - \frac{K}{g^2b^2} \tag{2.33}$$

The coefficients A and δ are evaluated from the boundary conditions at $\theta = \chi$. They are $r = \infty$ and $\dot{r} = -g$. Using Eq. (2.28) in (2.33) allows the values of A and δ to be found as shown below.

$$A = \frac{(1 + K^2/g^4b^2)^{1/2}}{b} \qquad \tan\delta = \frac{g^2b}{K}$$

The total angular deflection χ may be obtained by letting $r \to \infty$ in Eq. (2.33). One solution is $\chi = \pi$. The other solution is $\chi = \pi - 2\delta$.

Therefore,

$$\tan \frac{\chi}{2} = \cot \delta = \frac{K}{g^2 b} \tag{2.34}$$

Note that all χ's are possible since we have a large range of possible impact parameters. What we are now looking for, however, is the *distribution* of angular deflections. We are interested in obtaining $n(\chi)\, d\chi$, which is the number density of particles deflected through an angle χ. For the spherically symmetric force assumed here, the dependence on ε can be neglected. Then the number scattered into solid angle $d\Omega$ at angle χ is, from (2.20)

$$n_0\, d\varepsilon \frac{d\sigma}{d\Omega} = \frac{n_0\, b}{\sin \chi} \frac{db}{d\chi}\, d\varepsilon \tag{2.35}$$

n_0 is the incident particle density in particles per unit incident *area*.

Now since

$$\tan \frac{\chi}{2} = \frac{K}{g^2 b}$$

then

$$b = \frac{K}{g^2[\tan(\chi/2)]}$$

We can differentiate this expression to obtain $db/d\chi$ and substitute for this and b in Eq. (2.35). The result is

$$n_0 \frac{d\sigma}{d\Omega} = \frac{n_0 K^2}{4g^4 \sin^4(\chi/2)} \tag{2.36}$$

We can integrate Eq. (2.36) over all values of χ by use of the relation

$$d\Omega = 2\pi \sin \chi\, d\chi \tag{2.37}$$

where we have already integrated over ε.

We now proceed to a discussion of the effects observed when a large number of colliding particles are considered. One of the first things to be noted is that, in a plasma, it is very unlikely that the density will be the same over all space. The result, when collisions are considered, is a *diffusion* of particles from regions of higher density to regions of lower density. This free diffusion process is one of the important effects that appear when "collective" phenomena are considered.

E. DIFFUSION AND MOBILITY

Classical free diffusion, like classical collision theory, is an event that does not really occur, especially in plasmas, but often is a process that may be closely approximated in practice. Since the development is quite straightforward, it will be advantageous to consider it at this point.

We shall first concern ourselves with a weakly ionized plasma in which the majority of the collisional processes occur between plasma particles (here defined as ions and electrons) and neutral atoms. No long-range electrical forces are therefore involved in this type of diffusion process.

Figure 2.8 shows a one-dimensional model of a system used to develop diffusion concepts. Let us consider the behavior of particles moving along the x axis of Fig. 2.8 as they collide with scattering centers (neutral atoms). If the density of plasma particles is expressed as n per cubic meter where n is a function of position along the x axis, then

$$n(x)\frac{v}{l} = n(x)\nu \tag{2.38}$$

is the number of collisions per cubic meter per second occurring at a point x along the axis, using the notation previously defined for the mean free path and collision frequency. We will now assume that the velocity distribution of particles is isotropic and that a particle is equally likely to be scattered in any one of six possible coordinate directions after it undergoes a collision. The six directions refer to the positive and negative x, y, and z directions. Since the collisions can be assumed to take place in an incremental volume, we can now presume that, on the average, one sixth of all the collisions occurring at point x will result in a directed motion toward the observer, who is to be stationed at the origin (point O).

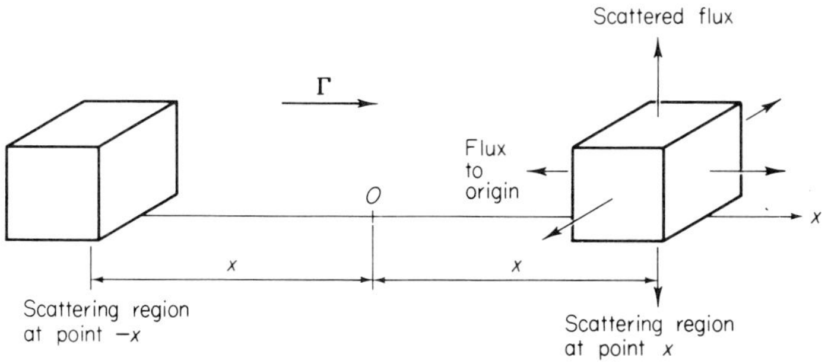

Fig. 2.8. Diffusion diagram.

Therefore, the number of collisions per cubic meter per second made at point x that are directed toward the origin is $(v/6l)n(x)$.

We now recall that the probability that a particle will traverse a distance x without making a collision is $e^{-x/l}$. We are now able to compute the net particle flux passing the origin produced in the region between x and $x + dx$ where $x > 0$ as

$$d\Gamma_{0x} = -\frac{v}{6l} n(x) e^{-x/l}\, dx \qquad \text{if} \quad x > 0 \tag{2.39}$$

Γ is defined as the *particle current* and is assumed positive for movement in the $+x$ direction. The particle current passing the origin from points $x < 0$ can be found in much the same way. It is

$$d\Gamma_{0x} = \frac{v}{6l} n(x) e^{-|x|/l}\, dx \qquad \text{if} \quad x < 0$$

The net particle flux passing the origin may then be obtained by integrating over all x, which yields

$$\Gamma_{0x} = \frac{v}{6l}\left[\int_{-\infty}^{0} n(x) e^{-|x|/l}\, dx - \int_{0}^{\infty} n(x) e^{-x/l}\, dx\right] \tag{2.40}$$

We shall now consider small variations in $n(x)$ so that it may be expanded in a Taylor series about point O, as

$$n(x) = n(0) + x\frac{dn}{dx} + \cdots$$

We truncate the Taylor series after the second term. The substitution of this series into Eq. (2.40) results in

$$\Gamma_{0x} = \frac{v}{6l}\left[\int_{-\infty}^{0} n(0) e^{-|x|/l}\, dx + \int_{-\infty}^{0} x \left.\frac{dn}{dx}\right|_{x=0} e^{-|x|/l}\, dx - \int_{0}^{\infty} n(0) e^{-x/l}\, dx \right.$$
$$\left. - \int_{0}^{\infty} x \left.\frac{dn}{dx}\right|_{x=0} e^{-x/l}\, dx\right] \tag{2.41}$$

for the net particle current. The first and third terms in (2.41) will cancel, leaving

$$\Gamma_{0x} = \frac{v}{6l}\left.\frac{dn}{dx}\right|_{x=0}\left[\int_{-\infty}^{0} x e^{-|x|/l}\, dx - \int_{0}^{\infty} x e^{-x/l}\, dx\right]$$

or, after completing the two integrals,

$$\Gamma_{0x} = -\frac{1}{3} lv \left.\frac{dn}{dx}\right|_{x=0} = -D\left.\frac{dn}{dx}\right|_{x=0} \tag{2.42}$$

Here we have set $D = \frac{1}{3}lv$, where D is called the *free diffusion coefficient.*

Since the origin is arbitrary in this problem, we may generalize Eq. (2.42) to find the particle current at any point x by evaluating the free diffusion coefficient and dn/dx at the point x, that is,

$$\Gamma = -D\frac{dn}{dx} \tag{2.43}$$

The free diffusion coefficient D is dependent on the values of l and v for whichever species of particle and kinds of interaction are being considered. For example, we see that the coefficients

$$D_i = \tfrac{1}{3}l_i v_i \tag{2.44}$$

for the ions and

$$D_e = \tfrac{1}{3}l_e v_e \tag{2.45}$$

for the electrons may be different. A similar expression may, of course, be written for the neutral atoms. The mean free paths for momentum transfer are often equal for ions and electrons but their velocities may be quite different. Usually $v_e > v_i$, which means that $D_e > D_i$. Remember that the diffusion process is such that particles seem to "slide down" density gradients, that is, from regions of higher particle density to regions of lower density.

The presence of a uniform dc magnetic field will tend to inhibit the diffusion process of charged particles across the magnetic field lines. We can obtain some idea of the reasons for this effect by considering the following development.

First, it is well known that a charged particle orbits about magnetic field lines. Figure 2.9 shows an orbiting particle in a plane perpendicular to a magnetic field line. We note that the arguments of the preceding development still apply regarding the probability of a particle making a collision, but they

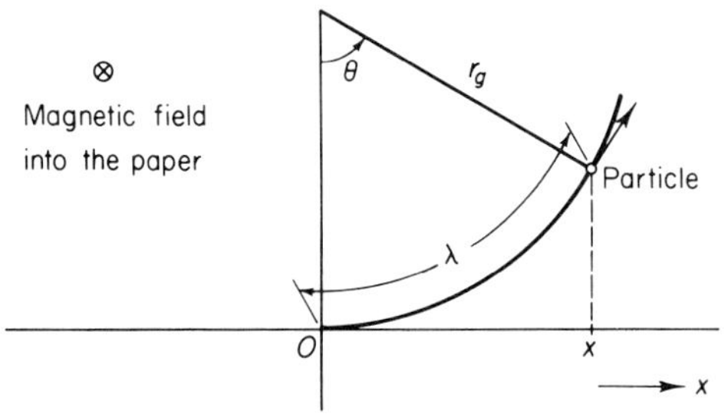

Fig. 2.9. Orbiting particle path.

must apply to the *arc length* rather than the linear distance x as before. The resultant particle current must then be (at the origin)

$$\Gamma_{0x} = \frac{v}{6l}\left[\int_{-\infty}^{0} n(\lambda)e^{-|\lambda|/l}\,d\lambda - \int_{0}^{\infty} n(\lambda)e^{-\lambda/l}\,d\lambda\right] \tag{2.46}$$

where λ is the arc length, defined as

$$\lambda = r_g \theta$$

The linear distance x is then $r_g \sin\theta$. Note that arc length can go from $-\infty$ to $+\infty$ even though, in principle, a particle retraces the same path over and over.

We now apply a Taylor series expansion to n, as before, and obtain

$$n(\lambda) \approx n(0) + x\left.\frac{dn}{dx}\right|_{x=0}$$

Note that the variation of n is *still* in the x direction, even though the particles move in the θ direction. If we assume that r_g, the radius of gyration, remains constant for a given particle (as long as it makes small angle collisions), then the particle current becomes

$$\Gamma_{0x} = -\frac{vr_g^2}{3l}\left.\frac{dn}{dx}\right|_{x=0}\int_0^{\infty}\sin\theta\,\exp\left(-r_g\frac{\theta}{l}\right)d\theta \tag{2.47}$$

after making the appropriate substitutions into Eq. (2.40). Integrating Eq. (2.47), we obtain for the net particle current

$$\Gamma_{0x} = -\frac{vr_g^2}{3l}\left.\frac{dn}{dx}\right|_{x=0}\frac{1}{1+(r_g^2/l^2)}$$

or

$$\Gamma_{0x} = -\frac{lv}{3}\frac{1}{1+(l^2/r_g^2)}\left.\frac{dn}{dx}\right|_{x=0} \tag{2.48}$$

Again, the origin is arbitrary, so the expression (2.48) may be generalized to any point x as in Eq. (2.43).

This looks very much like the expression for free diffusion in the absence of a magnetic field except for the multiplicative constant

$$\frac{1}{1+(l^2/r_g^2)} \tag{2.49}$$

We can then define a magnetic field diffusion coefficient as follows:

$$D_{\text{magfield}} = \frac{D_{\text{no mag field}}}{1+(l^2/r_g^2)} \tag{2.50}$$

It should be noted here that this coefficient $D_{\text{mag field}}$ is only valid for diffusion across the field lines, and therefore we equate

$$D_\perp = D_{\text{mag field}} \qquad D_\parallel = D_{\text{no mag field}}$$

This last expression is true because the path traveled by the particle parallel to the field is unaffected by the magnetic field. The resultant effects of the magnetic field on particle flux are, therefore, an inhibition of free diffusion *across* the magnetic field lines, but have essentially no effect on the free diffusion *along* the field lines.

If the density of charged particles increases, the process of free diffusion begins to be subverted by another type of diffusion process that exists solely for charged particles. This process is caused by a space-charge buildup when the diffusion rates of ions and electrons are unequal. The space charge tends to slow down the faster-diffusing particles and speed up the slower particles. This process, called *ambipolar diffusion*, will be treated in the following passage.

We define the *mobility* of a particle as the ratio of its velocity to an applied electric field; that is,

$$\mu = \frac{v}{E} \tag{2.51}$$

We will first consider an electron. The force $q_e E$ that is exerted by the field on the electron can be set equal to the rate of momentum change between collisions *if* it is assumed that the charged particle loses *all* of its ordered momentum (momentum obtained from the field) in making a collision. In a sense, this assumption was tacitly made in the definition of the momentum transfer collision frequency ν_m.

The result is then

$$q_e E = (m_e v)\left(\frac{v}{l}\right) = \frac{m_e v^2}{l} \tag{2.52}$$

Equation (2.52) assumes that a momentum of mv is gained by the electron between collisions and lost *immediately* and completely after each collision.

We can then solve for v/E as follows, from (2.52):

$$\mu_e = \frac{v_e}{E} = \frac{q_e l_e}{m_e v_e} \tag{2.53}$$

A similar development yields a mobility expression for the ions defined as

$$\mu_i = \frac{v_i}{E} = \frac{q_i l_i}{m_i v_i} \tag{2.54}$$

In general, $\mu_i \neq \mu_e$.

The net particle current resulting from ambipolar diffusion must be the *sum* of the free diffusion and the mobility particle currents. Separate particle currents may be defined for ions and electrons. The total ion particle current is

$$\Gamma_i = -D_i \frac{dn_i}{dx} + \mu_i E n_i \tag{2.55}$$

since the mobility particle current is just the density of particles times the velocity. Now, for the electrons, the total particle current is

$$\Gamma_e = -D_e \frac{dn_e}{dx} - \mu_e E n_e \tag{2.56}$$

The electronic charge is of negative sign (and we assume that the mobility will be a positive number), and therefore the sign of the mobility term in (2.56) must be reversed from that in (2.55).

Ambipolar diffusion will result in the particle *drift velocities* being equal. The *particle currents* will not be equal, in general, unless ions and electrons carry the same magnitude of charge. However, since the net particle current is essentially nv_{net}, we may divide Eqs. (2.55) and (2.56) by their respective densities and then set them equal as shown in Eq. (2.57):

$$v = -\frac{D_i}{n_i}\frac{dn_i}{dx} + \mu_i E = -\frac{D_e}{n_e}\frac{dn_e}{dx} - \mu_e E \tag{2.57}$$

If the densities of ions and electrons are now approximately equal (which is the tendency in a plasma) and we assume the density gradients of both species are approximately equal, we can rewrite (2.57) as

$$v = -\frac{D_i}{n}\frac{dn}{dx} + \mu_i E = -\frac{D_e}{n}\frac{dn}{dx} - \mu_e E \tag{2.58}$$

We can eliminate E and solve for v, namely,

$$v = -\left(\frac{\mu_e D_i + \mu_i D_e}{\mu_i + \mu_e}\right)\frac{1}{n}\frac{dn}{dx} \tag{2.59}$$

The resultant particle current nv is then

$$\Gamma = -\left(\frac{\mu_e D_i + \mu_i D_e}{\mu_i + \mu_e}\right)\frac{dn}{dx} \tag{2.60}$$

This looks in form like the diffusion particle current obtained for free diffusion previously. We can define, therefore, an *ambipolar* diffusion coefficient which is the *same for both electrons and ions*, namely,

$$D_{\text{amb}} = \frac{\mu_e D_i + \mu_i D_e}{\mu_i + \mu_e} \tag{2.61}$$

The usual value of D_{amb} is nearly equal to D_i, because the mass of the ion is much larger than the mass of the electron. This implies that electrons tend to be slowed greatly in their paths while ions tend to be speeded up only slightly.

We can also define an ambipolar diffusion coefficient in the presence of a magnetic field. In order to obtain this we note that the particle mobilities in a dc magnetic field will be adjusted by the factor $1/[1 + (l^2/r_g{}^2)]$ as in Eq. (2.50). This is obtained by exactly the same method as was the magnetic field diffusion previously, by using *arc length*, rather than linear distance (see Problem 5). So

$$\mu_\perp = \frac{\mu_\parallel}{1 + (l^2/r_g{}^2)}$$

for each species of particle. Again, the symbol $\perp$ means the direction perpendicular to the magnetic field lines and the symbol $\parallel$ refers to the parallel-to-the-field direction. Recall that the radius of gyration of a particle is proportional to its mass. Therefore, the perpendicular reduction factor for the mobilities and free diffusion coefficients in this case is not as great for ions as for electrons. Now in free diffusion, ions will tend to diffuse more rapidly across the field lines than electrons. But, for ambipolar diffusion, since

$$D_{i\perp} > D_{e\perp} \qquad \text{and} \qquad \mu_{i\perp} > \mu_{e\perp}$$

we have

$$D_{amb\perp} \approx 2D_{e\perp}$$

So, we have also slowed down the ion diffusion across the field lines. Considering the parallel direction, we find that

$$D_{amb\parallel} \approx 2D_{i\parallel}$$

which is not as much of a change as in the perpendicular case. Ambipolar diffusion demonstrates a very important attribute of plasmas; that is, they tend to be electrically neutral. If a perturbation of one species of charge develops, the space-charge fields often become strong enough to drag the other species along and thus tend to preserve charge neutrality.

When we consider diffusion in a magnetic field of finite dimensions (that is, the field lines terminate on a conducting wall), it may be possible for electrons to diffuse more rapidly along the path shown in Fig. 2.10 than to diffuse across the field lines.

The conducting end wall permits a possible path for the electrons in going from A to B to be along the field lines, where their rate of diffusion is faster, and across the conducting plate, which acts as a "short circuit." This may

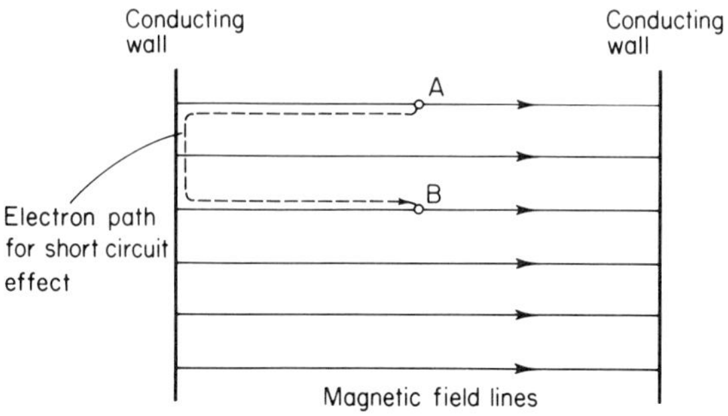

Fig. 2.10. The "short-circuit" diffusion process.

result in an effective experimentally measured diffusion coefficient that is anomalously large. It should be noted here that travel of electrons from the plasma along the plate and back again is a complicated effect and may not occur spontaneously. Such short-circuit effects often tend to "damp out" plasma instabilities by quickly eliminating localized density fluctuations before they grow cataclysmically. Often the placement of a conducting end plate in a magnetically confined plasma device will result in this type of "line-tying" stabilization of an otherwise unstable plasma.

We can also consider the time-dependent diffusion effects. Figure 2.11 shows an incremental volume element of plasma. If we suppose that particles

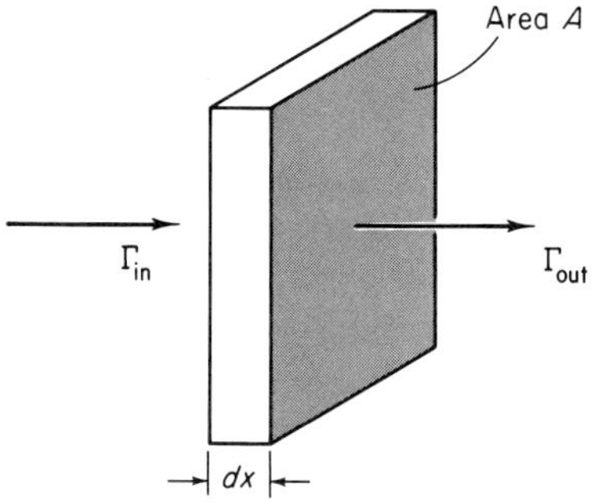

Fig. 2.11. Plasma slab.

may be stored or removed from this incremental volume, then the net change of particle storage must be equal to the difference in flow of particles through the element. This net flow (in an element of time dt), assuming diffusion, is

$$A\,dt\left[-D\left.\frac{dn}{dx}\right|_{x=x}+D\left.\frac{dn}{dx}\right|_{x=x+dx}\right]=dn\,A\,dx \tag{2.62}$$

Dividing both sides of Eq. (2.62) by $A\,dx\,dt$, we arrive at

$$\frac{D\left[\left.\frac{dn}{dx}\right|_{x+dx} - \left.\frac{dn}{dx}\right|_{x}\right]}{dx} = \frac{dn}{dt} \tag{2.63}$$

Passing to the limit as $dx \to 0$ gives

$$D\frac{\partial^2 n}{\partial x^2} = \frac{\partial n}{\partial t} \tag{2.64}$$

This is the *time-dependent* diffusion equation and it can be utilized in the solution of boundary value problems. In many cases the standard method of separation of variables may be utilized for this purpose.

In three dimensions, Eq. (2.64) becomes

$$D\,\nabla_r{}^2 n = \frac{\partial n}{\partial t} \tag{2.65}$$

The separation constants in the solutions of Eqs. (2.64) and (2.65) will be characteristic lengths or times, often called *diffusion lengths* and *diffusion times*, respectively.

When we have considered diffusion in a magnetic field, we have noted that the electrons diffuse at a different rate than the ions, although both rates are inhibited from their zero magnetic field rates. The resultant space-charge fields can then produce a magnetic field ambipolar diffusion, which tends to inhibit the faster-moving species.

However, it has been found that in measuring the diffusion rates of plasma across the magnetic field lines, the rate is often faster than that predicted by the classical theory. The reason for this is thought to be a turbulence or instability developed by the unequal rates of diffusion across the magnetic field. An empirical formula developed by Bohm gives the perpendicular diffusion rate as

$$D_{\text{Bohm}} = \frac{10^{-4} k T}{16 q_e B} \tag{2.66}$$

where k is Boltzmann's constant and T is the plasma temperature. Very often, this diffusion coefficient (and associated diffusion rate) is used for comparison with experimental measurements.

This concludes the elementary introduction to collisional phenomena. They will never be far away from all the material in the following chapters, for the interactions between particles are perhaps one of the most complicated yet necessary parts of plasma behavior. If there were no collisions, there would be less ionization, less turbulent mixing and heating of a plasma, and less damping of waves that may be propagating through the plasma. On the other hand, there would also be less deionization, less loss of energy, and less attenuation of the waves.

It is often possible to design experiments so that the plasma appears to be either "collision dominated" or "collisionless" *if* the collision cross sections and particle densities are known, since they determine the collision times.

We now proceed to a detailed discussion of how charged particles move when they are *not* making a collision. We will consider the plasma particles as single "isolated" charged particles with the only forces acting on them being those resulting from "nonstochastic" continuous fields, such as externally applied electric, magnetic, and gravitational fields.

SUGGESTED READING

Artsimovich, L. A., "Controlled Thermonuclear Reaction," Oliver & Boyd, Edinburgh and London, 1964. This book presents a good exposition of Bohm diffusion, as well as a broad outline of thermonuclear research methods.

Barnett, C. F., Gauster, W. B., and Ray, J. A., Oak Ridge Natl. Lab. Rep. No. ORNL-3113 (revised), 1964. A good collection of cross-section measurements is given.

Bohm, D., *in* "The Characteristics of Electrical Discharges in Magnetic Fields" (A. Guthrie and R. K. Wakerling, eds.), McGraw-Hill, New York, 1949.

Brown, S. C., "Basic Data of Plasma Physics," Wiley, New York, 1959. This is a good collection of experimental data on cross sections.

Chapman, S., and Cowling, T. G., "The Mathematical Theory of Non-Uniform Gases," Cambridge Univ. Press, London and New York, 1960. An excellent theoretical treatment of collisional processes is presented.

Kennard, E. H., "Kinetic Theory of Gases," McGraw-Hill, New York, 1938.

MacDaniel, E. W., "Collision Phenomena in Ionized Gases," Wiley, New York, 1964. This work is an even more complete collection of collisional data.

Mott, N. F., and Massey, H. S. W., "Theory of Atomic Collisions," 3rd ed., Oxford Univ. Press, London and New York, 1965.

Rose, D. J., and Clark, M., "Plasmas and Controlled Fusion," M.I.T. Press, Cambridge, Massachusetts, 1961.

Spitzer, L., Jr., "Physics of Fully Ionized Gases," 2nd ed., Wiley (Interscience), New York, 1962.

Problems

1 Locate, in the Suggested Reading for this chapter or elsewhere, some cross-section data. An example is sketched in the figure on p. 37. In particular, examine electron–electron, electron–ion, and electron–neutral collisions. Which cross sections seem to have the highest magnitudes? Why?

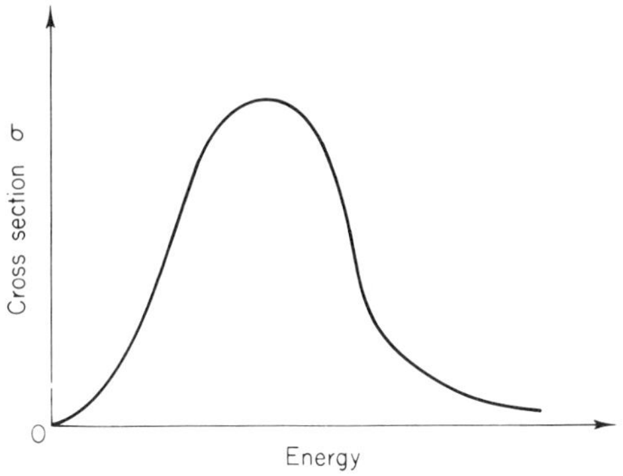

Explain why the cross section shown above is low for *both* high and low energies. Is this the case for the cross sections found in the references? What are some typical values for σ?

2 If Eq. (2.33) is evaluated at the appropriate boundary conditions, obtain the expressions for the coefficients A and δ.

3 What is the particle current in the radial direction as a function of radius if a beam of electrons of uniform velocity v is moving in the z direction? The radial density profile is $AJ_0(\alpha r)$, where J_0 is the Bessel function of the first kind of order zero and α is obtained by noting that the beam has zero density at $r = R$. What should be the end result of this process if no means were available to replace the electrons diffused out of the beam, if the electrons could only move as far as a wall of radius R?

4 What is the particle current if a dc magnetic field B is directed along the z axis in Problem 3?

5 Show that the mobility of ions and electrons in the direction perpendicular to a dc magnetic field is reduced by the factor $1/[1 + (l^2/r_g{}^2)]$ from the mobility in the absence of the dc magnetic field.

6 Solve the time-dependent diffusion equation

$$D\,\nabla_r{}^2 n = \frac{\partial n}{\partial t}$$

for $n(\mathbf{r}, t)$, assuming the following initial conditions.

(a) No axial z variation.

(b) The radial density profile is $n_0 = NJ_0(\alpha r)$.

(c) The density goes to zero at $r = a$.

7 Solve Problem 6 if the axial and radial density variations at $t = 0$ are

$$n_0 = NJ_0(\alpha r)\cos\left(\frac{\pi z}{2L}\right)$$

The density goes to zero at $r = a$ and at $z = \pm L$.

8 Compute the free diffusion coefficient for electrons of 10 eV energy in a helium plasma of low ionization. The ambient gas pressure is 10^{-4} Torr. Hint: To do this, the collision cross section, the velocity of the electrons, and the density of target particles must be known. (Since we have low ionization, the most likely collisions are between electrons and neutrals.) Use appropriate reference material.

9 Suppose that an electron is acted upon by a viscous force cv where v is the velocity of the electron, and is also acted upon by an alternating electric field $E_0 \cos(\omega t)$ pointing in the x direction. Find the steady state velocity of the electron and the ratio of the magnitude of this velocity to that of the electric field (mobility).

10 The atmosphere is assumed to have a density that varies as

$$n = n_0 \exp(-r/\gamma)$$

where r is the distance from the earth's *surface*. For a collision cross section σ, find the distance r from where a particle may completely escape the atmosphere with a probability of $\frac{1}{2}$ if it is moving radially outward at high speed.

11 Assume that the free diffusion coefficients of ions and electrons are known (normally, electrons diffuse faster than ions).

(a) What would be an approximate ratio for $D_{\text{electrons}}/D_{\text{ions}}$?

(b) Then compute the diffusion coefficients across magnetic field lines. What is the ratio of $D_{\perp\,\text{electrons}}/D_{\perp\,\text{ions}}$? Which diffuses faster?

(c) Do parts (a) and (b) for the ambipolar diffusion coefficients.

12 Examine Fig. 2.3. If $\mathbf{g}_{21}$ and $\hat{k}$ are known, in both magnitude and direction, find $\mathbf{g}'_{21}$.

13 Prove that, by using Eq. (2.22), we may determine the motion of particles 1 and 2 by knowing the reduced mass and vector $\mathbf{r}_{12}$ as a function of time.

3 The Motion of Isolated Charged Particles

A. THE MOTION OF A CHARGED PARTICLE IN A MAGNETIC FIELD

We shall now begin the study of the motion of isolated charged particles in electric, magnetic, and gravitational fields. In many situations (low plasma density and low background gas pressure) the collision frequency is low enough so that over the time of interest for a particular problem, the single-particle approach has significant application.

The fundamental equation of motion of a single charged particle in an electric, magnetic, and gravitational field is

$$\mathbf{F} = m\frac{d\mathbf{v}}{dt} = q(\mathbf{E} + \mathbf{v} \times \mathbf{B}) + m\mathbf{g} \tag{3.1}$$

In Eq. (3.1), **E** is the electric field seen at the particle, **B** is the magnetic field present, and **g** is the gravitational acceleration vector. Since **B** is always perpendicular to **v**, the magnetic field can do no work on the charge. Hence, the energy of a charged particle is unaffected by the magnetic field alone. However, if the magnetic field should change rapidly in time, an induced electric field may produce a change in energy of the particle. The conditions under which any or all of the three terms in Eq. (3.1) must be considered depend upon the magnitudes of the fields and the time of interaction. We shall be interested in steady state solutions to Eq. (3.1), and we will include the effects of time-varying fields of constant amplitude.

The solution to Eq. (3.1) will yield much information. We shall be able to obtain the velocity as a function of an applied electric field (the mobility). This result is especially interesting when a dc magnetic field is included. For the first time, we will observe analytically that the mobility is a function of direction with respect to the magnetic field, showing that a plasma becomes *anisotropic* in a magnetic field.

Subsequently, we will generalize the single-particle results to a collection of particles, each of which may also be considered isolated. The results obtained are elegant and very useful for many applications.

If both gravity and electric field effects can be neglected, then Eq. (3.1) tells us that if there is a component of particle velocity in the direction of the magnetic field, the particle will move in a helical path. The motion perpendicular to the field lines will be a circle if the magnetic field is uniform. The diameter of the circle (radius of gyration) and the frequency of rotation (cyclotron frequency) may be determined by equating the centripetal rate of momentum change (centrifugal force) and the centripetal force in Eq. (3.1) to obtain the equilibrium condition on the particle motion. This relation is

$$\frac{mv_\perp^{\,2}}{r_g} = qv_\perp B = mr_g\omega_c^{\,2} \tag{3.2}$$

Equation (3.2) is the component of Eq. (3.1) in the r direction written in cylindrical coordinates; $v_\perp$ is the component of particle velocity in the direction perpendicular to the magnetic field lines, r_g is the radius of gyration, and ω_c is the cyclotron frequency. Then

$$r_g = \frac{mv_\perp}{qB} \quad \text{and} \quad \omega_c = \frac{qB}{m}$$

Note that r_g, which is shown in Fig. 2.9, is proportional to the mass and perpendicular component of particle velocity. The cyclotron frequency is proportional to the magnetic field strength and inversely proportional to the mass. This means, for example, that in a plasma composed of ions and electrons, in general

$$r_{g_i} > r_{g_e} \qquad \omega_{c_i} < \omega_{c_e}$$

The smaller the radius of gyration, the greater the centrifugal force on the particle. This means that for no force, the radius of gyration is infinite (straight-line motion). Note also that the direction of rotation of ions and electrons about the field lines is not the same. Figure 3.1 shows the different path directions for ions and electrons for a fixed direction of magnetic field. The motion in the direction parallel to the magnetic field line is unchanged and remains $v_\parallel$.

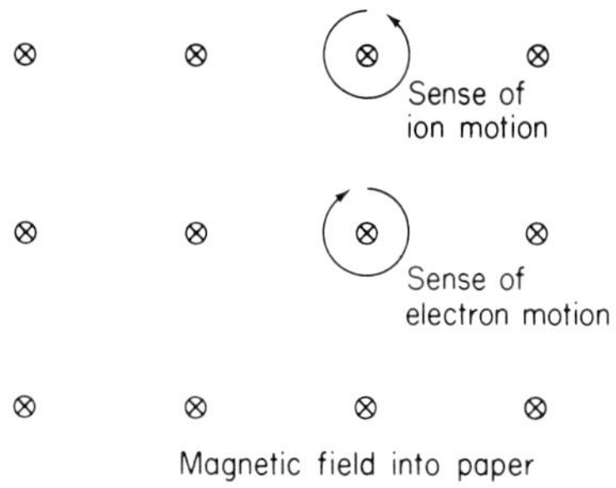

Fig. 3.1. Ion and electron paths in a uniform magnetic field. Note that the diameters of the orbits may be different.

B. CROSSED ELECTRIC AND MAGNETIC FIELDS

We now assume a large dc magnetic field and introduce perpendicular dc electric and gravitational fields as small variations on the magnetic field motion. These electric and gravitational fields will be assumed to be uniform in space.

Figure 3.2 shows the appropriate conditions for a dc electric field. At first glance it would appear that an orbiting charged particle could just move in the same direction as the electric field vector. However, this is not the case. If we remember that the radius of gyration of a particle is inversely proportional to the centrifugal force, as shown in Eq. (3.2), we can examine Fig. 3.3 and note the changes in the radius of gyration at the top and bottom of the orbit.

This figure is drawn for a positively charged particle, so that the electric field force tends to "cancel" some of the magnetic force at the top of the orbit but "aids" the force at the bottom of it. Therefore, the radius of

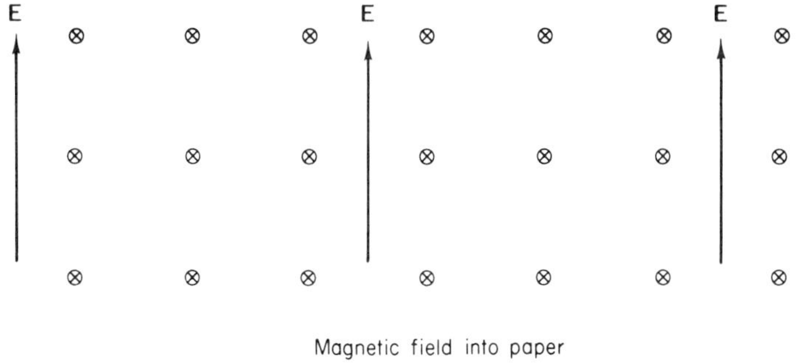

Fig. 3.2. Electric and magnetic field directions.

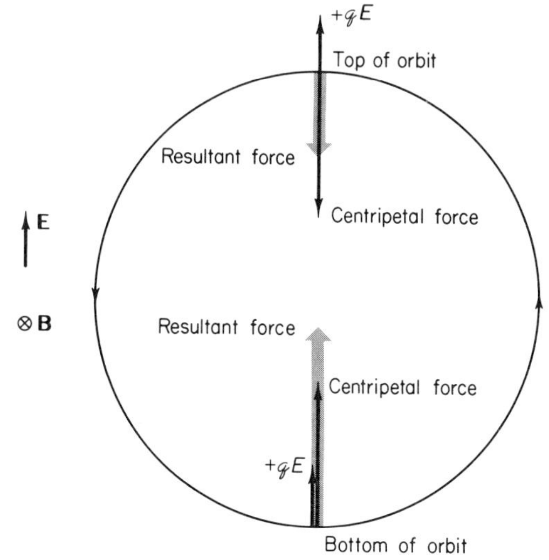

Fig. 3.3. Diagram of the forces experienced by a positively charged particle in crossed electric and magnetic fields.

gyration is larger at the top of the orbit and smaller at the bottom of it, since it is inversely proportional to the net force. A sketch of the path of the particle's motion under this condition is shown in Fig. 3.4. The motion is a particle *drift* to the left in this figure. Under the same conditions, a negatively charged particle will have a large radius of gyration at the *bottom* of its orbit and a small one at the top, since the electric field force is oppositely directed. Figure 3.5 shows the result of this type of particle motion under the same condition as that which produced Fig. 3.4.

Note that *both* species of particles drift to the left. The magnitudes of the drift speeds are the same. This can be shown by Eq. (3.1) if a steady state condition (zero force) and zero gravitation are included. Equation (3.1) will

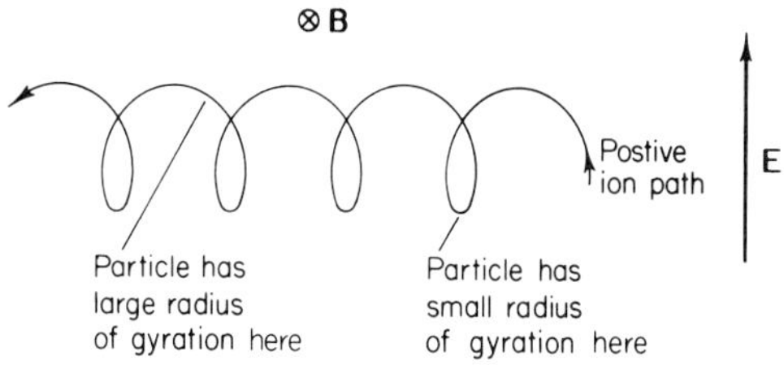

Fig. 3.4. Electric field drift of a positively charged particle in a dc magnetic field.

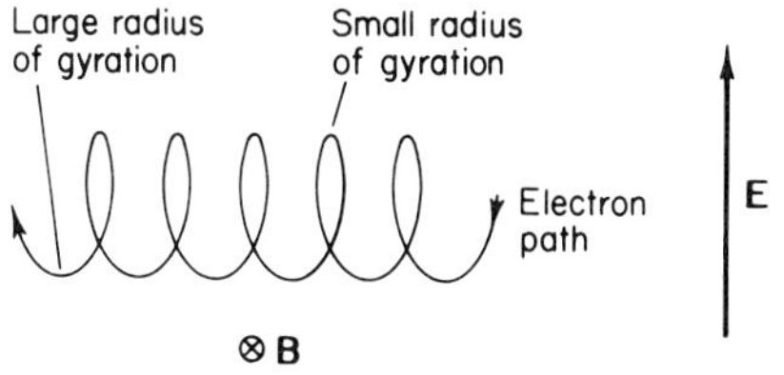

Fig. 3.5. Electric field drift of a negatively charged particle in a dc magnetic field.

then have two components, assuming **B** is in the z direction. They are

$$\frac{\partial v_x}{\partial t} = q \frac{v_y B_z}{m} \tag{3.3}$$

and

$$\frac{\partial v_y}{\partial t} = \frac{q}{m} E_y - q \frac{v_x B_z}{m} \tag{3.4}$$

if **E** has only a y component. The steady state solution ($\partial/\partial t = 0$) is then $v_x = E_y/B_z$ and $v_y = 0$. Therefore the drift velocity is independent of the magnitude and the sign of the charge and the mass. Equation (3.5)

$$\mathbf{v}_d = \frac{\mathbf{E} \times \mathbf{B}}{B^2} \tag{3.5}$$

gives the vector formulation for the drift velocity.

Note that this last equation tells us that as $\mathbf{B} \to 0$ the drift velocity approaches infinity. It is to be expected that the charged particles would move *along* the electric field lines when $\mathbf{B} = 0$. This matter can be resolved by noting that the radius of gyration of the charged particles also goes to infinity as $\mathbf{B} \to 0$. Hence, a particle, moving with a velocity in the electric field direction (accelerated from zero by the electric field) which is perpendicular to the plane of **B** will continue moving in that direction if $\mathbf{B} = 0$.

C. GRAVITATIONAL FIELDS

We now consider gravitational effects. In this case, gravity will be assumed to be acting, as shown in Fig. 3.6, in a downward direction. The force of gravity is independent of the sign of the charge. Figure 3.6 shows the results of a gravitational drift for both electrons and positively charged ions, assuming the gravitational force is small compared to the $\mathbf{v} \times \mathbf{B}$ force. The direction of the drift may be determined by constructing a figure similar to Fig. 3.3 for the gravitational and magnetic forces.

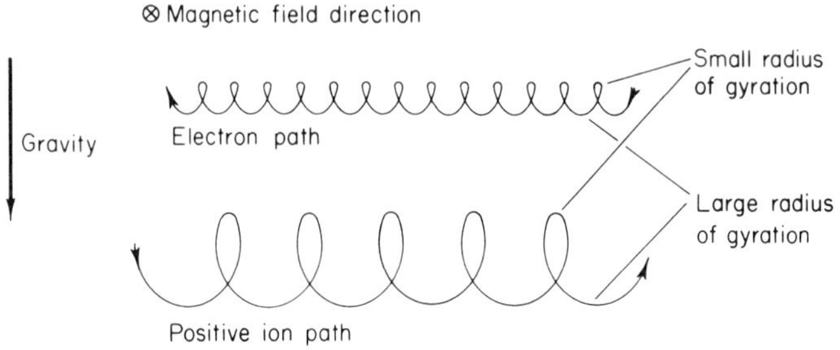

Fig. 3.6. Gravitational drifts of positively and negatively charged particles in a dc magnetic field.

Note that the gravitational drift produces a *charge separation*. The drift direction is still perpendicular to both the gravitational and magnetic fields. The velocity may be determined in the same way as for Eq. (3.4) and will result in the following vector expression.

$$\mathbf{v}_{d_{\text{grav}}} = \frac{m}{q}\frac{\mathbf{g} \times \mathbf{B}}{B^2} \tag{3.6}$$

This velocity depends upon the sign of the charge, and therefore results in the charge separation mentioned above.

The next type of drift to be considered is one due to the nonuniformity of the magnetic field, a result that must be present in all real experiments and devices.

D. MAGNETIC FIELD GRADIENTS—MAGNETIC MIRRORS AND CUSPS

The radius of gyration (mv/qB) is inversely proportional to the strength of the magnetic field; that is, the weaker the field, the larger the value of r_g. Figure 3.7 shows a model in which the gradient of **B** (its direction of variation,

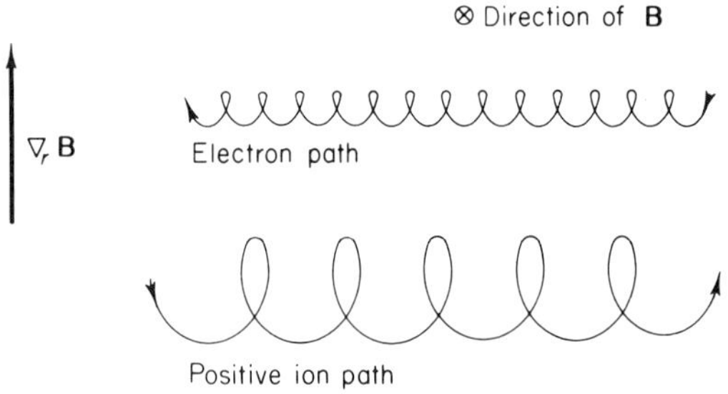

Fig. 3.7. Particle drifts in a nonuniform magnetic field.

not the direction of the field) goes upward; **B** increases as one moves upward. As the particles move to the top of their orbits, their radii of gyration decrease; as they move to the bottom, their radii increase. Therefore a drift occurs, as shown in Fig. 3.7. Note here that the direction of the particle drifts also produces a net charge separation. This is often a problem in many types of plasma devices, since a magnetic field produced by real magnets, be they permanent magnets or electromagnets, is always at least slightly nonuniform.

The orbits of the particles have been examined as yet only on a plane perpendicular to the magnetic field lines. The particle behavior in the direction parallel to the field lines can be immediately ascertained to be that which would exist without the presence of the magnetic field for both the electric and gravitational fields by examining the component of Eq. (3.1) in the magnetic field direction. A sketch of the general path of an electron for this class of fields is shown in three dimensions in Fig. 3.8. Here the particle drifts

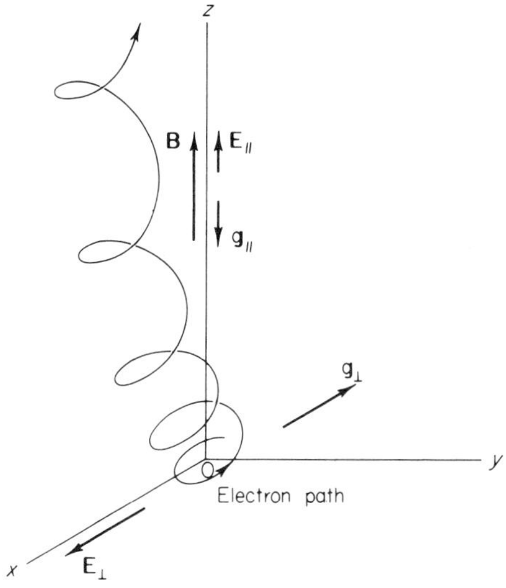

Fig. 3.8. Three-dimensional electron path.

in a direction perpendicular to **E** and **B**, but also has a component of motion along the *z* axis. Figure 3.8 also shows a gravitational field vector. If this vector is broken up into two components $\mathbf{g}_\perp$ and $\mathbf{g}_\parallel$, the charged particle will exhibit a $\mathbf{g} \times \mathbf{B}$ drift due to $\mathbf{g}_\perp$. $\mathbf{g}_\parallel$ and $\mathbf{E}_\parallel$ will accelerate the particle in the direction parallel to **B** as though the magnetic field were not present.

The conditions of the particle motion when the magnetic field has a gradient *in* the direction of the field is not quite so ascertainable. However, by means of one major assumption, we can observe a very interesting behavior of these charged particles when they move in a magnetic field of this type. The assumption to be made is that the magnetic moment of the orbiting particle is constant. The conditions under which this is so will be given in Chapter 5. The magnetic moment is the product of the current developed by the motion of the particle times the area swept out by its orbit. The magnetic moment to be considered here is not that of the "spin" of the particle, but only the result of a charge flowing around a circular path. The magnetic moment is, then, explicitly

$$\mu_m = IA = \left(\frac{q\omega_c}{2\pi}\right)(\pi r_g{}^2) \tag{3.7}$$

The current is $(q\omega_c/2\pi)$. If we now substitute the results of Eq. (3.2), we observe that μ_m can be evaluated as

$$\mu_m = \frac{1}{2}\frac{mv_\perp{}^2}{B} \tag{3.8}$$

This result tells us that if B increases, $v_\perp$ must also increase to keep the magnetic moment constant. However, we now have a problem. Since the magnetic force from a constant magnetic field on a charged particle is always perpendicular to the direction of motion of the particle, then it can do no work and therefore cannot change the energy of the particle. If $v_\perp$ should change, there must be an increase in kinetic energy of the particle in the perpendicular direction. The source of this energy can only be from the kinetic energy in the parallel direction. Hence, as the charged particle spirals into a region of higher magnetic field, the parallel component of kinetic energy decreases. However, since the parallel component of kinetic energy determines the fact that the particle moves in the direction of increasing magnetic field, there may be a point where the particle cannot move any farther, and, in fact, can only begin moving in the reverse direction.

At first glance, it would seem that the particle could rest indefinitely at its reflection point, since there seems to be no force to push it back toward the region of weaker field.

However, note that a gradient of the magnetic field exists. It can be seen that a force due to $\nabla_r B$ will *accelerate* the particle back into the lower field region (see Problem 4). The total kinetic energy

$$\tfrac{1}{2}mv_\parallel{}^2 + \tfrac{1}{2}mv_\perp{}^2$$

is conserved if there are no other forces acting on the particle, and the values of $v_\perp$ and $v_\parallel$ will then be continuously adjusted to correspond to the constancy of the magnetic moment. If the magnetic field increases indefinitely, all particles will eventually be reflected ($\tfrac{1}{2}mv_\parallel{}^2 = 0$). However, for real systems, the magnetic field will not increase indefinitely, but will reach a maximum value and then, most likely, decrease. The result is, then, that if the particle has enough parallel kinetic energy to carry it through the maximum value of the field, it will then not be reflected, but continue in the same direction. A device such as that just described, in which some but not all particles are reflected, is called a *magnetic mirror*. We now turn to a further quantitative development of such a system.

Consider a magnetic field configuration described as shown in Fig. 3.9.

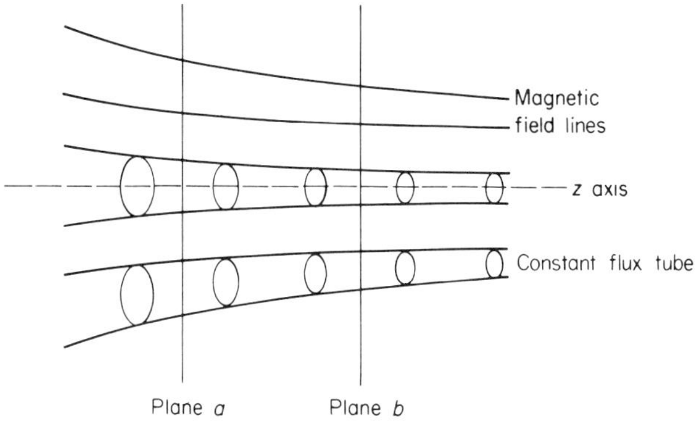

Fig. 3.9. Magnetic mirror.

Since the magnetic moment is assumed constant and, as shown in Eq. (3.7), is proportional to the square of the radius of gyration r_g and to the magnetic field ($\omega_c = qB/m$), we can say that

$$r_g{}^2 B = \text{constant} \tag{3.9}$$

Now, $r_g{}^2 B$ represents the magnetic flux through the orbit circle if $\mathbf{B}$ is essentially constant over the circle diameter. Therefore, the conclusion is that the particle orbits around a tube of constant flux. Determination of the configuration of such tubes of constant flux would then give the general paths of the orbiting particles.

We next turn to the problem of determining the point of reflection of a given particle. Since total kinetic energy is conserved, we may note that at two points, a and b, along a magnetic mirror field like that in Fig. 3.9,

$$W_{\parallel a} + W_{\perp a} = W_{\parallel b} + W_{\perp b} \tag{3.10}$$

where W is the kinetic energy, and the symbols $\parallel$ and $\perp$ refer to the coordinate directions parallel to and perpendicular to $\mathbf{B}$, respectively. But, due to the constancy of μ_m,

$$\frac{W_{\perp a}}{B_a} = \frac{W_{\perp b}}{B_b} \tag{3.11}$$

Therefore, we may solve for $W_{\parallel b}$ as a function of the other energies and the values of the magnetic field. It is

$$W_{\parallel b} = W_{\parallel a} - W_{\perp a} \left(\frac{B_b}{B_a} - 1\right) \tag{3.12}$$

If $W_{\parallel b}$ becomes zero, then the particle is reflected at point b back into the weaker field region. This reflection condition can be evaluated in terms of the energies at point a to be

$$\frac{W_{\parallel a}}{W_{\perp a}} = \frac{B_b}{B_a} - 1 \tag{3.13}$$

If B_b is the highest value of field, and B_a some reference value (perhaps that in the midplane of some magnetic coil device), then Eq. (3.13) gives the reflection condition as a ratio of parallel to perpendicular kinetic energies at the midplane. Those particles whose energy ratio is greater than this value will escape from the mirror. Geometrically, this condition may be represented as shown in Fig. 3.10. Particles whose velocity vectors at the midplane fall in

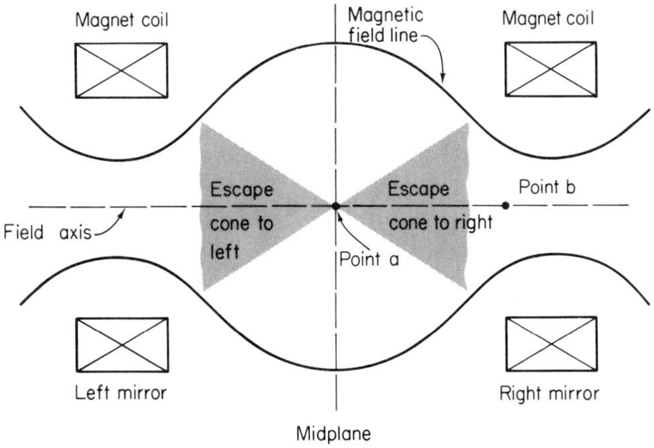

Fig. 3.10. Magnetic bottle confinement device ("mirror machine").

the shaded region will escape from the mirror system. Figure 3.10 depicts a magnetic "bottle" configuration in which mirror fields are placed on both ends of a cylindrically symmetric device. The field lines shown here are not the magnitude of the field, but a representation of a single field line. The intensity of the field is represented by the number of lines per unit area. The field is maximum in amplitude at the "throats" of each mirror, located directly under each coil, and reaches a local minimum at the midplane. Plasma devices with this geometry are utilized to provide a measure of inhibition of axial particle loss. However, these field configurations are plagued by a difficulty that occurs when the plasma density and/or energy increases: instabilities. This means that under certain conditions the plasma may suddenly fly apart or decay away more rapidly than would be predicted by the classical diffusion and recombination theories.

The ratio of the maximum to the minimum value of magnetic field along the axis in this type of device is called the *mirror ratio*. It is defined as

$$R = \frac{B_{\max}}{B_{\min}} \tag{3.14}$$

For confinement, if point a is at the midplane, we must have

$$\frac{W_{\parallel a}}{W_{\perp a}} \leq R - 1 \tag{3.15}$$

as can be seen by using the reflection condition (3.13). But since

$$W_{\parallel a} = W - W_{\perp a}$$

Eq. (3.14) may be rewritten as

$$\frac{W}{W_{\perp a}} \leq R \tag{3.16}$$

We can also define an angle θ such that

$$\cot \theta = \frac{v_{\parallel a}}{v_{\perp a}} \tag{3.17}$$

Now, therefore

$$\frac{W_{\parallel a}}{W_{\perp a}} = \left(\frac{v_{\parallel a}}{v_{\perp a}}\right)^2 = \cot^2 \theta \tag{3.18}$$

We now have the angle θ in terms of the energy ratio. For angle θ greater than a value $\theta = \theta_{\max}$, particles are confined. For $\theta < \theta_{\max}$ the particles escape. $\theta_{\max}$ can then be determined as

$$\cot^2 \theta_{\max} = R - 1$$

$\theta_{\max}$ is the boundary of the *loss cones* of the system, which are also shown in Fig. 3.10.

Another point to note here is that orbiting charged particles of either sign generate their own magnetic field, which is in *opposition* to the applied magnetic field. This can be noted by utilizing the "right-hand rule" from electromagnetism on the orbiting motion. The result of this opposing field is that a plasma tends to be diamagnetic. This means that the net magnetic field inside a plasma is less than what it would be if the plasma was not present. Later we shall see that measurement of the diamagnetism may provide a measurement of the plasma temperature or density. In vector form, to show this diamagnetism the magnetic moment may be written as

$$\boldsymbol{\mu}_m = \tfrac{1}{2}q(\mathbf{r}_g \times \mathbf{v}_\perp)$$

One difficulty with a mirror machine as depicted in Fig. 3.10 is that initially we wished to use the dc magnetic field to inhibit perpendicular diffusion losses as well as to reflect particles from the ends of the device. The amount of inhibition of such diffusion loss is proportional to the strength of the magnetic field. If a plasma is placed inside the mirror region, even some of this slight diffusion across the field lines will result in plasma being transported to a region where the magnetic field is weaker. The result will be a smaller inhibiting effect and the diffusion process may be enhanced, causing even higher losses and/or instabilities.

A possible magnetic field configuration to improve upon this point is shown in Fig. 3.11. This configuration is essentially the mirror coil geometry

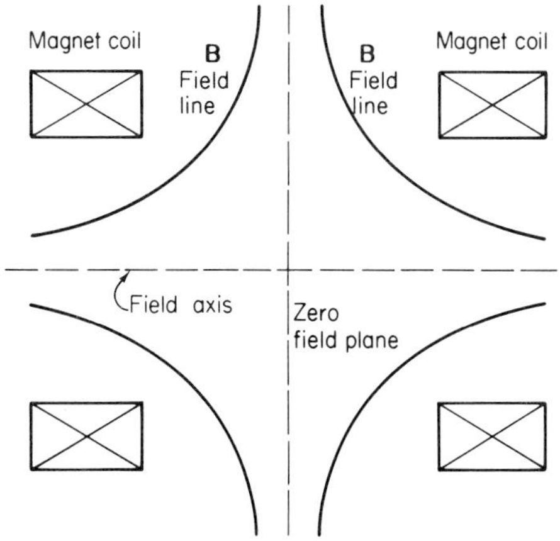

Fig. 3.11. Cusped magnetic fields.

of Fig. 3.10 with the current in one of the coils reversed. It is as though we had arranged to have two North poles of a magnet face each other, rather than a North–South confrontation as in the mirror field. The result is a field "zero" in the midplane of the device. The magnetic field increases everywhere out from this plane and therefore the inhibitory effects of the magnetic field should increase in all directions from the center, except along the zero field plane itself.

There are two difficulties with this scheme. The first is that there are some paths where the particles can still escape (along the midplane-outward). Second, for a zero value of a magnetic field, it is obvious that the magnetic moment can no longer be considered constant. The result would be that the reflection properties of the mirror fields in the throats of each coil would not be as effective.

A scheme that permits a nonzero but minimum magnetic field (a magnetic well) can be made by combining a mirror and a cusp field. This Ioffe bar system is shown in Fig. 3.12.

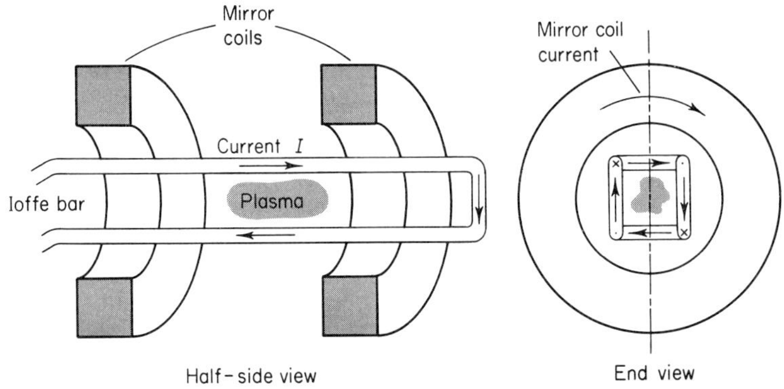

Fig. 3.12. Combined mirror and cusp fields (Ioffe system).

The field is produced by a set of mirror coils and a group of "bars" that conduct high currents. The bars produce a cusped field geometry which is superimposed on the mirror field. The resultant field configuration (minimum B) has a nonzero minimum, which will result in improved confinement characteristics.

In addition, systems such as that shown in Fig. 3.11 may be "stuffed" with a conductor along the axis of the device. Current flowing through this bar produces an azimuthal magnetic field around the bar and thus removes the field zero.

The Ioffe bar containment system may also be made in a toroidal configuration. The bars are bent around to form closed hoops. Such *toroidal multipoles* have excellent confinement characteristics. An even better confinement characteristic can be obtained if the hoops producing the multipole fields are made superconducting. They can then be levitated to eliminate plasma losses caused by collisions with the hoop supports.

E. THE EFFECTS OF AC ELECTRIC FIELDS

The foregoing aspects of particle motion do not include the effects of ac electric fields. Such fields are extremely commonplace in plasmas. Whenever electromagnetic radiation interacts with a plasma, plasma waves or instabilities appear, and whenever the effects of neighboring charges must be considered, an ac or a fluctuating electric field is often the result.

We shall assume that the dc magnetic field is sufficiently strong to outweigh the effects of any ac magnetic field associated with the fluctuating electric field. If the ac magnetic field is to be included, however, it may be obtained in terms of the electric field from Maxwell's equations.

We will now investigate the behavior of charged particles in the presence of a dc magnetic field and an ac electric field. Figure 3.13 shows the coordinate geometry used for this problem.

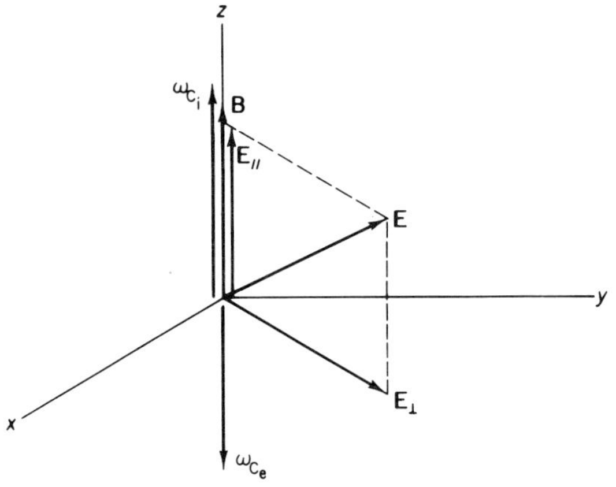

Fig. 3.13. Coordinate system for charged particle motion in an ac electric field.

Let us consider that the charged particle is an electron. We shall break up the electric field into two components, $\mathbf{E}_{\parallel}$ and $\mathbf{E}_{\perp}$. Their directions are related to the dc magnetic field direction that is shown here as being parallel to the z axis. The dc magnetic field is assumed to be so large that the induced ac magnetic fields, from the ac electric field, can be neglected. The nonrelativistic equation of motion of the electron is then

$$m_e \frac{d\mathbf{v}}{dt} = q_e\mathbf{E}_{\parallel} + q_e\mathbf{E}_{\perp} + q_e(\mathbf{v} \times \mathbf{B}) \tag{3.19}$$

We can solve this equation directly for its complete solution, if initial conditions on velocity and electric field are known. Let us now assume, first, that $\mathbf{v}$, the velocity of the electron, may be broken up into two components, as shown in Eq. (3.20).

$$\mathbf{v} = \mathbf{v}_1 + \mathbf{v}_2 \tag{3.20}$$

We shall let $\mathbf{v}_2$ be the motion of the electron in the presence of the dc magnetic field alone. It must therefore be a solution to the following equation:

$$m_e \frac{d\mathbf{v}_2}{dt} = q_e(\mathbf{v}_2 \times \mathbf{B}) \tag{3.21}$$

$\mathbf{v}_2$ will be zero if there is no random thermal motion, since Eq. (3.21) provides for no mechanism to make $\mathbf{v}_2$ other than zero, except for initial conditions. We may express $\mathbf{v}_2$ in component directions as

$$\mathbf{v}_2 = \mathbf{v}_{2\parallel} - (\boldsymbol{\omega}_c \times \mathbf{r}_g) \tag{3.22}$$

$\mathbf{r}_g$ is a vector pointing to the particle from the center of the orbit whose magnitude is the radius of gyration. A complete discussion of this solution was given earlier in the chapter.

We shall define $\boldsymbol{\omega}_c$ for the electrons and the ions to be $q\mathbf{B}/m$. The sign of the charge is carried with $\boldsymbol{\omega}_c$ by q. For ions, $\boldsymbol{\omega}_c$ points in the same direction as $\mathbf{B}$. $\mathbf{r}_g$ is, again, in the direction at any instant of the particle from the center of the orbit.

On the other hand, $\mathbf{v}_1$ is the velocity of the particle when the electric field *is* present. It is a *superposition* of this velocity upon the one described by Eq. (3.22) that gives the net particle motion. Therefore, it must be such that $\mathbf{v}_1$ is zero when $\mathbf{E}$ is zero. This separation, defined in Eq. (3.20), is known as the *guiding center* approximation.

We may restate this approximation. It is essentially the assumption that the complete motion of the particle may be made by superimposing the *gyration* at the cyclotron frequency $\mathbf{v}_2$ of the particle over the motion of the center of the orbit $\mathbf{v}_1$. We may compute the motion of the guiding center by use of the particle momentum equation (3.1) with the addition of forces dependent upon the nonuniformities in the electric and magnetic fields if they vary across the orbit diameter (see Problem 4).

Obviously, if the particle started up from zero velocity, the guiding center solution is *exactly* the complete solution, since the guiding center and the actual position of the particle coincide. As the particle gains energy and begins to spiral around the lines of force, the approximation may become less and less valid, depending upon nonuniformities in the electric and magnetic fields that may appear as the orbit diameter increases. The approximation is always valid if there is no random orbital motion, since the momentum equation may be solved exactly. Recall again that it is *also* always valid if there are no non-uniformities in the electric and magnetic fields, regardless of the particle's velocities.

The drift velocities previously defined are the drifts of the guiding centers in fields that do not change in time. For the following development, we compute the velocity of the guiding center when it is subjected to a time-changing electric field. It will be seen that "orbital" motion of the guiding center does develop, but again, this motion must be superimposed on the original "random" orbital motion before the complete path of the particle is known. As the frequency of the electric field becomes comparable to the cyclotron frequency, the guiding center approximation becomes less valid, although we may still use an expansion for the particle's motion like Eq. (3.20). In this case, $\mathbf{v}_1$ and $\mathbf{v}_2$ are velocities of the particle itself, but only $\mathbf{v}_1$ is dependent upon the ac electric field.

We now define a rectangular set of coordinates, shown in Fig. 3.14, which can be used to describe the motion of $\mathbf{v}_1$. They are superimposed over the ones of Fig. 3.13. The three coordinate vectors, for electrons, are $\mathbf{E}_\parallel$, $\mathbf{E}_\perp$, and $\mathbf{E}_\perp \times \boldsymbol{\omega}_{c_e}$. Note that they are all mutually perpendicular, but they are *not*

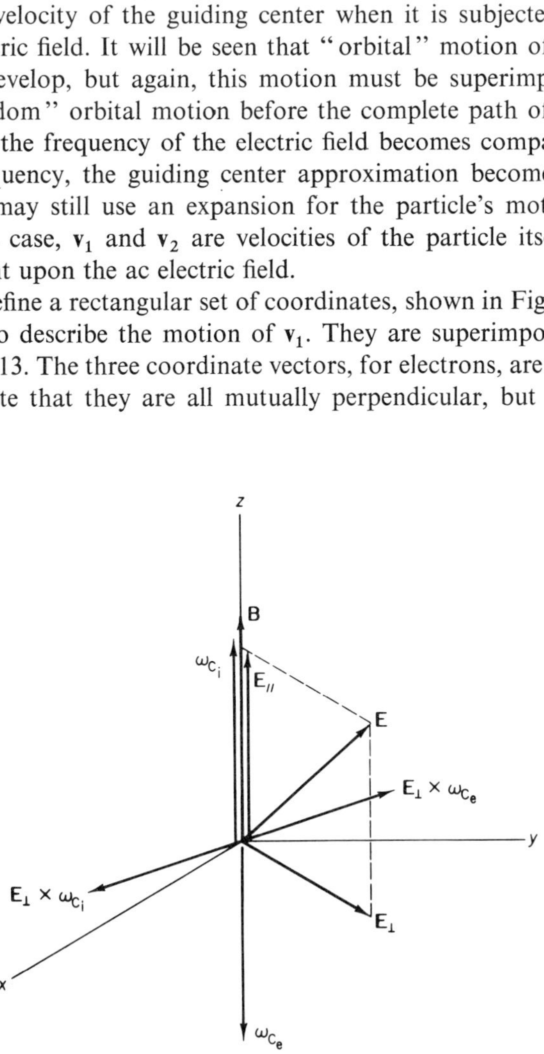

Fig. 3.14. Right-handed coordinates $\mathbf{E}_\parallel$, $\mathbf{E}_\perp$, and $\mathbf{E}_\perp \times \boldsymbol{\omega}_c$.

unit vectors. For electrons, these vectors form a right-handed set of coordinates, since

$$\frac{\mathbf{E}_{\parallel} \times \mathbf{E}_{\perp}}{|E_{\parallel}||E_{\perp}|} = \frac{\mathbf{E}_{\perp} \times \boldsymbol{\omega}_{c_e}}{|E_{\perp}|\,|\omega_{c_e}|} \tag{3.23}$$

For ions the vectors must be

$$\mathbf{E}_{\parallel}, \qquad \mathbf{E}_{\perp}, \qquad \text{and} \qquad -(\mathbf{E}_{\perp} \times \boldsymbol{\omega}_{c_i})$$

to ensure right-handedness.

We will now follow the electrons. We can determine the velocity $\mathbf{v}_1$, which must obey the equation

$$m_e \frac{d\mathbf{v}_1}{dt} = q_e(\mathbf{E} + \mathbf{v}_1 \times \mathbf{B}) \tag{3.24}$$

$\mathbf{v}_1$ is a linear function of $\mathbf{E}$ and has components in the three directions ($\mathbf{E}_{\parallel}$, $\mathbf{E}_{\perp}$, and $\mathbf{E}_{\perp} \times \boldsymbol{\omega}_{c_e}$), defined by the new set of coordinates. Therefore

$$\mathbf{v}_1 = a_0 \mathbf{E}_{\parallel} + a_1 \mathbf{E}_{\perp} + a_2(\mathbf{E}_{\perp} \times \boldsymbol{\omega}_{c_e}) \tag{3.25}$$

The coefficients a_0, a_1, and a_2 are found by solving for the steady state solution of Eq. (3.24) with an electric field present. The field may vary in time, but only sinusoidally. This causes little loss in generality, since a Fourier representation of any periodic signal may be obtained. Note that a_0, a_1, and a_2 may be functions of time.

To consider the most general case, we shall let $\mathbf{E}$ vary as the real part of $\mathbf{E}_0 e^{-j\omega t}$. Since $\mathbf{v}_1$ is a linear function of $\mathbf{E}$, it will also vary as $e^{-j\omega t}$. There are three cases of interest:

(1) $\omega = 0$;
(2) $\omega \neq 0$, $\omega \neq \omega_c$;
(3) $\omega = \omega_c$.

We have removed the subscript for the electrons, since the three cases above will also apply to the ions. Let us consider the steady state solutions to Eq. (3.24). We will not consider the general solutions to Eq. (3.24) because we are interested in long-time behavior. Although there is no damping in this equation, all real plasmas exhibit some damping, and the steady state solution should be that which remains after a long time has elapsed. The general solutions, if any, will appear as solutions to Eq. (3.21), which determines $\mathbf{v}_2$. The result then will be the *complete* solution to Eq. (3.19), $\mathbf{v}_1 + \mathbf{v}_2$, if we assume the single particle model holds.

Equation (3.24) is a vector equation and it will result in three component equations: $\mathbf{E}_{\parallel}$ components, $\mathbf{E}_{\perp}$ components, and $\mathbf{E}_{\perp} \times \boldsymbol{\omega}_{c_e}$ components.

We now write Eq. (3.24) with (3.25) substituted in after dividing through by m_e. This is

$$
\begin{aligned}
&e^{-j\omega t}[\dot{a}_0 \mathbf{E}_{\parallel 0} - j\omega a_0 \mathbf{E}_{\parallel 0} + \dot{a}_1 \mathbf{E}_{\perp 0} - j\omega a_1 \mathbf{E}_{\perp 0} \\
&\quad + \dot{a}_2(\mathbf{E}_{\perp 0} \times \boldsymbol{\omega}_{c_e}) - j\omega a_2(\mathbf{E}_{\perp 0} \times \boldsymbol{\omega}_{c_e})] \\
&= e^{-j\omega t}\left[\frac{q_e}{m_e}\mathbf{E}_{\parallel 0} + \frac{q_e}{m_e}\mathbf{E}_{\perp 0} + \frac{q_e}{m_e}\left(a_0(\mathbf{E}_{\parallel 0} \times \mathbf{B})\right.\right. \\
&\quad \left.\left. + a_1(\mathbf{E}_{\perp 0} \times \mathbf{B}) + a_2(\mathbf{E}_{\perp 0} \times \boldsymbol{\omega}_{c_e}) \times \mathbf{B}\right)\right]
\end{aligned} \tag{3.26}
$$

Note that $(q_e/m_e)\mathbf{B} = \boldsymbol{\omega}_{c_e}$, since the sign of the charge is carried by q_e, so the right-hand side of (3.26) becomes

$$
\begin{aligned}
&e^{-j\omega t}\left[\frac{q_e}{m_e}\mathbf{E}_{\parallel 0} + \frac{q_e}{m_e}\mathbf{E}_{\perp 0} + a_0(\mathbf{E}_{\parallel 0} \times \boldsymbol{\omega}_{c_e})\right. \\
&\qquad\qquad \left. + a_1(\mathbf{E}_{\perp 0} \times \boldsymbol{\omega}_{c_e}) + a_2(\mathbf{E}_{\perp 0} \times \boldsymbol{\omega}_{c_e}) \times \boldsymbol{\omega}_{c_e}\right]
\end{aligned}
$$

Using the vector diagram of Fig. 3.14 we may now simplify the foregoing expression to yield

$$
e^{-j\omega t}\left[\frac{q_e}{m_e}\mathbf{E}_{\parallel 0} + \frac{q_e}{m_e}\mathbf{E}_{\perp 0} + a_1(\mathbf{E}_{\perp 0} \times \boldsymbol{\omega}_{c_e}) - a_2 \omega_{c_e}^2 \mathbf{E}_{\perp 0}\right]
$$

Note that $\mathbf{E}_{\parallel 0} \times \boldsymbol{\omega}_{c_e} \equiv 0$. We can obtain three component equations from (3.26):

$E_{\parallel 0}$ Component

$$
\dot{a}_0 \mathbf{E}_{\parallel 0} - j\omega a_0 \mathbf{E}_{\parallel 0} = \frac{q_e}{m_e}\mathbf{E}_{\parallel 0} \qquad \text{or} \qquad \dot{a}_0 - j\omega a_0 = \frac{q_e}{m_e} \tag{3.27}
$$

$E_{\perp 0}$ Component

$$
\dot{a}_1 \mathbf{E}_{\perp 0} - j\omega a_1 \mathbf{E}_{\perp 0} = \frac{q_e}{m_e}\mathbf{E}_{\perp 0} - a_2 \omega_{c_e}^2 \mathbf{E}_{\perp 0} \qquad \text{or} \qquad \dot{a}_1 - j\omega a_1 + a_2 \omega_{c_e}^2 = \frac{q_e}{m_e} \tag{3.28}
$$

$\mathbf{E}_{\perp 0} \times \boldsymbol{\omega}_{c_e}$ *Component*

$$
\dot{a}_2(\mathbf{E}_{\perp 0} \times \boldsymbol{\omega}_{c_e}) - j\omega a_2(\mathbf{E}_{\perp 0} \times \boldsymbol{\omega}_{c_e}) = a_1(\mathbf{E}_{\perp 0} \times \boldsymbol{\omega}_{c_e}) \qquad \text{or} \qquad \dot{a}_2 - j\omega a_2 = a_1 \tag{3.29}
$$

The dot refers to differentiation of the coefficient with respect to time. We will now obtain the steady state solutions to these equations for the three ranges of ω.

We must add a word for the ions. Note that when considering the ions, the third vector must be $-(\mathbf{E}_\perp \times \boldsymbol{\omega}_{c_i})$. What this means is that the solutions for a_2 for ion motion must be the negative of what they are for the electrons. This is the only difference in the two methods of solution. It is very important to remember that q, wherever it appears, *must* change sign, including in ω_{c_e} or ω_{c_i}. From now on, therefore, we will use ω_c without an additional subscript and note the cases for the ions.

Case 1, $\omega = 0$ (dc electric field)

The solutions for a_0, a_1, and a_2 are

$$a_0 = \frac{q}{m} t \qquad a_1 = 0 \qquad a_2 = \begin{cases} \dfrac{q}{m\omega_c^2} & \\ -\dfrac{q}{m\omega_c^2} & \text{for ions} \end{cases}$$

The motion of a particle under these conditions results in a constantly increasing velocity in the $E_\parallel$ direction, no component of motion in the $E_\perp$ direction, and a constant (drift) velocity in the $\mathbf{E}_\perp \times \boldsymbol{\omega}_c$ direction. Therefore we may use $\mathbf{E}_\perp \times \boldsymbol{\omega}_{c_e}$ or $-(\mathbf{E}_\perp \times \boldsymbol{\omega}_{c_i})$ interchangeably since the sign of the coefficient a_2 takes care of the opposite vector direction and the $\mathbf{E} \times \mathbf{B}$ drifts remain the same for ions and electrons. The resultant velocity $\mathbf{v}_1$ is

$$\mathbf{v}_1 = \frac{q}{m} t\mathbf{E}_\parallel \pm \frac{q}{m\omega_c^2} (\mathbf{E}_\perp \times \boldsymbol{\omega}_c) \qquad \text{(minus sign for ions)} \tag{3.30}$$

Upon examining this velocity expression, we note that these results were obtained previously by the radius of gyration arguments. There is a constantly increasing velocity in the $\mathbf{E}_\parallel$ direction (parallel to $\mathbf{B}$) and a *constant* drift velocity in the direction perpendicular to both $\mathbf{E}_\perp$ and $\mathbf{B}$. Its magnitude is then seen to be equal to $E_\perp/B$, which is the previous result.

We can investigate the solutions of Eqs. (3.27), (3.28), and (3.29) for

Case 2, $\omega \neq 0$, $\omega \neq \omega_c$

The results for the steady state solution are

$$a_0 = -\frac{q}{j\omega m} \qquad a_1 = -\frac{j\omega q}{m} \frac{1}{(\omega_c^2 - \omega^2)} \qquad a_2 = \begin{cases} \dfrac{q}{m} \dfrac{1}{(\omega_c^2 - \omega^2)} & \\ -\dfrac{q}{m} \dfrac{1}{(\omega_c^2 - \omega^2)} & \text{for ions} \end{cases}$$

The result is that the velocity $\mathbf{v}_1$ becomes, for Case 2,

$$\mathbf{v}_1 = \left[-\frac{q}{jm\omega} \mathbf{E}_{\parallel 0} - \frac{j\omega q}{m} \frac{1}{(\omega_c^2 - \omega^2)} \mathbf{E}_{\perp 0} \pm \frac{q}{m} \frac{1}{(\omega_c^2 - \omega^2)} (\mathbf{E}_{\perp 0} \times \boldsymbol{\omega}_c) \right] e^{-j\omega t} \tag{3.31}$$

(Again, the minus sign is used for ions.) The subscript 0 for the **E**'s is defined as the Fourier amplitude. The velocity in all three coordinate directions is constant in magnitude, but varies sinusoidally in time. This means that the time average velocity is zero for each of the $\mathbf{v}_1$ components. Converting back to trigonometric functions, by taking the real part of (3.31), we obtain

$$\mathbf{v}_1 = \left[\frac{q}{\omega m} \mathbf{E}_{\parallel 0} \sin(\omega t) + \frac{q\omega}{m} \frac{1}{(\omega_c^2 - \omega^2)} \mathbf{E}_{\perp 0} \sin(\omega t) \right.$$
$$\left. \pm \frac{q}{m} \frac{1}{(\omega_c^2 - \omega^2)} (\mathbf{E}_{\perp 0} \times \boldsymbol{\omega}_c) \cos(\omega t) \right] \tag{3.32}$$

(The minus sign is used for ions.)

Case 3, $\omega = \omega_c$

Case 3, the condition when $\omega = \omega_c$, is known as electron (or ion) cyclotron resonance. The solutions of the three equations become

$$a_0 = -\frac{q}{j\omega m} \qquad a_1 = -\frac{q}{j2\omega_c m}(1 - j\omega_c t) \qquad a_2 = \begin{cases} -\dfrac{qt}{j2\omega_c m} & \\ \dfrac{qt}{j2\omega_c m} & \text{for ions} \end{cases}$$

Here the velocity is constant in magnitude in the direction parallel to the field lines, but the magnitude increases linearly with time in the $\mathbf{E}_\perp$ and $\mathbf{E}_\perp \times \boldsymbol{\omega}_c$ directions. The velocity $\mathbf{v}_1$ for Case 3 would then be

$$\mathbf{v}_1 = \left[-\frac{q}{j\omega_c m} \mathbf{E}_{\parallel 0} - \frac{q}{j2\omega_c m}(1 - j\omega_c t)\mathbf{E}_{\perp 0} \mp \frac{qt}{j2\omega_c m} (\mathbf{E}_{\perp 0} \times \boldsymbol{\omega}_c) \right] e^{-j\omega t} \tag{3.33}$$

with the plus sign used for ions. Converting to trigonometric form, we get

$$\mathbf{v}_1 = \frac{q}{\omega_c m} \mathbf{E}_{\parallel 0} \sin(\omega_c t) + \frac{q}{2\omega_c m} \mathbf{E}_{\perp 0} \sin(\omega_c t) + \frac{q\omega_c t}{2\omega_c m} \mathbf{E}_{\perp 0} \cos(\omega_c t)$$
$$\pm \frac{qt}{2\omega_c m} (\mathbf{E}_{\perp 0} \times \boldsymbol{\omega}_c) \sin(\omega_c t) \tag{3.34}$$

where the minus sign is used for ions. The velocity in the plane perpendicular to $\mathbf{B}$ will increase indefinitely with time as long as the nonrelativistic, noncollisional, single-particle model holds. Obviously, in a laboratory plasma, there must exist some limitation on the behavior of the plasma since the plasma particles eventually become relativistic and go off of the resonance condition due to their increased mass, collide with the vacuum chamber walls, travel to a region where the magnetic field changes, make collisions with other particles, or are subjected to a combination of all of these effects. Also, if $\mathbf{v}_2$ is directed oppositely to $\mathbf{v}_1$, there will result a net *deceleration* of the particle, rather than an acceleration.

Note that we have now established a relationship, in these three cases, between $\mathbf{v}_1$ and the electric field $\mathbf{E}$. The ratio of $\mathbf{v}_1$ to $\mathbf{E}$ has been defined as the *mobility*. Therefore, in the plane perpendicular to the magnetic field, a velocity in one coordinate direction may be obtained from an electric field in another direction, as shown by the nonzero coefficients of the $\mathbf{E}_\perp \times \boldsymbol{\omega}_c$ terms in Eqs. (3.30), (3.32), and (3.34). The result must be that the mobility in general is a tensor. We could develop this by expressing $\mathbf{E}$ in terms of its components in rectangular coordinates. If $\mathbf{B}$ is parallel to the z axis, then the representation for $\mathbf{E}$ in rectangular coordinates is

$$\mathbf{E}_\parallel = E_z\,\hat{a}_z \qquad \mathbf{E}_\perp = E_x\,\hat{a}_x + E_y\,\hat{a}_y \qquad \text{and} \qquad \boldsymbol{\omega}_c = \omega_c\,\hat{a}_z$$

The expression for $\mathbf{v}_1$ is then

$$\mathbf{v}_1 = a_0\,E_z\,\hat{a}_z + a_1(E_x\,\hat{a}_x + E_y\,\hat{a}_y) \pm a_2[(E_x\,\hat{a}_x + E_y\,\hat{a}_y) \times \omega_c\,\hat{a}_z]$$

(The minus sign is used for the ions.) We can now determine the x, y, and z components of $\mathbf{v}_1$ to be

$$v_{1x} = a_1 E_x \pm a_2\,\omega_c\,E_y \tag{3.35}$$

$$v_{1y} = \mp\,a_2\,\omega_c\,E_x + a_1 E_y \tag{3.36}$$

$$v_{1z} = a_0\,E_z \tag{3.37}$$

with the lower signs for the ions. This is expressible in matrix notation as

$$\begin{bmatrix} v_{1x} \\ v_{1y} \\ v_{1z} \end{bmatrix} = \begin{bmatrix} a_1 & \pm a_2\,\omega_c & 0 \\ \mp\,a_2\,\omega_c & a_1 & 0 \\ 0 & 0 & a_0 \end{bmatrix} \times \begin{bmatrix} E_x \\ E_y \\ E_z \end{bmatrix} \tag{3.38}$$

(The lower signs are for the ions.) The off-diagonal terms of the matrix are called Hall effect terms from the similar behavior observed in solid state work. Note that this can result in, for example, a velocity in the x direction due to an electric field in the y direction. It should also be noted that for ions, the off-diagonal terms change sign. Each subsequent mobility tensor, therefore, may be simply changed to show the ion behavior by changing the signs of the off-diagonal terms and q. Remember that in an *isotropic* medium, $\mathbf{v}$ is always directly proportional to $\mathbf{E}$ and the proportionality constant is μ, the mobility, which is a scalar in this case. Equation (3.38) shows a similar relation for an *anisotropic* material (the plasma in a magnetic field) and the tensor shown is called the mobility tensor.

If Case 2 is assumed to be the most general, we can then obtain the mobility tensor by substitution of a_0, a_1, and a_2 into Eq. (3.38) as shown in Eq. (3.39).

$$\mu = \begin{bmatrix} \dfrac{-j\omega q}{m}\dfrac{1}{(\omega_c^2-\omega^2)} & \dfrac{\omega_c q}{m(\omega_c^2-\omega^2)} & 0 \\ \dfrac{-\omega_c q}{m(\omega_c^2-\omega^2)} & \dfrac{-j\omega q}{m}\dfrac{1}{(\omega_c^2-\omega^2)} & 0 \\ 0 & 0 & \dfrac{-q}{j\omega m} \end{bmatrix} \tag{3.39}$$

We can factor out $-q/j\omega m$ from all terms of Eq. (3.39) and arrive at

$$\mu = \frac{-q}{j\omega m}\begin{bmatrix} \dfrac{\omega^2}{\omega^2-\omega_c^2} & \dfrac{j\omega_c\omega}{\omega^2-\omega_c^2} & 0 \\ \dfrac{-j\omega_c\omega}{\omega^2-\omega_c^2} & \dfrac{\omega^2}{\omega^2-\omega_c^2} & 0 \\ 0 & 0 & 1 \end{bmatrix} \tag{3.40}$$

for the mobility tensor.

Up to this point, we have only followed the equation of motion for a single electron or ion. Since a plasma is composed of many particles, we would like to obtain the collected effects of electric and magnetic fields on all the particles. We will obtain these effects by considering the electric current produced by all the particles. A simple straightforward approximation to the collective electric current behavior is to assume that the current of N particles is N times the electric current of a single one. Thus, the single-particle electric current would be

$$I_1 = qv$$

The electric current of N particles would then be

$$I_N = Nqv \tag{3.41}$$

The collective electric current density vector is then

$$\mathbf{J}_E = \mathcal{q} n \mathbf{v} \tag{3.42}$$

where n is the particle density. So, in terms of the mobility,

$$\mathbf{J}_E = \mathcal{q} n \boldsymbol{\mu} \cdot \mathbf{E} = \boldsymbol{\sigma} \cdot \mathbf{E} \tag{3.43}$$

We have now defined a *conductivity* tensor such that

$$\boldsymbol{\sigma} = \mathcal{q} n \boldsymbol{\mu} \tag{3.44}$$

We may convert Eq. (3.40) into a conductivity tensor as

$$\boldsymbol{\sigma} = \frac{jn\mathcal{q}^2}{\omega m} \begin{bmatrix} \dfrac{\omega^2}{\omega^2 - \omega_c{}^2} & \dfrac{j\omega_c\omega}{\omega^2 - \omega_c{}^2} & 0 \\ \dfrac{-j\omega_c\omega}{\omega^2 - \omega_c{}^2} & \dfrac{\omega^2}{\omega^2 - \omega_c{}^2} & 0 \\ 0 & 0 & 1 \end{bmatrix} \tag{3.45}$$

The conductivity of the plasma when the magnetic field is removed is a scalar, and is the constant that multiplies Eq. (3.45). Note also that the conductivity *along* the magnetic field lines is the same as though the field were not present. At this juncture, we note that if the rule to obtain the ion effects, that is, changing the sign of ω_c and $\mathcal{q}$ is followed, no net changes are noted in the signs of any diagonal terms in (3.45). This makes sense, since the ions and electrons making up the plasma act as two conducting fluids that are in parallel, so their respective conductivities would add directly. The off-diagonal terms *do* change sign, however, which implies that the Hall effects of ions and electrons tend to cancel.

We now apply a collisional process to this problem. Since only a collisionless single-particle model was first assumed, and then a generalization to the mobility of a collection of particles was made in Eq. (3.41) by merely multiplying the single-particle mobility by the density, we ought to at least consider the interaction between particles in terms of a momentum transfer collision frequency, to see if we might be able to improve the model.

We can now examine the momentum conservation equation with this effect. It becomes, for $\mathbf{v}_1$,

$$m\frac{d\mathbf{v}_1}{dt} + m\mathbf{v}_1\nu_m = \mathcal{q}(\mathbf{E} + \mathbf{v}_1 \times \mathbf{B}) \tag{3.46}$$

The inclusion of the second term on the left-hand side of Eq. (3.46) is permitted if we assume that the particle *loses* all of its ordered momentum after each collision, that is, ν_m times per second.

If an $e^{-j\omega t}$ dependence is assumed, then the left-hand side of Eq. (3.46) becomes

$$-j\omega m\mathbf{v}_1\left(1-\frac{\nu_m}{j\omega}\right)=-j\omega m\mathbf{v}_1\left(\frac{\omega+j\nu_m}{\omega}\right) \tag{3.47}$$

The result tells us that if we replace m by $m[(\omega+j\nu_m)/\omega]$ in Eq. (3.45) we should obtain the conductivity tensor including the effects of collisions in the form shown in Eq. (3.48).

$$\sigma=\frac{jnq^2}{(\omega+j\nu_m)m}\begin{bmatrix}\dfrac{(\omega+j\nu_m)^2}{(\omega+j\nu_m)^2-\omega_c^{\,2}} & \dfrac{j\omega_c(\omega+j\nu_m)}{(\omega+j\nu_m)^2-\omega_c^{\,2}} & 0\\[2ex] \dfrac{-j\omega_c(\omega+j\nu_m)}{(\omega+j\nu_m)^2-\omega_c^{\,2}} & \dfrac{(\omega+j\nu_m)^2}{(\omega+j\nu_m)^2-\omega_c^{\,2}} & 0\\[2ex] 0 & 0 & 1\end{bmatrix} \tag{3.48}$$

Note that since m appears in the expression for ω_c, the substitution must be made there as well.

If $\omega=0$, we can obtain a nonsingular value for the conductivity tensor, since the velocity in the z direction will not now approach infinity as time gets large. When $\omega=0$, σ is then

$$\sigma=\frac{nq^2}{\nu_m m}\begin{bmatrix}\dfrac{\nu_m^{\,2}}{\omega_c^{\,2}+\nu_m^{\,2}} & \dfrac{\omega_c\nu_m}{\omega_c^{\,2}+\nu_m^{\,2}} & 0\\[2ex] \dfrac{-\omega_c\nu_m}{\omega_c^{\,2}+\nu_m^{\,2}} & \dfrac{\nu_m^{\,2}}{\omega_c^{\,2}+\nu_m^{\,2}} & 0\\[2ex] 0 & 0 & 1\end{bmatrix} \tag{3.49}$$

Note here that the off-diagonal terms *will* again change signs depending upon the sign of the charge. The result with both $\omega_c=0$ and $\omega=0$ is then

$$\sigma=\frac{nq^2}{m\nu_m} \tag{3.50}$$

again a scalar quantity. The conductivity representation is often used in Maxwell's equations, where it appears as

$$\nabla_r \times \mathbf{H} = \mathbf{J} + \frac{\partial \mathbf{D}}{\partial t} = \sigma \cdot \mathbf{E} + \frac{\partial \mathbf{D}}{\partial t} = \sigma \cdot \mathbf{E} + \varepsilon_0 \frac{\partial \mathbf{E}}{\partial t} \tag{3.51}$$

The only effect of the electrons and ions is found in the conductivity term of Eq. (3.51). The permittivity is still that of free space. However, a form that is completely equivalent, in which Eq. (3.51) is written with zero conductivity and a permittivity different from that of free space, is often used. With the $e^{-j\omega t}$ notation this is a relatively straightforward situation. Equation (3.51) becomes

$$\nabla_r \times \mathbf{H} = \sigma \cdot \mathbf{E} - j\omega\varepsilon_0 \mathbf{E} = (\sigma \cdot - j\omega\varepsilon_0 \cdot)\mathbf{E} = j\omega\varepsilon_0\left(\mathbf{1} - \frac{\sigma}{j\omega\varepsilon_0}\right) \cdot \mathbf{E} \tag{3.52}$$

where $\mathbf{1}$ is the unit tensor

$$\begin{bmatrix} 1 & 0 & 0 \\ 0 & 1 & 0 \\ 0 & 0 & 1 \end{bmatrix}$$

and ε_0 is

$$\begin{bmatrix} \varepsilon_0 & 0 & 0 \\ 0 & \varepsilon_0 & 0 \\ 0 & 0 & \varepsilon_0 \end{bmatrix}$$

The quantity in the parentheses in Eq. (3.52) is the relative permittivity of the plasma and Eq. (3.51) may now be modified to present this "dielectric constant." To do this, we first compute $\sigma/j\omega\varepsilon_0$ to be

$$\frac{\sigma}{j\omega\varepsilon_0} = \frac{nq^2}{m\varepsilon_0\,\omega(\omega + j\nu_m)} \begin{bmatrix} \dfrac{(\omega + j\nu_m)^2}{(\omega + j\nu_m)^2 - \omega_c^2} & \dfrac{j\omega_c(\omega + j\nu_m)}{(\omega + j\nu_m)^2 - \omega_c^2} & 0 \\ \dfrac{-j\omega_c(\omega + j\nu_m)}{(\omega + j\nu_m)^2 - \omega_c^2} & \dfrac{(\omega + j\nu_m)^2}{(\omega + j\nu_m)^2 - \omega_c^2} & 0 \\ 0 & 0 & 1 \end{bmatrix} \tag{3.53}$$

The result for $\varepsilon/\varepsilon_0$ using Eq. (3.53) in (3.52) is then given by Eq. (3.54) (see p. 64).

$$\frac{\boldsymbol{\varepsilon}}{\varepsilon_0} = \boldsymbol{\kappa} = \begin{bmatrix} 1 - \dfrac{nq^2(\omega + j\nu_m)}{m\omega\varepsilon_0[(\omega + j\nu_m)^2 - \omega_c{}^2]} & \dfrac{-j\omega_c nq^2}{m\omega\varepsilon_0[(\omega + j\nu_m)^2 - \omega_c{}^2]} & 0 \\ \dfrac{j\omega_c nq^2}{m\omega\varepsilon_0[(\omega + j\nu_m)^2 - \omega_c{}^2]} & 1 - \dfrac{nq^2(\omega + j\nu_m)}{m\omega\varepsilon_0[(\omega + j\nu_m)^2 - \omega_c{}^2]} & 0 \\ 0 & 0 & 1 - \dfrac{nq^2}{m\omega\varepsilon_0(\omega + j\nu_m)} \end{bmatrix} = \begin{cases} \text{cold plasma} \\ \text{dielectric constant} \end{cases} \tag{3.54}$$

We may simplify Eq. (3.54) by introducing a substitution. Let

$$\omega_p{}^2 = \frac{nq^2}{m\varepsilon_0}$$

Then Eq. (3.54) becomes

$$\frac{\varepsilon}{\varepsilon_0} = \mathrm{K} = \begin{bmatrix} 1 - \dfrac{\omega_p{}^2}{(\omega + j\nu_m)^2 - \omega_c{}^2} & \dfrac{-j\omega_c\,\omega_p{}^2}{(\omega + j\nu_m)^2 - \omega_c{}^2} & 0 \\ \dfrac{j\omega_c\,\omega_p{}^2}{(\omega + j\nu_m)^2 - \omega_c{}^2} & 1 - \dfrac{\omega_p{}^2}{(\omega + j\nu_m)^2 - \omega_c{}^2} & 0 \\ 0 & 0 & 1 - \dfrac{\omega_p{}^2}{(\omega + j\nu_m)^2} \end{bmatrix} \tag{3.55}$$

We call ω_p the *plasma frequency* for either electrons or ions, depending upon which species is considered. This frequency is a very fundamental characteristic of all plasmas, and its significance will appear in later chapters; note that it is independent of the sign of the charge.

Similar calculations for the ion motion result in the *identical* form for the ion dielectric constant. The effects of the ions may be combined directly with those of the electrons by assuming conductivities in parallel to give a net plasma conductivity and thence a net permittivity. This last expression [Eq. (3.55)] is valid for all three cases of ω, since the singularities for $\omega = 0$ or $\omega = \omega_c$ are damped out by the collisional effects.

The single-particle model developed in this chapter is valid under many conditions when the "cooperative" nature of plasma interactions is not applicable. Further extension of validity of plasma behavior now requires a statistical approach.

Suggested Reading

Allis, W. P., Motion of Ions and Electrons, *in* "Handbuch der Physik" (S. Flügge, ed.), Vol. 21, Springer, Berlin, 1956. This is an outstanding treatment, in English, of particle motions.

Chandrasekhar, S., "Plasma Physics," Univ. of Chicago Press, Chicago, 1960.

Delcroix, J. L., "Introduction to the Theory of Ionized Gases," Wiley (Interscience), New York, 1960.

Linhart, J. G., "Plasma Physics," North-Holland Publ., Amsterdam, 1960.

Longmire, C. L., "Elementary Plasma Physics," Wiley (Interscience), New York, 1963.

Northrup, T. G., "The Adiabatic Motion of Charged Particles," Wiley (Interscience), New York, 1963.

Schmidt, G., "Physics of High Temperature Plasmas—An Introduction," Academic Press, New York, 1966. This book gives a wide-ranging coverage with much material in depth.
Spitzer, L., Jr., "Physics of Fully Ionized Gases," 2nd ed., Wiley (Interscience), New York, 1962.

Problems

1 Calculate the following quantities:

(a) ω_c, r_g, and μ_m for
(1) an electron of 1 keV energy,
(2) an electron of 1 MeV energy,
in a 3 kG magnetic field.
(b) Do part (a) for a hydrogen ion.
(c) Do part (a) for a deuterium ion. (Deuterium is an isotope of hydrogen containing 1 neutron and 1 proton in its nucleus.) Order the magnetic moments, radii of gyration, and cyclotron frequencies for all three parts together.

2 Compute the points of reflection in a magnetic mirror field of an electron whose total kinetic energy is 1 keV. The dc magnetic field is assumed to increase linearly in both directions from $z = 0$ with the following relation:

$$B_z = B_{0_z}(1 + \alpha|z|)$$

Compute the reflection for angles that the electron makes with the dc magnetic field at the midplane of

(a) 5°, (b) 20°, (c) 45°, (d) 67°.

3 In many cases it is necessary to consider not only the steady state responses of charged particles, but also transient responses.
(a) Solve the equation of motion of a charged particle in a uniform ac electric field $E_0 \sin \omega t$, pointing in the x direction, that is switched on at $t = 0$. Assume that the particle begins at rest. Plot the motion and velocity as functions of time. Does the velocity increase indefinitely without limit? Assume $\mathbf{B} \equiv 0$.
(b) Do part (a) again, but assume $E = E_0 \cos \omega t$. Can the guiding center approximation be used for these cases?

4 Suppose that a particle moves in a dc magnetic field that is pointing in the z direction. The magnetic field varies with x (not z). Calculate the drift of a charged particle in this $\nabla_r B$ field if $\mathbf{B} = B_0(1 + \beta x^2)\,\hat{a}_z$. (Note: The expression for guiding center drifts in the presence of magnetic field gradients has not been derived. The vector equation of motion may be solved instead, however.) Compare the drift velocity obtained in this manner with the results obtained by using the theoretical expression for the force on a guiding center due to $\nabla_r B$, which is

$$\mathbf{F}_{\nabla B} = -\mu_m \nabla_r B$$

Show that when the gradient is perpendicular to **B** the drift velocity is a constant of value

$$\mathbf{v}_{\nabla B} = -\frac{1}{q}\frac{\mu_m}{B}\frac{\nabla_r(B^2/2) \times \mathbf{B}}{B^2}$$

What is the drift when the gradient is in the direction of **B**? Compute the drift if $\mathbf{B} = B_0(1 + \beta z^2)\,\hat{a}_z$. In both cases, assume β is small.

5 We have previously shown that if the magnetic moment can be assumed constant, a particle, orbiting in a dc magnetic field, encloses a tube of constant flux. Suppose that a mirror field increases *slowly* in time. What will happen to a particle that is confined between the magnetic mirrors?

6 *Numerical Problem.* Determine the velocity of a charged particle under the conditions of cyclotron resonance. Include the effects of relativity by noting that the mass of the particle changes as

$$m = \frac{m_0}{[1 - (v^2/c^2)]^{1/2}}$$

where m_0 is the "rest" mass and c is the velocity of light. Assume an electron in a dc magnetic field such that the electron has a cyclotron frequency of 10^{10} Hz. The electric field strength will be uniform in space of value $10^4 \cos(\omega_c t)$ V/m pointing in the x direction. The dc magnetic field is in the z direction. Assume that the particle begins from rest. The equation of motion is to be solved by "finite differences." That is, we may use the equation

$$m\frac{d\mathbf{v}}{dt} = q(\mathbf{E} + \mathbf{v} \times \mathbf{B})$$

to solve for the motion. However, after breaking it up into components, we may approximate the time derivative by, for example, for the x component of velocity,

$$\frac{dv_x}{dt} \cong \frac{v_{x,t+\Delta t} - v_{x,t}}{\Delta t}$$

where $v_{x,t+\Delta t}$ = the x component of velocity at time $t + \Delta t$. This is the unknown quantity. $v_{x,t}$ is the known velocity. Set the time step Δt to, say,

$\frac{1}{20}$th of a cyclotron period. The result is that the two component momentum equations become algebraic, perhaps of the form

$$\frac{m_0}{\{1 - [(v_{x,t}^2 + v_{y,t}^2)/c^2]\}^{1/2}} \frac{v_{x,t+\Delta t} - v_{x,t}}{\Delta t} = qE_{x,t} + qv_{y,t}B_z$$

for the x component, and a similar one for the y component. The two component equations may be solved simultaneously for $v_{x,t+\Delta t}$ and $v_{y,t+\Delta t}$. The process may be repeated using the newly found values of v_x and v_y for the known velocities and solving again. You may also wish to compare these results to the approximation where the term $qv_{y,t}B_z$ becomes $q(v_{y,t+\Delta t} + v_{y,t})/2$ and similarly for the term $-qv_{x,t}B_z$ in the other momentum equation. Plot the velocity as a function of time for five cyclotron periods.

7 Find the complete steady state plus transient solution for the equation of motion of a charged particle in a uniform dc magnetic field **B** pointing in the z direction, starting from rest with an electric field

$$\mathbf{E} = E_x \sin \omega t\, \hat{a}_x + E_z \cos \omega t\, \hat{a}_z$$

8 Determine the complete cold plasma conductivity and permittivity tensors for a plasma considering the effects of ions and electrons. Using hydrogen, and a magnetic field of 1000 G, which terms dominate at a frequency of 10 Hz? At a frequency of 10^8 Hz?

9 Suppose that a charged particle orbits in a uniform magnetic field. Then suppose that the field increases linearly with time fast enough so that energy is not conserved. Find the rate of increase of the particle's energy if the field varies as

$$B = B_0(1 + \gamma t)$$

10 (a) Find the motion of a charged particle in the field of an infinitely long straight current-carrying wire. Assure that the particle has both parallel and perpendicular components of velocity. Use the guiding center approximation, if applicable.

(b) Find the motion of the particle if it is also subjected to a uniform dc electric field in the direction parallel to the wire.

11 Calculate the gravitational drift velocity of hydrogen ions and electrons in the earth's gravitational and magnetic fields ($\mathbf{B}_{\text{earth}} \approx 0.3$ G).

12 Show that the vector formulation for the magnetic moment of a charged particle

$$\boldsymbol{\mu}_m = \tfrac{1}{2}q(\mathbf{r}_g \times \mathbf{v}_\perp)$$

produces a magnetic field in opposition to the dc field regardless of the sign of the charge. Sketch this diamagnetic field centered about the particle.

4 The Beginnings of Collective Phenomena—Plasma Statistical Mechanics

A. FUNDAMENTAL DEFINITIONS

We now begin a section where some of the difficulties and approximations caused by the presence of extremely large numbers of particles make themselves felt. We must recall that plasmas are not composed of two or three charged particles but trillions of them, all moving and interacting together. We cannot ever hope to follow them all around their respective paths, even if we could obtain the exact equations to be solved.

The statistical treatment to be covered here is needed *because* we cannot predict the *exact* behavior of every particle in the plasma. The statistical method will enable us to predict what happens to a collection of particles, *on the average*, subject to certain conditions which will become evident throughout this chapter.

The results of the development in this chapter will be a special function, called a *distribution function*. This function will yield, when the plasma is in equilibrium (time invariance of the total energy and number of particles in the plasma) a representation of the plasma. The representation will be the number of particles making up the plasma in given incremental ranges of velocity and position. From this function, collective averages and the behavior of the plasma may be developed. If the plasma is not in equilibrium, other methods must be used to obtain the distribution function.

We need to define the following terms to begin our work on collective phenomena.

1. *Stationary coordinates*. Coordinates that can be derived by a transformation, which does not involve time, from the coordinates corresponding to an unaccelerated set of Cartesian axes are called stationary coordinates. For

example, cylindrical and spherical coordinates are time-invariant transformations of Cartesian coordinates. Therefore, they come under this classification, along with translated and rotated (not rotating) coordinates. The number of coordinates must be sufficient for a complete specification of the position, momentum, and internal and configuration of the system.

2. *Holonomic System.* Each coordinate of a holonomic system can be varied *independently* without violating any constraints, such as conservation of energy or magnetic moment.

3. *Nonholonomic System.* Some or all of the coordinates of this type of system are connected or constrained, so as not to permit independent motion.

4. *Degrees of Freedom.* The number of independent variations that can be made in the coordinates is called the number of degrees of freedom of a system. It may be assumed for nonquantum-mechanical systems that we can observe the momentum and position for each degree of freedom needed for the specification of the system. This specification of the momenta and positions alone should be sufficient to describe each state of the system instantaneously.

5. *System.* The physical object of interest. It may be composed of elements. Note that the elements themselves *may* have degrees of freedom as well.

6. *Ensemble.* A collection of systems distributed over a range of possible states. The ensemble is *not real*, but is merely a grouping of all possible states that the system can be in during the complete range of the time of interest. All of these states making up the ensemble are assumed to be in existence at the *same time*, so that the ensemble is a *collection of systems*. Note that the different states of the system *could* be ascertained by watching the system and recording its states as time progresses. It is obvious, then, that the system may appear in nearly the same state (to within an incremental range) more than once, as time progresses. For example, a ball may be in the same space, moving with the same velocity, but at two different times. We must then examine each state, and record the number of times each state is reached by the system. This should then give us the "probability" for the system to appear in a given state. The ensemble theory, to be described, permits us to find this probability function *without* waiting for the system to progress through its various states.

7. *Phase Space.* For r degrees of freedom, we can construct a space of $2r$ dimensions, with $2r$ rectangular axes, one axis for the position and one for the velocity (or momentum) coordinates of each degree of freedom.

8. *Representative Point.* The instantaneous state of any system may be regarded by noting the position of a representative point in the phase space (phase point) corresponding to a state of the system.

Systems in an ensemble do not and cannot react with each other. A gas or a plasma is *not* an ensemble. *It is a system.* It is composed of a large number of similar particles colliding with each other. These particles can be called the *elements* of the system. In using statistical mechanics to describe the behavior of a plasma, we will consider a collection of *systems*, each of which is the plasma in a given state. This is the ensemble described previously. Figure 4.1 shows the relations between an ensemble and systems and elements.

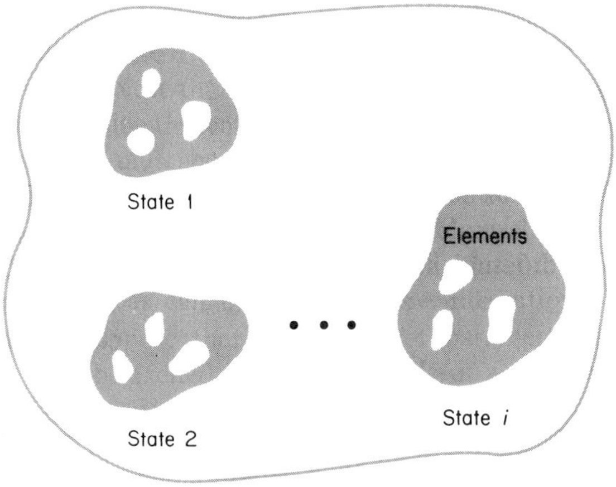

Fig. 4.1. An ensemble, showing systems and elements.

We are now interested in the problem of developing the mathematical expressions describing the different states of the system. This will permit us to examine the behavior of the *ensemble as a whole*, since the ensemble is composed of the system in many different states. Assume that a system is composed of N elements, each one of which has a mass m_k, position coordinates $x_k, y_k, z_k, \ldots$, and velocity coordinates $v_{xk}, v_{yk}, v_{zk}, \ldots$, where

$$v_{xk} = \frac{dx_k}{dt} = \dot{x}_k \tag{4.1}$$

There is one position and velocity coordinate for each degree of freedom of the element. We will assume r degrees of freedom for each element and therefore there will be Nr degrees of freedom for the system. We can then write an expression for the *Lagrangian* of each element as

$$L_k = T_k - \mathscr{V}_k = \tfrac{1}{2}m_k(\dot{x}_k^{\,2} + \dot{y}_k^{\,2} + \dot{z}_k^{\,2}) - \mathscr{V}_k$$

T_k is the kinetic energy of the kth element and $\mathscr{V}_k$ is its potential energy. The *Lagrangian* of the system L is then the sum over all elements k of the following

$$L = T - \mathscr{V} = \sum_{k=1}^{N} (\tfrac{1}{2}m_k(\dot{x}_k^2 + \dot{y}_k^2 + \dot{z}_k^2)) - \mathscr{V} \tag{4.2}$$

T is the kinetic energy of the system. $\mathscr{V}$ is the potential energy of the system. It will be assumed to depend only upon the position and time. $\mathscr{V}$ can be written as

$$\mathscr{V} = \mathscr{V}(x, y, z, \ldots, x_k, y_k, z_k, \ldots, t) \tag{4.3}$$

$\mathscr{V}$, then, is a function of *all* the position coordinates and the time. Lagrange's equations may now be written for each element as shown below. These can be shown to be equivalent to Newton's equations, which are written to the right.

Lagrangian equations	Newtonian equations	
$\dfrac{d}{dt}(m_k \dot{x}_k) + \dfrac{\partial \mathscr{V}}{\partial x_k} = 0$	$F_{xk} = m_k \ddot{x}_k$	(4.4)
$\dfrac{d}{dt}(m_k \dot{y}_k) + \dfrac{\partial \mathscr{V}}{\partial y_k} = 0$	$F_{yk} = m_k \ddot{y}_k$	(4.5)
$\dfrac{d}{dt}(m_k \dot{z}_k) + \dfrac{\partial \mathscr{V}}{\partial z_k} = 0$	$F_{zk} = m_k \ddot{z}_k$	(4.6)

We must now make sure that we will be working in *generalized coordinates.* Generalized coordinates become necessary if we are to ensure that the coordinates of the system and its elements are to be holonomic. That is, each coordinate must be capable of being varied independently of the other. Sometimes, various *physical constraints* in the system or the elements prohibit this independent variation and we must *transform* the old coordinates into a set of generalized coordinates where coupling does not exist between the generalized coordinates. We may obtain generalized coordinates when the coupling between the actual coordinates is expressible as an algebraic equation.

If there is coupling we can eliminate one or more of the coordinates by using the constraint and its stated relation between the coordinates. In addition, we may introduce $(r - c)$ *new* coordinates instead, where the element (or system) originally had r degrees of freedom and c constraints. These new coordinates are called generalized coordinates. Note that the constraints are external. That is, they do not result from a solution to the equation of motion, but are imposed upon the motion by some outside means.

Let us now consider an example of a problem where we can easily find the generalized coordinates, and show how we may eliminate one of the ordinary

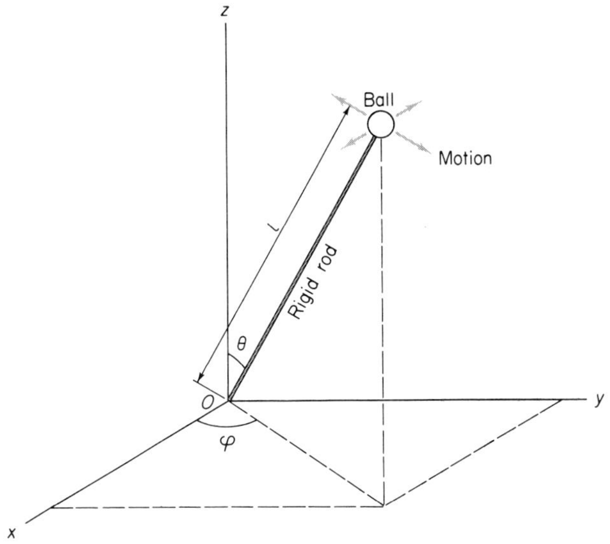

Fig. 4.2. Ball–rod system.

coordinates. The problem will be that of a ball, suspended on a rigid rod, which is revolving about the opposite end of the rod as shown in Fig. 4.2. In other words, the ball and rod are fixed together but can revolve about point O.

If it were not fixed to the rod, the ball would have three degrees of freedom. Its coordinates are x, y, z, and v_x, v_y, v_z. Now, however, the motion is constrained, so the system is nonholonomic. This means that there is coupling between the coordinates. We may express the two constraints for this problem in equation form as

$$x^2 + y^2 + z^2 = l^2 \qquad \text{and} \qquad v_z z + v_x x + v_y y = 0 \tag{4.7}$$

These constraints are that the length l is constant and the radial velocity is zero. We may solve for, say, z and v_z and eliminate these coordinates from the equation of motion by substitution. The remaining four coordinates can be generalized coordinates for the problem.

If we transform to spherical coordinates, the problem is much simpler to visualize. The coordinates are now r, θ, and φ and v_r, v_θ, and v_φ. Now the physical constraints show that $r =$ constant and $v_r = 0$. Therefore these two coordinates are eliminated immediately and the remaining four are another set of generalized coordinates for this problem. Note that we can transform back to Cartesian coordinates using only the four generalized spherical coordinates and the constraints if we desire.

As has been shown, these are not necessarily unique sets of generalized coordinates, but it is usually possible to find some set that will be more useful than another. Not only has the number of variables been reduced, but the system has been made holonomic. Note again that by using the constraints in reverse, we may find *all* of the original coordinates.

A generalized position coordinate may be represented as q_{ik} and a generalized *momentum* coordinate may be defined as p_{ik} from the relation

$$p_{ik} = \frac{\partial L}{\partial \dot{q}_{ik}} \tag{4.8}$$

The Lagrangian for the system may be rewritten in terms of the generalized coordinates as

$$L = T - \mathscr{V} = \sum_{k=1}^{N} \sum_{i=1}^{r} [\tfrac{1}{2} m_k \dot{q}_{ik}^2] - \mathscr{V}(q_1 \cdots q_{Nr}, t) \tag{4.9}$$

The *Hamiltonian* can now be defined for the system as

$$H = \sum_{k=1}^{N} \sum_{i=1}^{r} p_{ik} \dot{q}_{ik} - L$$

where

$$H = H(q_1 \cdots q_{Nr}, p_1 \cdots p_{Nr}, t)$$

in general. We may eliminate the subscript k and use one summation set of generalized coordinates with $2Nr$ axes. Recall again that r is the number of degrees of freedom of each element and Nr the number of degrees of freedom of the system as shown below. H is a function of position, momentum, and time. The total differential dH of H is written as

$$dH = \sum_{i=1}^{Nr} p_i \, d\dot{q}_i + \sum_{i=1}^{Nr} \dot{q}_i \, dp_i - \sum_{i=1}^{Nr} \frac{\partial L}{\partial q_i} dq_i - \sum_{i=1}^{Nr} \frac{\partial L}{\partial \dot{q}_i} d\dot{q}_i - \frac{\partial L}{\partial t} dt$$

Since

$$p_i = \frac{\partial L}{\partial \dot{q}_i}$$

we may cancel the first and fourth terms on the right of the equation above and it therefore becomes

$$dH = \sum_{i=1}^{Nr} \dot{q}_i \, dp_i - \sum_{i=1}^{Nr} \frac{\partial L}{\partial q_i} dq_i - \frac{\partial L}{\partial t} dt \tag{4.10}$$

The partial differentials for Eq. (4.10) are then

$$\frac{\partial H}{\partial p_i} = \dot{q}_i, \qquad \frac{\partial H}{\partial q_i} = -\frac{\partial L}{\partial q_i}, \qquad \text{and} \qquad \frac{\partial H}{\partial t} = -\frac{\partial L}{\partial t} \tag{4.11}$$

which in turn produce, by use of (4.11) together with

$$\frac{d}{dt}\left(\frac{\partial L}{\partial \dot{q}_i}\right) - \frac{\partial L}{\partial q_i} = 0 \qquad \text{(Lagrange's equations)}$$

the two Hamiltonian equations

$$\dot{q}_i = \frac{\partial H}{\partial p_i} \qquad \text{and} \qquad \dot{p}_i = -\frac{\partial H}{\partial q_i} \tag{4.12}$$

Any quantity F that depends upon q_i and p_i may be differentiated with respect to time as follows:

$$\frac{dF}{dt} = \frac{\partial F}{\partial t} + \sum_{i=1}^{Nr} \left[\frac{\partial F}{\partial q_i}\frac{dq_i}{dt} + \frac{\partial F}{\partial p_i}\frac{dp_i}{dt}\right] \tag{4.13}$$

or

$$\frac{dF}{dt} = \frac{\partial F}{\partial t} + \sum_{i=1}^{Nr} \left[\frac{\partial F}{\partial q_i}\frac{\partial H}{\partial p_i} - \frac{\partial H}{\partial q_i}\frac{\partial F}{\partial p_i}\right] \tag{4.14}$$

by using Hamilton's equations (4.12). The right-hand term of Eq. (4.14) is called a *Poisson bracket*. It may be defined in a symbolic notation as

$$[M, N] = \sum_{i=1}^{Nr} \left[\frac{\partial M}{\partial q_i}\frac{\partial N}{\partial p_i} - \frac{\partial N}{\partial q_i}\frac{\partial M}{\partial q_i}\right] \tag{4.15}$$

Therefore, Eq. (4.14) now becomes

$$\frac{dF}{dt} = \frac{\partial F}{\partial t} + [F, H] \tag{4.16}$$

B. CONSERVATION LAWS FOR SYSTEMS

In conservative systems, H is not a function of time, so that

$$\frac{dH}{dt} = \frac{\partial H}{\partial t} + [H, H] \equiv 0 \tag{4.17}$$

H can be thought of as the total energy of the system in this case.

If we assume that the elements are acting under a force derivable from a potential, and if we neglect internal energy states of the elements, we may write H for the system as

$$H = \sum_{k=1}^{N} \left[\frac{1}{2m_k}(p_{xk}^2 + p_{yk}^2 + p_{zk}^2) + \mathscr{V}(x_k, y_k, z_k)\right] \tag{4.18}$$

where

$$p_{xk} = m_k \dot{x}_k \qquad p_{yk} = m_k \dot{y}_k \qquad \text{and} \qquad p_{zk} = m_k \dot{z}_k$$

We will now apply the results above to some physical quantities of the system.

The *linear momentum* in the x direction of the system is

$$P_x = \sum_{k=1}^{N} p_{xk} \tag{4.19}$$

The *angular momentum* (moment of momentum) about the z axis for the system is

$$M_z = \sum_{k=1}^{N} (x_k p_{yk} - y_k p_{zk}) \tag{4.20}$$

We may write similar expressions for the linear and angular momenta in the other coordinate directions.

The rate of change of any of these just-mentioned quantities with respect to time is

$$\frac{dF}{dt} = [F, H] \tag{4.21}$$

since they do not depend explicitly on time.

Therefore, for the system, we may write, using the two-summation notation,

$$\frac{dP_x}{dt} = [P_x, H] = \sum_{k=1}^{N} \sum_{i=1}^{r} \left[\frac{\partial p_{xk}}{\partial q_{ik}} \frac{\partial H}{\partial p_{ik}} - \frac{\partial H}{\partial q_{ik}} \frac{\partial p_{xk}}{\partial p_{ik}}\right]$$

Now, returning to Cartesian coordinates, we get, for the foregoing equation,

$$\frac{dP_x}{dt} = -\sum_{k=1}^{N} \frac{\partial H}{\partial x_k} = -\sum_{k=1}^{N} \frac{\partial \mathscr{V}}{\partial x_k} \tag{4.22}$$

because $\partial p_{xk}/\partial q_{ik} = 0$ always and

$$\frac{\partial p_{xk}}{\partial p_{ik}} = \begin{cases} 0 & \text{if} \quad i \neq x \\ 1 & \text{if} \quad i = x \end{cases}$$

In addition, from the definition of H, the only term that now remains in Eq. (4.22) will be the one involving $\mathscr{V}$.

For an isolated system, $\mathscr{V}$ depends only upon the relative positions of the particles in the system and would remain constant even if all the particles were displaced by the same amount. Each element in the system might change its potential energy but the net result for the system must be no net change. Therefore

$$\frac{dP_x}{dt} = 0$$

that is, the time rate of change of momentum for the system is zero. This means that

$$P_x = \text{constant} \qquad P_y = \text{constant} \qquad \text{and} \qquad P_z = \text{constant}$$

for the system.

The same type of analysis may be followed with regard to the angular momentum. We may write the time rate of change of M_z as in Cartesian coordinates as

$$\frac{dM_z}{dt} = \sum_{k=1}^{N} \left[\frac{1}{m_k} (p_{yk}\, p_{xk} - p_{xk}\, p_{yk}) \right] + \sum_{k=1}^{N} \left[\frac{\partial \mathscr{V}}{\partial x_k} y_k - \frac{\partial \mathscr{V}}{\partial y_k} x_k \right]$$

If polar coordinates are used ($x_k = r_k \cos \varphi_k$ and $y_k = r_k \sin \varphi_k$) and if we use Hamilton's equations, then it can be seen that $\partial \mathscr{V}/\partial r_k$ has no moment arm through the origin about the z axis, and the resultant remaining term is then just

$$\frac{dM_z}{dt} = - \sum_{k=1}^{N} \frac{1}{r_k} \frac{\partial (r_k \mathscr{V})}{\partial \varphi_k} \tag{4.23}$$

For a single system, a rotation about the z axis produces no change in $\mathscr{V}$ and therefore

$$\frac{dM_z}{dt} = 0$$

or, in other words

$$M_x = \text{constant} \qquad M_y = \text{constant} \qquad \text{and} \qquad M_z = \text{constant}$$

for the system.

C. DENSITY AND DISTRIBUTION FUNCTIONS AND AVERAGES

In using ensembles in statistical mechanics, however, it is to be noted that *there is no need to maintain distinctions between the individual systems* in the ensemble because only the *number* of systems that would be found in the different states that correspond to different regions in phase space is of interest. This is what we need to determine the probability distribution.

At this point, a brief recapitulation of the aims of this development is in order. We are interested in learning the probability distribution of states of a system. We may find this in one of two ways.

First, we may observe the system as it changes its state and record the number of times it passes through each state. The relation between the number of times the system reaches a given state and all the states of the system is the exact probability distribution of the states of the system, but requires a long time period before the distribution can be obtained to any degree of accuracy.

However, instead, we may use ensemble theory. In this case, we determine a probability distribution by *counting*, *at a single instant of time*, the *number* of systems in the ensemble that are in each state.

The two methods are *not* the same, but their results, given enough representative points, *are equivalent*. This has never been proven to be so, but no contradictions have as yet been found.

Note that, in actuality, no system can really ever be in exactly the same state as it once was. However, we will "count" systems to be in the same state if they lie within the same incremental volume in phase space.

The number of systems in an ensemble should be large enough so that the numbers in different states can be regarded as continuously changing from one point to the next. It is therefore possible to specify a quantity ρ which can be thought of as a state density function. ρ specifies the density with which representative points are scattered over the phase space. We therefore write ρ as

$$\rho = \rho(q_1 \cdots q_{Nr}, p_1 \cdots p_{Nr}, t) \qquad \text{or} \qquad \rho = \rho(q, p, t)$$

as an abbreviation. By use of the density function ρ, we may obtain the number of systems $\Delta\nu$ that would be found at time t to have coordinates lying in any selected infinitesimal range of coordinates (incremental "volume"), that is,

$$\Delta q_1 \cdots \Delta q_{Nr} \, \Delta p_1 \cdots \Delta p_{Nr}$$

in accordance with the equation

$$\Delta\nu = \rho(q, p, t) \, \Delta q_1 \cdots \Delta q_{Nr} \, \Delta p_1 \cdots \Delta p_{Nr} \tag{4.24}$$

Recall again that $\Delta\nu$ is exactly the number of systems lying inside the incremental volume.

The total number of systems ν used in the ensemble can be found by integrating ρ over all phase space. Thus

$$\nu = \int \rho(q, p, t) \, dq_1 \cdots dq_{Nr} \, dp_1 \cdots dp_{Nr} \tag{4.25}$$

To determine the probability of systems occurring within the infinitesimal boundaries, it is only necessary to *normalize* ρ as shown.

$$\text{Prob} = \frac{\rho(q, p, t)}{v} = \frac{\rho(q, p, t)}{\int \rho(q, p, t)\, dq_1 \cdots dq_{Nr}\, dp_1 \cdots dp_{Nr}} \tag{4.26}$$

This normalized function is sometimes known as a *distribution function*. The integral of ρ/v over all phase space is exactly 1.

For any quantity $F(q, p)$, which may be a function of all $2Nr$ coordinates, we may determine its *mean value* for all the systems in an ensemble at a given instant of time to be

$$\langle F \rangle = \frac{1}{v} \int F(q, p)\rho(q, p, t)\, dq_1 \cdots dq_{Nr}\, dp_1 \cdots dp_{Nr}$$

This is called the *ensemble average* value of the quantity F.

This *ensemble average* of the quantity F is not the same as the *system time average* of the quantity F. The latter average is the value of a variable, which may be a function of all the coordinates of all the elements in the system, averaged over time, as the system passes through its various states, whereas the ensemble average is the average of a variable associated with each system in the ensemble at a given instant of time. It will turn out that the time average will, given enough time, approach the ensemble average. Again, this is unproven, but no contradictions have yet been found.

We may now write, for comparison, the mathematical expressions for the two averaging processes.

Time Average

$$\bar{F} = \frac{1}{T} \int_{-T/2}^{+T/2} F(q, p, t)\, dt$$

where T is the interval of time over which the average is taken. T is often passed to the limit of ∞.

Ensemble Average

$$\langle F \rangle = \frac{1}{v} \int F(q, p, t)\rho(q, p, t)\, dq_1 \cdots dp_{Nr}$$

(t is fixed in this averaging process.) Note that q and p may be functions of time, so that the values of the coordinates change with time in the time-averaging process. Also, of course, if the ensemble average is taken at another value of time, these coordinates may change.

D. LIOUVILLE'S THEOREM

We now develop a fundamental theorem of statistical mechanics. Consider a differential volume in phase space

$$\Delta q_1 \cdots \Delta q_{Nr} \, \Delta p_1 \cdots \Delta p_{Nr}$$

The number of systems in this volume are

$$\rho \, \Delta q_1 \cdots \Delta q_{Nr} \, \Delta p_1 \cdots \Delta p_{Nr}$$

If the two faces in this incremental volume perpendicular to the q_1 axis are considered, they can be thought of as being at the coordinates

$$q_1 \qquad \text{and} \qquad q_1 + \Delta q_1$$

The number of systems that cross the face at q_1 into the incremental volume per unit time is

$$\rho \dot{q}_1 \, \Delta q_2 \cdots \Delta q_{Nr} \, \Delta p_1 \cdots \Delta p_{Nr} \tag{4.27}$$

By expanding in a Taylor series we can determine the number of systems leaving the opposite face at $q_1 + \Delta q_1$ per unit time. This flux is

$$\left(\rho + \frac{\partial \rho}{\partial q_1} \Delta q_1\right)\left(\dot{q}_1 + \frac{\partial \dot{q}_1}{\partial q_1} \Delta q_1\right)(\Delta q_2 \cdots \Delta q_{Nr} \, \Delta p_1 \cdots \Delta p_{Nr}) \tag{4.28}$$

neglecting higher-order differentials. The number of systems remaining in the volume due to motion in the q_1 direction is therefore the difference between (4.27) and (4.28), or

$$-\left[\rho \frac{\partial \dot{q}_1}{\partial q_1} + \dot{q}_1 \frac{\partial \rho}{\partial q_1}\right] \Delta q_1 \cdots \Delta q_{Nr} \, \Delta p_1 \cdots \Delta p_{Nr} \tag{4.29}$$

neglecting terms of the order of $(\Delta q_1)^2$. Figure 4.3 shows a diagram for the process leading to Eq. (4.29).

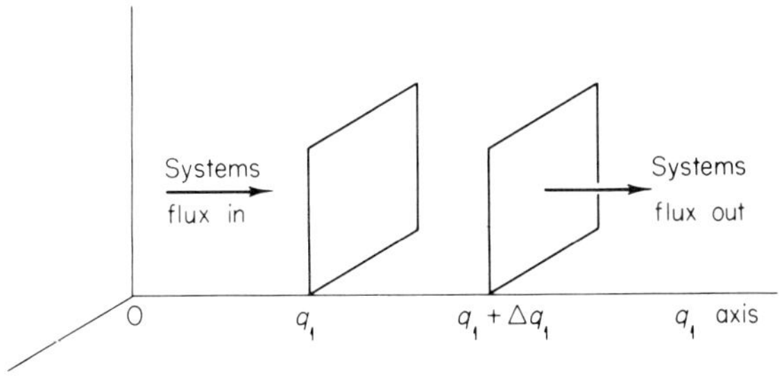

Fig. 4.3. Flux of systems through two parallel faces of an incremental volume.

Extending this last result over all $2Nr$ coordinates *and momenta* and summing gives the net rate of accumulation of systems inside the incremental volume. This is

$$\frac{\partial(\Delta v)}{\partial t} = -\sum_{i=1}^{Nr}\left[\rho\left(\frac{\partial \dot{q}_i}{\partial q_i} + \frac{\partial \dot{p}_i}{\partial p_i}\right) + \left(\frac{\partial \rho}{\partial q_i}\dot{q}_i + \frac{\partial \rho}{\partial p_i}\dot{p}_i\right)\right]\Delta q_1 \cdots \Delta p_{Nr} \quad (4.30)$$

The terms $\partial \dot{p}_i/\partial p_i$ and $\dot{p}_i\, \partial\rho/\partial p_i$ come from the system flux relations in the momentum coordinate directions. Remember that

$$\dot{q}_i = \frac{\partial H}{\partial p_i} \quad (4.31)$$

and

$$\dot{p}_i = -\frac{\partial H}{\partial q_i} \quad (4.32)$$

(Hamilton's equations). From these two equations we may then obtain the following relation.

$$\frac{\partial \dot{q}_i}{\partial q_i} = -\frac{\partial \dot{p}_i}{\partial p_i}$$

This can be found by differentiating (4.31) with respect to q_i and (4.32) with respect to p_i and setting the results equal to each other.

The foregoing shows that the term in the first parenthesis on the right-hand side of Eq. (4.30) is identically equal to zero. That is,

$$\sum_{i=1}^{Nr}\frac{\partial \dot{q}_i}{\partial q_i} + \frac{\partial \dot{p}_i}{\partial p_i} = 0$$

If Eq. (4.30) is divided by $\Delta q_1 \cdots \Delta p_{Nr}$ we then get

$$\frac{\partial \rho}{\partial t} = -\sum_{i=1}^{Nr}\left(\frac{\partial \rho}{\partial q_i}\dot{q}_i + \frac{\partial \rho}{\partial p_i}\dot{p}_i\right) \quad (4.33)$$

Since $\rho = \Delta v/(\Delta q_1 \cdots \Delta p_{Nr})$, we may rewrite Equation (4.33) to be

$$\frac{\partial \rho}{\partial t} = -\sum_{i=1}^{Nr}\left(\frac{\partial \rho}{\partial q_i}\frac{\partial H}{\partial p_i} - \frac{\partial \rho}{\partial p_i}\frac{\partial H}{\partial q_i}\right) = -[\rho, H] \quad (4.34)$$

with the aid of Hamilton's equations. Equation (4.34) is known as *Liouville's theorem*.

We may write Eq. (4.34) in terms of a total derivative, since

$$\frac{d\rho}{dt} = \frac{\partial \rho}{\partial t} + [\rho, H] \quad (4.35)$$

We have seen that, in Eq. (4.34), the two terms on the right-hand side of Eq. (4.35) were equal and opposite. Then we may say

$$\frac{d\rho}{dt} = 0 \tag{4.36}$$

Equation (4.36) is the commonly used formal statement of Liouville's theorem. This theorem states that the complete rate of change with time, considering explicit and implicit variations, at a given point of phase space, of a density function ρ for an ensemble of systems is given by an expression involving the explicit rate of change with time and an additional function of the coordinates and momenta.

Note that the density $\rho(q, p, t)$ depends upon the values of q, p, $\dot{q}$, and $\dot{p}$, which are expressions for the components of ρ in the coordinates through which a representative point would move through phase space.

Liouville's theorem is also a statement that systems in an ensemble are *conserved*; this statement is quite similar in form to the more familiar expression for conservation of charge in electromagnetics. In conservation of charge we write

$$\frac{\partial \rho_E}{\partial t} + \nabla_r \cdot \mathbf{J}_E = 0$$

where ρ_E is the charge density, $\nabla_r \cdot$ is the spatial divergence operator, and $\mathbf{J}_E$ is the electric current density vector. This is equivalent to

$$\begin{aligned}\frac{d\rho_E}{dt} &= \frac{\partial \rho_E}{\partial t} + \frac{\partial \rho_E}{\partial x}\frac{dx}{dt} + \frac{\partial \rho_E}{\partial y}\frac{dy}{dt} + \frac{\partial \rho_E}{dz}\frac{dz}{dt} \\ &= \frac{\partial \rho_E}{\partial t} + \nabla_r \cdot \mathbf{J}_E = 0\end{aligned}$$

where $\mathbf{J}_E = \rho_E \mathbf{v}$. This is true if $\nabla_r \cdot \mathbf{v} = 0$ and there are no sources.

This looks exactly like the formulation in Eq. (4.35). Here the *total* derivative of ρ_E is zero, and this must be true everywhere in position space. Hence ρ_E is conserved. Note that the implicit time differentiation is made along the path (trajectory) of the charge motion.

In other words, in order for the charge to increase or decrease at a given point, it must move in or out along a trajectory. However, since the total derivative of ρ_E is zero along all trajectories, it is impossible to change the charge at one point without inversely changing it at some other point.

In Liouville's theorem we are conserving systems in phase space (which has both position and momentum coordinates) but the idea is exactly the same. The theorem means that the total rate of change of ρ along the trajectory of the system is zero. The trajectory of the system is obtained from the solution to Hamilton's equations for the system.

We will now fit the notions developed in this chapter to a plasma. Recall that a plasma is the *system* that we are interested in, and each of its particles tends to interact with every other particle.

E. THE MICROCANONICAL ENSEMBLE

We shall consider an *equilibrium* condition, in which the *total energy* of the plasma is conserved. We shall construct a particular type of ensemble, called a *microcanonical ensemble*, to aid this development.

Since the energy is assumed to be conserved, it is defined as a *constant of the motion*. This means that, while all the elements of the plasma move on their respective trajectories, the total energy E of the plasma remains constant. We can now produce the microcanonical ensemble, which is the special ensemble required for the plasma. This ensemble is described most easily by a particular region of position and momentum space such that, if the system (plasma) is located anywhere in the region, the energy of the plasma will always be between E and $E + dE$. (Note again that we assume that the system will never have *exactly* the same energy as it changes its state, so we must allow for a "spread" in E.

For example, let us suppose that the system is composed of two free particles in linear motion. The total energy is then

$$E = \frac{P_1{}^2}{2m_1} + \frac{P_2{}^2}{2m_2} \tag{4.37}$$

where P_1 and P_2 are the momenta of the two particles, respectively. We may now examine the phase space for this system. The set of momentum coordinates for the system is shown in Fig. 4.4. The shaded region is the only *allowed location* for all possible *states of the system*. That is, any point in the shaded region will specify the two values of P_1 and P_2 that will make the total energy be between E and $E + dE$. The particles may be *anywhere* in linear *position* space as long as the total energy is not a function of position, which is the case when they are free uncharged particles. Therefore, for this example, the phase space is the shaded area in Fig. 4.4 plus all of position space.

If there are a large number of particles, a similar region of phase space may be constructed, although it is difficult to visualize in more than two dimensions. The only restriction is that the total energy be between E and $E + dE$.

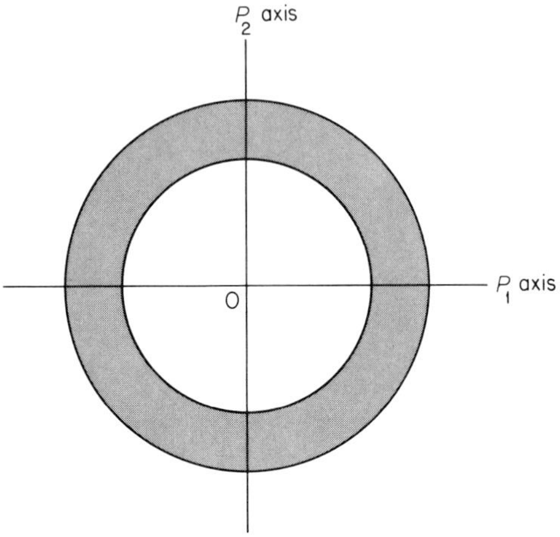

Fig. 4.4. The momentum part of the phase space for two free particles in linear motion.

We now make the fundamental postulate of statistical mechanics:

Any state of the system that is specified by a point in phase space, subject to such restrictions as conservation of energy, and so forth, is equally likely to occur.

In other words, if we were to divide the phase space for the microcanonical ensemble into volumes of equal dimensions, the probability of finding the system in *any* of these volumes of phase space would be equally likely to occur. The fact that the probability is assumed constant is what is most important. Note that the conservation of energy restriction requires that the system be in a particular volume of phase space but within this volume the probability is constant.

F. DEVELOPMENT OF THE DISTRIBUTION LAWS

The next step is building up the representation of the states of the individual elements making up the plasma. That is, we wish to know where the *elements* are in phase space. Once we can determine the possible states of the system from the states of its elements, we will be in a position to find the probability distribution for the elements. Remember again that the elements will be the plasma particles and the system will be the plasma.

We can determine the state of a single element to be defined by the values of the following phase space coordinates.

$$q_1, \ldots, q_r; p_1, \ldots, p_r$$

where r = number of degrees of freedom of each element.

Therefore, $f = Nr$ (the total number of degrees of freedom of the system, since it is to be composed of N elements).

Let $q_1 \cdots q_3$ and $p_1 \cdots p_3$ refer to the position and momentum of the center of gravity of the element and allow the rest of the coordinates to describe only its internal configuration. The set of phase space coordinates would then be

$$x, y, z; q_4, \ldots, q_r; p_x, p_y, p_z; p_4, \ldots, p_r$$

This result is not required for further work, but it is instructive to see how the external and internal coordinates may be separated. We can call these $2r$ axes for describing the *element phase space coordinates* μ space (element space). We now divide up the μ space into a collection of equal incremental regions

$$\Delta V_\mu = \Delta q_1 \cdots \Delta q_r \, \Delta p_1 \cdots \Delta p_r$$

all having the same "volume" in phase space. We call these *cells* in μ space. Each cell can be numbered 1, 2, 3, ..., i, ... We can then specify the state of any element within the range desired by giving its particular cell number in its μ space. Recall that there is a separate set of μ-space coordinates for each element in the system. We can permit more than one element to be in the same cell in μ space if no external constraints are violated. What this means is that these elements are located in the same numbered cell in their particular set of μ-space coordinates. Note that there will be an energy E_i associated with each cell as well, since by knowing the coordinates we may compute the energy of each element. This is not the total energy of the system, but we may compute that as follows. If we define the number of elements (particles) that are in each cell of μ space as n_i, then each particle has energy E_i. The total energy of the system is then

$$E = \sum_{i=1}^{\infty} n_i E_i \tag{4.38}$$

assuming an "infinite" number of cells.

If the velocities of the elements are small compared to that of light, all of the N sets of $2r$ coordinates and momenta for the element become together $2Nr$ coordinates and momenta, which can describe the *system* itself. The state of the whole system can then be regarded as determined by the individual states of its component elements.

We can now construct the γ *space*, or phase space for the *entire system*, which has been used previously. The exact location of a single representative point in γ space ($2Nr$ coordinates) will then give the exact location of each element in μ space and therefore an exact specification of the state of the whole system. This γ space is the phase space needed to describe the system in the microcanonical ensemble, since from γ space we may determine the total energy, or other constants of the motion directly. For example, the coordinates previously described for the two particle system are the γ space for that system. Each point in γ space also has an energy of the system associated with it.

We are now interested in the ranges of γ space. We will therefore divide the γ space into cells as follows.

$$\Delta V_\gamma = (\Delta q_1 \cdots \Delta p_r)_1 (\Delta q_1 \cdots \Delta p_r)_2 \cdots (\Delta q_1 \cdots \Delta p_r)_N$$

The product is taken up to N elements. This is the product of all the volumes of each of the μ space cells in which a particle appears. We may also number these cells. Recall again that if more than one element is in the same cell of μ space, it is permitted as long as any other restrictions do not prohibit it. If this is so, the expression for ΔV_γ may have factors that are the same.

By then specifying the cell number in γ space, we can obtain the state of the system in the range of that cell. Again, remember that γ space is equivalent to the phase space used for ensemble theory. It gives the state of a system at each point. The function ρ gives the density distribution of *systems* in γ space.

It is important to note that a number of these cells in γ space would correspond to the *same condition of the system from its macroscopic properties*, since from a macroscopic viewpoint *it is assumed to make no difference* which particular elements of the N similar ones are taken as lying in specified cells of μ space.

If, for example, the system was specified by a region ΔV_γ having such a character that each of the N elements is assigned to a different cell in its μ space, then it is evident that $N!$ different regions in the γ space all correspond to the same gross observational properties. Remember that for a given state of the system we will specify the numbers of elements in each corresponding cell in its μ *space* to be

$$n_1, n_2, n_3, \ldots, n_i, \ldots$$

without specifying just which elements are used. Then there will be a total of

$$G = \frac{N!}{n_1!\, n_2!\, n_3! \cdots n_i! \cdots} \tag{4.39}$$

different incremental volumes of γ space which would all correspond to the same macroscopic condition. In other words, these are the different microscopic conditions that can exist without our knowing *which* of the elements are in *which* cell and are independent of the order of selection. N, again, is the total number of elements in the system. This can be shown as follows.

Starting with the first cell (1) in μ space with n_1 molecules, there are exactly N ways of choosing the first element from a "pot" of N elements that compose the system. There are $(N - 1)$ ways of choosing the second element, and so on, so that the number of ways of choosing n_1 elements is

$$G_{n_1} = N(N-1)(N-2)\cdots(N-n_1+1) = \frac{N!}{(N-n_1)!} \tag{4.40}$$

In this last expression, we have counted as a different arrangement each separate sequence in which the first n_1 particles could have been selected. However, we need to know only the number of elements that are in cell (1), not the order in which they have been removed from the "pot." Since there are $n_1!$ different ways of ordering these n_1 elements, then we must divide (4.40) by $n_1!$. Then

$$G_1 = \frac{N!}{n_1!(N-n_1)!}$$

is the number of ways of placing n_1 molecules in cell (1), independently of the order in which they are placed. G_2 is found in the same way and is given as

$$G_2 = \frac{(N-n_1)!}{n_2!(N-n_1-n_2)!}$$

since there are only $N - n_1$ molecules left to put in cell (2).

The total number of ways of arranging the elements in all of the cells in μ space is then just given by the product of all the G factors

$$G = G_1 G_2 G_3 \cdots G_i \cdots$$

which is

$$G = \frac{N!(N-n_1)!(N-n_1-n_2)!\cdots(N-n_1\cdots-n_{N-i})!}{n_1!(N-n_1)!n_2!(N-n_1-n_2)!\cdots n_i!(N-n_1-n_2\cdots-n_i)!\cdots}$$

or

$$G = \frac{N!}{n_1!\,n_2!\,n_3!\cdots n_i!\cdots} \tag{4.41}$$

since $0!$ is 1. Recall, then, that this expression is the total number of states that the elements can have that correspond to the *same* macroscopic state of the *system*.

Inside each cell in μ space there may be a number (ε_i) of discrete energy levels. These may restrict the energies that each *element* may have, but the total energy of the system must still be between E and $E + \Delta E$. To determine the various probability distributions, three kinds of systems will be considered. They are composed of elements, which we will now call *particles*. The three systems are

(a) identical, but distinguishable, particles;
(b) identical, indistinguishable particles of $\frac{1}{2}$ integral spin;
(c) identical, indistinguishable particles of integral spin.

We now investigate the three cases and find the macroscopic state of the system that is most probable. This state is the one that has the greatest number of arrangements of the particles corresponding to it.

In case (a), the exclusion principles does not apply, and therefore each particle is equally likely to be in any energy level. That is, there is always room for another particle, no matter how many are already present. There are ε_i ways the first particle can be placed in cell i in μ space, ε_i ways for the second particle, and so on. The total number of possibilities, considering the energy levels, is therefore

$$\varepsilon_i^{n_i}$$

This means that the G term must now be

$$G_a = N! \prod_{i=1}^{k} \frac{\varepsilon_i^{n_i}}{n_i!} \qquad \text{for case (a)} \tag{4.42}$$

if we have k cells (k may approach ∞). In case (b), the indistinguishability of the particles prevents us from knowing which ones of the N particles are placed in a cell in μ space, so the only distinguishing feature of a given microscopic distribution is which of the ε_i energy levels are occupied by the n_i molecules. The exclusion principle requires that not more than one particle be in each energy level.

If we first imagine the particles to be distinguishable, the first particle can be put in any one of the ε_i levels. For each of these choices, because of the exclusion principle, the second particle can only be put into $\varepsilon_i - 1$ levels, and so on. It must therefore be true that $\varepsilon_i \geq n_i$ for all cells.

The total number of arrangements of n_i particles in cell i is therefore

$$G_i = \varepsilon_i(\varepsilon_i - 1) \cdots (\varepsilon_i - n_i + 1) = \frac{\varepsilon_i!}{(\varepsilon_i - n_i)!} \qquad \text{for case (b)}$$

However, the indistinguishability of the particles now requires that we not count the various possible permutations of the particles among themselves, so we must again divide by $n_i!$, as was done previously. This results in

$$G_i = \frac{\varepsilon_i!}{n_i!(\varepsilon_i - n_i)!} \tag{4.43}$$

For the total number of microscopic distributions (possible states of the particles) that can lead to a given *macroscopic* state (state of the system), we now obtain, by taking the product of all of the appropriate terms like (4.43), the following.

$$G_b = \prod_{i=1}^{k} G_i = \prod_{i=1}^{k} \frac{\varepsilon_i!}{n_i!(\varepsilon_i - n_i)!} \qquad \text{for case (b)} \tag{4.44}$$

For case (c), the indistinguishability of the particles again prevents us from knowing which of the N particles have been placed in each cell in μ space, but now the exclusion principle does not act to limit the population of a given energy level.

Consider the following pictorial device. Suppose the ith cell consists of $n_i + \varepsilon_i - 1$ spaces, into which particles or partitions between energy levels may be placed. $\varepsilon_i - 1$ partitions break down into ε_i energy levels as shown.

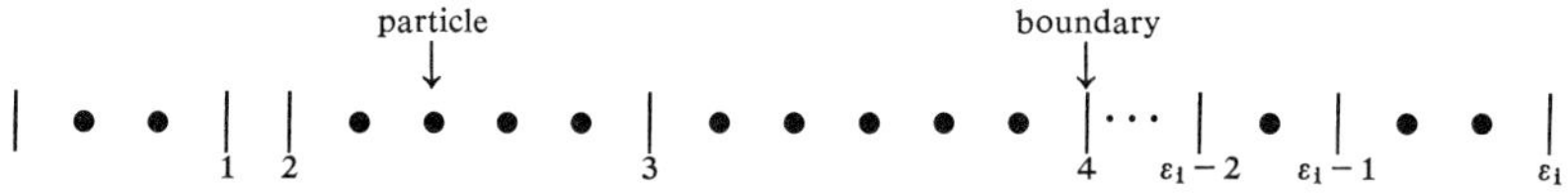

To arrange n_i particles in the ε_i energy levels we can now see, if we use the pictorial device, that the number of ways of taking the first particle is $n_i + \varepsilon_i - 1$ and the number of ways for the second particle is $n_i + \varepsilon_i - 2$, and so on. The total number of ways for n_i particles is then

$$G_i = (n_i + \varepsilon_i - 1)(n_i + \varepsilon_i - 2) \cdots n_i = \frac{(n_i + \varepsilon_i - 1)!}{(\varepsilon_i - 1)!}$$

We must divide by $n_i!$ for the same reasons as before, to eliminate the ways in which the order of taking the particles matters, so G_i is now

$$G_i = \frac{(n_i + \varepsilon_i - 1)!}{n_i!(\varepsilon_i - 1)!}$$

The resultant total number of ways for a microscopic distribution to be the same as a macroscopic distribution for case (c) is then

$$G_c = \prod_{i=1}^{k} \frac{(n_i + \varepsilon_i - 1)}{n_i!(\varepsilon_i - 1)!} \tag{4.45}$$

since we have taken the product for k cells.

We now have a set of three numbers, corresponding to the number of possible microscopic states possible for the three cases stated. We wish to find the probability distribution of the states. To do this we multiply each state by the *probability* of finding the system in that state. From the discussion previously developed, we have assumed that the probability of a given microscopic state is the same for all states, and the result is that the G factors are simply multiplied by a constant to obtain the probability, or

$$\text{Prob} = G \cdot \text{constant}$$

This gives us the probability of finding the system in a state specified by the numbers n_i of elements in their respective cells in μ space, i.

We now recall that there will be some *microscopic* states that correspond to the same *macroscopic* state. The most probable macroscopic state is therefore the one that has the most microscopic states corresponding to it, subject to these two auxiliary conditions:

$$\sum_{i=1}^{\infty} n_i = N = \text{constant} \tag{4.46}$$

conservation of particles, and

$$\sum_{i=1}^{\infty} E_i n_i = E = \text{constant} \tag{4.47}$$

conservation of energy. E_i is the energy of the particles in the ith cell in μ space. We have also extended the number of cells to infinity instead of k. The values of the three G (or probability) functions are then dependent on the numbers of the particles in each cell in μ space.

We are usually dealing with such large numbers of particles, and even with such large numbers of energy levels in a single cell, that we may consider the various summation expressions above to be continuous functions of continuous variables and thus treat the problem by using calculus. For convenience, we now change the problem to the equivalent one of maximizing the logarithm of $G(n_i)$ (times a constant) where for the three cases considered the logarithm of G is

Case (a)

$$\ln G(n_i) = \ln N! + \sum_{i=1}^{\infty} [n_i \ln \varepsilon_i - \ln n_i!] \tag{4.48}$$

Case (b)

$$\ln G(n_i) = \sum_{i=1}^{\infty} [\ln \varepsilon_i! - \ln n_i! - \ln(\varepsilon_i - n_i)!] \tag{4.49}$$

Case (c)

$$\ln G(n_i) = \sum_{i=1}^{\infty} [\ln(n_i + \varepsilon_i - 1)! - \ln n_i! - \ln(\varepsilon_i - 1)!] \tag{4.50}$$

We must now maximize the value of ln G with respect to all small variations of the various μ-space cell populations that can occur while still satisfying the auxiliary conditions. That is, if each n_i changes by a small amount δn_i, we must have, for the change in ln G

$$\delta(\ln G) = 0 \tag{4.51}$$

for any small values of δn_i that satisfy both of the following relations.

$$\delta N = \sum_{i=1}^{\infty} \delta n_i = 0 \tag{4.52}$$

$$\delta E = \sum_{i=1}^{\infty} E_i \, \delta n_i = 0 \tag{4.53}$$

Equation (4.51) shows us that we are at an extremum. We use Lagrange's method of undetermined multipliers in order to utilize conditions (4.52) and (4.53). This results in the following variational equation.

$$\delta(\ln G) - \alpha \, \delta N - \beta \, \delta E = 0 \tag{4.54}$$

α and β are constants to be determined. If we now look at Eqs. (4.51), (4.52), and (4.53), we see that if they are all satisfied, then Eq. (4.54) will automatically be satisfied.

Case (a) will be considered first. Equation (4.51) for this case is obtained from Eq. (4.48). It is

$$\delta(\ln G) = \sum_{i=1}^{\infty} (\ln \varepsilon_i - \ln n_i) \, \delta n_i \tag{4.55}$$

using the approximation $\delta(\ln n_i!) \approx \ln n_i \, \delta n_i$, which is true if $\delta n_i \ll n_i$.

Putting this result in Eq. (4.54), we obtain, for Case (a)

$$\sum_{i=1}^{\infty} (\ln \varepsilon_i - \ln n_i - \alpha - \beta E_i) \, \delta n_i = 0 \tag{4.56}$$

The two conditions on conservation of particles and energy can result in two undertermined coefficients δn_i. We can obtain values of α and β to make each term in the series (4.56) be zero if we realize that the δn_i's are arbitrary.

The result is that

$$\ln \varepsilon_i - \ln n_i - \alpha - \beta E_i = 0 \tag{4.57}$$

for *all* values of *i*. We can find similar equations for cases (b) and (c). They are

$$-\ln n_i + \ln(\varepsilon_i - n_i) - \alpha - \beta E_i = 0 \qquad \text{for case (b)} \tag{4.58}$$

and

$$\ln(n_i + \varepsilon_i) - \ln n_i - \alpha - \beta E_i = 0 \qquad \text{for case (c)} \tag{4.59}$$

Solving (4.57), (4.58), and (4.59) for n_i, we obtain the desired forms of the distribution functions for the three cases:

$$n_i = \frac{\varepsilon_i}{\exp(\alpha + \beta E_i)} \qquad \text{case (a)} \tag{4.60}$$

$$n_i = \frac{\varepsilon_i}{\exp(\alpha + \beta E_i) + 1} \qquad \text{case (b)} \tag{4.61}$$

$$n_i = \frac{\varepsilon_i}{\exp(\alpha + \beta E_i) - 1} \qquad \text{case (c)} \tag{4.62}$$

Equations (4.60), (4.61), and (4.62) are known respectively as Maxwell–Boltzmann, Fermi–Dirac, and Bose–Einstein distributions.

The constant multipliers α and β may be evaluated from the conditions of conservation of particles and energy for the three systems of particles. We shall be primarily concerned with case (a), the Maxwell–Boltzmann distribution. It will turn out, however, that the constant β will have the same form for all three cases:

$$\beta = \frac{1}{kT} \tag{4.63}$$

where T is the absolute temperature and k is Boltzmann's constant (see Problem 6). The three laws then become

$$n_i = \varepsilon_i \exp\left(-\alpha - \frac{E_i}{kT}\right) \qquad \text{Maxwell–Boltzmann} \tag{4.64}$$

$$n_i = \frac{\varepsilon_i}{\exp[\alpha + (E_i/kT)] + 1} \qquad \text{Fermi–Dirac} \tag{4.65}$$

$$n_i = \frac{\varepsilon_i}{\exp[\alpha + (E_i/kT)] - 1} \qquad \text{Bose–Einstein} \tag{4.66}$$

Plots of the three functions are shown in Fig. 4.5. The constant e^{α} may be used as a normalization constant if the number of particles is conserved.

It should be noted that for high values of E_i, the form of all three distribution functions becomes essentially that of Eq. (4.64). In other words, all of these distribution functions tend to look Maxwellian in nature for high

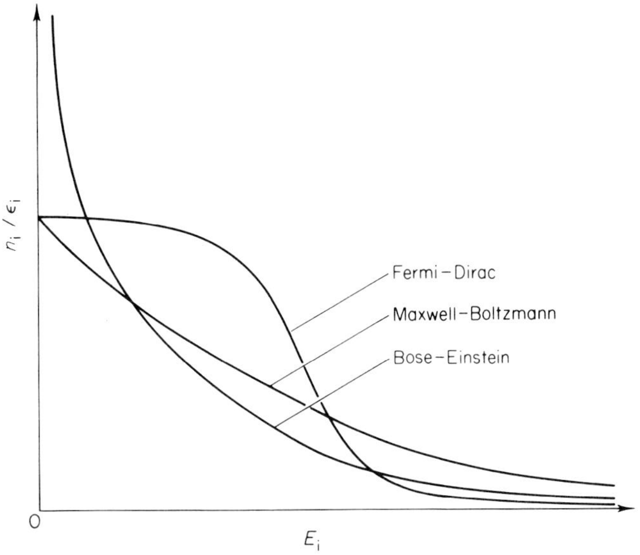

Fig. 4.5. The energy distributions of Maxwell–Boltzmann, Fermi–Dirac, and Bose–Einstein statistics.

energies. Note that if there is only one energy level in each cell, then in the functions (4.64), (4.65), or (4.66), ε_i becomes unity. Often the spacing between cells (in energy) is so small that only one energy level is possible.

G. VELOCITY, SPEED, AND ENERGY DISTRIBUTION FUNCTIONS

Much of plasma work begins with some kind of basic and *equilibrium* distribution function. In many cases, the Maxwellian distribution, or a variant of it, is assumed to specify the state of a system of plasma particles, both ions and electrons. The temperatures of these species of particles can be, and often are, quite different from each other. The form of the distribution function can be obtained from (4.64) by noting that we will henceforth use

$$f_N(q, p, t) = A(q, t) \exp\left(-\frac{E_i}{kT}\right) \tag{4.67}$$

for the function. We use the subscript N to show that this function is the energy distribution function for N particles (elements) that compose the plasma (system). E_i is the energy at the ith cell in μ space and in general is of the form

$$E_i = \frac{p_i^{\ 2}}{2m} + \mathscr{V}_i(q, t) = \frac{1}{2}m(v_{i_x}^2 + v_{i_y}^2 + v_{i_z}^2) + \mathscr{V}_i(x_i, y_i, z_i, t)$$

if we do not consider the internal energy or interactions between the particles. By integrating Eq. (4.67) over all velocity space and passing to the limit where the energy spacing between cells is infinitesimally small, we can find that

$$A(q, t) = n(q, t)\left(\frac{m}{2\pi kT}\right)^{3/2} \tag{4.68}$$

where $n(q, t)$ is the density function in *position* space. The value of the integral of (4.68) over all position space is set equal to N, the total number of particles. The potential energy comes out of the integral in (4.67) since only velocity space is integrated out, and we have assumed that it is made part of $A(q,t)$. We can use the ensemble averaging process to obtain the average kinetic energy as

$$\langle E_{kin}\rangle = \frac{m\langle v^2\rangle}{2} = \frac{3}{2}kT \tag{4.69}$$

The Maxwellian velocity distribution function is now

$$\begin{aligned}\delta N &= f_N(q, v, t)\, dv_x\, dv_y\, dv_z \\ &= n(q, t)\left(\frac{m}{2\pi kT}\right)^{3/2} \exp\left[-\frac{m(v_x^{\ 2} + v_y^{\ 2} + v_z^{\ 2})}{2kT}\right] dv_x\, dv_y\, dv_z \end{aligned} \tag{4.70}$$

where δN is the number of particles in a given incremental volume.

Equation (4.70) is the *velocity distribution*. That is, it gives the number of particles having *components of velocity* between v_x and $v_x + dv_x$, v_y and $v_y + dv_y$, and v_z and $v_z + dv_z$ at a given point in position space. The left-hand side of (4.70) has been set equal to a variable δN, which represents the number of particles in the range specified.

We may modify Eq. (4.70) in several ways to give us more information. If we assume *spherical* coordinates in velocity space, then Fig. 4.6 applies. The spherical volume element is then

$$v^2 \sin\theta\, dv\, d\theta\, d\varphi$$

Therefore, if we are interested in the *speed* distribution, Eq. (4.70) becomes

$$\begin{aligned}\delta N &= f_N(q, v, \varphi, \theta, t)v^2 \sin\theta\, dv\, d\theta\, d\varphi \\ &= n(q, t)\left(\frac{m}{2\pi kT}\right)^{3/2} \exp\left(\frac{-mv^2}{2kT}\right) v^2 \sin\theta\, dv\, d\theta\, d\varphi \end{aligned} \tag{4.71}$$

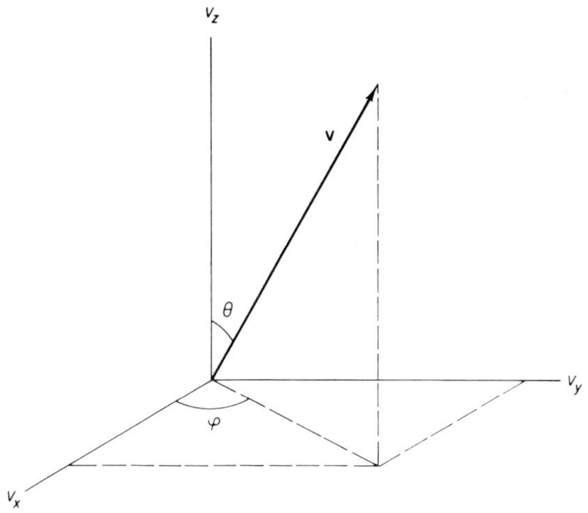

Fig. 4.6. Spherical velocity coordinates.

In Eq. (4.71) we have converted the function $f_N(q, v_x, v_y, v_z, t)$ to spherical coordinates. This gives the speed distribution (δN is now the number with *speed* between v and $v + dv$) in the direction defined by the angles θ and φ. We may integrate over these angles to obtain the speed distribution *regardless* of angle. The δN now is

$$\delta N = 4\pi f_N(q, v, t)v^2\,dv = n(q, t)\left(\frac{m}{2\pi kT}\right)^{3/2}\exp\left(\frac{-mv^2}{2kT}\right)4\pi v^2\,dv \quad (4.72)$$

If, finally, we wish the *energy* distribution, we may use (4.72) with the substitution $E = \frac{1}{2}mv^2$. Then the energy distribution is

$$\delta N = 2\pi f_N(q, E, t)E^{1/2}\,dE = n(q, t)\left(\frac{1}{\pi kT}\right)^{3/2}\exp\left(\frac{-E}{kT}\right)2\pi E^{1/2}\,dE \quad (4.73)$$

Equation (4.73) gives the number of particles whose energy is between E and $E + \Delta E$. We may plot these various distributions on the same set of coordinates for comparison, as shown in Fig. 4.7. We will compare Eqs. (4.70) and (4.72), which show the difference between the velocity and speed distributions. Recall that the *velocity* distribution is symmetric about the $v = 0$ axis, and extends from $v = -\infty$ to $v = +\infty$. Note that the average velocity will be zero, but the average speed is not zero!

There are often situations in a plasma when the distribution is not Maxwellian. However, the non-Maxwellian functions often look very much like Eq. (4.70) in form. For example, the average energy, and therefore the temperature, may be different in different directions. Such an anisotropic

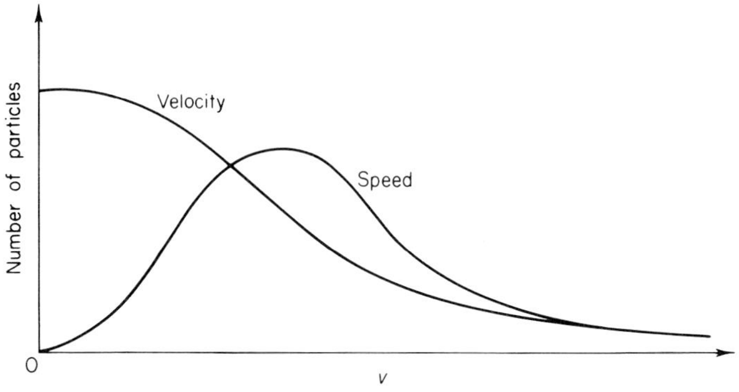

Fig. 4.7. Velocity and speed distributions.

velocity distribution may sometimes be written in terms of two "temperatures," T_1 and T_2. Perhaps T_1 is the "temperature" for the components of velocity in the x direction and the components of velocity in the y and z directions are at temperature T_2. The form of this anisotropic velocity distribution function could then be

$$\begin{aligned}\delta N &= f_N(q, v_x, v_y, v_z, t)\, dv_x\, dv_y\, dv_z \\ &= \frac{n(q, t)m^{3/2}}{(2\pi k)^{3/2} T_1^{1/2} T_2} \exp\left(-\frac{mv_x^2}{2kT_1} - \frac{m(v_y^2 + v_z^2)}{2kT_2} \right) dv_x\, dv_y\, dv_z\end{aligned} \tag{4.74}$$

The Maxwellian is often used even though the actual distribution function is not Maxwellian, especially if it is only *slightly* non-Maxwellian or if the actual form of the function is intractable or unmeasurable. It should be noted that the function described in Eq. (4.74) is not an equilibrium distribution, and is not derivable using the methods of this chapter. However, it is very useful as an approximation in many cases.

In any event, it is important to know the form of the distribution function, both from the viewpoint of determining the plasma temperature and also for developing theories on the nature of plasma behavior under *nonequilibrium* conditions. Very often, instabilities and waves may appear if the distribution function is made non-Maxwellian.

H. APPLICATION OF ENSEMBLE THEORY TO A PLASMA

The problem to be considered now is the following. We have permitted ourselves to consider a development for a particle (element) distribution function, and in doing so have *neglected* interactions. There is a great advantage in doing so. It will permit us to apply *ensemble theory* to a plasma. In this case the plasma can be considered as an "ensemble" and the particles are "systems." It is only this assumption that now permits us to say that we may write a Liouville theorem for the plasma.

Liouville's theorem was developed for the function ρ, which was the distribution function for systems (in γ space) in an ensemble. It is simply written as

$$\frac{df_N}{dt} = 0 \tag{4.75}$$

for the distribution function f_N, for the elements in the system, *assuming noninteraction of the elements.*

Note that in the function ρ, all the systems in an ensemble correspond to *possible* states of the same system, whereas in f_N, the coordinates of all the elements correspond to the *actual* states of all the elements at the time f_N is evaluated. The distribution function f_N is, however, still very similar to ρ. It tells us the distribution of all *elements* in the *system.* To repeat: We can assume that Liouville's theorem may *also* be applied to the element distribution function if interactions between particles are neglected. This is an approximation, because the ensemble theory used to develop Liouville's theorem assumed *noninteracting* systems, and the plasma particles do interact.

If interactions (collisions) are to be considered, then Liouville's theorem does not apply, but, as we shall see later, we can make a perturbation expansion of the distribution function that will permit us to extend the theory to include the effects of collisions.

It should be remembered that the f_N used in Liouville's theorem for a plasma is the distribution function for all elements in the system. We are now interested in obtaining a result that will enable us to predict the behavior of a single particle (element).

Suppose we now try to find the distribution function for a single particle. We note that f_N gives the number of particles per unit volume of phase space. It should now be possible to see that f_N can be written in terms of N sets of μ-space coordinates (that is, f_N is a function of all the individual coordinates of each particle), since a separate μ space could be defined for each particle. So, we may write f_N as

$$f_N = f_N[(q_1 \cdots p_r)_1(q_1 \cdots p_r)_2 \cdots] \tag{4.76}$$

By separating these out, we are utilizing the μ space previously defined. It is possible to use *either* notation interchangeably. Note that in the γ-space notation, the function f_N is only at a single point, while the function ρ may be spread out over this phase space. Thus, a single system distribution function may be obtained by integrating over all the coordinates and momenta *except* that for one system. Performing this integration results in a function

$$f = \int f_N (dq_1 \cdots dp_r)_2 (dq_1 \cdots dp_r)_3 \cdots \tag{4.47}$$

The function f is the *single-particle* distribution function. It will have a form similar to that of Eq. (4.70). We may now also apply this integration to Liouville's theorem to obtain

$$\int \frac{df_N}{dt}(dq_1 \cdots dp_r)_{2\cdots N} = \int \frac{\partial f_N}{\partial t}(dq_1 \cdots dp_r)_{2\cdots N} + \int [f_N, H](dq_1 \cdots dp_r)_{2\cdots N} \tag{4.78}$$

Equation (4.78) becomes, after completing some of the integration,

$$\frac{df}{dt} = \frac{\partial f}{\partial t} + \int [f_N, H](dq_1 \cdots dp_r)_{2\cdots N} \tag{4.79}$$

We must now investigate the second term on the right-hand of Eq. (4.79). The Hamiltonian H for the plasma as a whole is, if there are no interactions between particles, merely the sum of the Hamiltonians of the individual particles. If H is this sum, then (4.79) has no interaction terms present and we obtain

$$\frac{df}{dt} = \frac{\partial f}{\partial t} + [f, H_1] = 0 \tag{4.80}$$

where H_1 is the Hamiltonian of a single system. Equation (4.80) is known as the Vlasov or the collisionless Boltzmann equation. Note that the function f (without subscript) is a function *only* of the coordinates of a single particle at a given point in μ space. Its value is the *probability* of finding that particle in the particular incremental volume of μ space associated with that point. Therefore, the integral of f over all of μ space must be 1.

If interactions (collisions) between particles are considered, then there does exist a coupling between the coordinates of two particles, and the Boltzmann–Vlasov equation (4.80) may not apply. More will be said about this later.

The development of the function f_N or f has been the goal of this chapter. We will now examine some of its uses in plasma studies. In fact, throughout most of the rest of this book, the distribution function will be close at hand.

Suggested Reading

Goldstein, S., "Classical Mechanics," Addison-Wesley, Reading, Massachusetts, 1956. This book gives an excellent introduction to the Hamiltonian formulations, Hamilton–Jacobi theory, generalized coordinates, and more.

Kittel, C., "Elementary Statistical Physics," Wiley, New York, 1958. A clear, concise explanation of ensembles, systems, and so forth, is given.
Leighton, R. B., "Principles of Modern Physics," McGraw-Hill, New York, 1959.
Present, R. D., "Kinetic Theory of Gases," McGraw-Hill, New York, 1958.
Sears, F. W., "An Introduction to Thermodynamics, Kinetic Theory and Statistical Mechanics," Addison-Wesley, Reading, Massachusetts, 1953.
Tolman, R. C., "The Principles of Statistical Mechanics," Oxford Univ. Press, London and New York, 1962. This is a basic text in the field of statistical mechanics.

Problems

1 Consider a system that is a harmonic oscillator. That is, a system whose equation of motion is

$$m\frac{d^2x}{dt^2} + kx = 0$$

Show that, if energy is conserved, we may construct a phase space of two coordinates x and p_x, and that the path in phase space of the state of the system is an ellipse. Write an expression for the trajectory in phase space.

2 The Hamiltonian of a charged particle orbiting in a magnetic field is

$$H = \mathbf{p}\cdot\mathbf{v} - \tfrac{1}{2}mv^2 + q\,\varphi - q\mathbf{v}\cdot\mathbf{A}$$

or

$$H = \tfrac{1}{2}m(\mathbf{p} - q\mathbf{A})^2 - q\varphi$$

where φ is the scalar electric potential and $\mathbf{A}$ is the vector potential.

(a) Write Hamilton's equations for this particle.
(b) Show that the two definitions of H are identical.
(c) Show that the angular momentum of the particle in cylindrical coordinates about the magnetic field axis is

$$mr^2\dot{\theta} + qrA_\theta$$

where A_θ is the vector potential needed for a uniform magnetic field that is constant and points only in the z direction.

3 (a) Show that the root mean square velocity of an isotropic Maxwellian distribution is

$$v_{\text{rms}} = \left(\frac{3kT}{m}\right)^{1/2}$$

(b) Show that the most probable speed is

$$v_{\text{mps}} = \left(\frac{2kT}{m}\right)^{1/2}$$

(c) Calculate the total number of particles passing through a unit area per unit time for an isotropic uniform Maxwellian velocity distribution. Hint: Consider the flux in both directions.

4 What is the velocity distribution of uncharged particles escaping from a box in which there is an isotropic Maxwellian distribution of velocities? The particles escape through a very small hole in one face of the cubical box. In fact, the hole diameter is much smaller than the mean free path.

5 Find the average energy, speed, and velocity for particles characterized by Fermi–Dirac distribution.

6 In taking case (a) (Maxwell–Boltzmann statistics), we wish to evaluate the constant β as in Eq. (4.63). Let us assume an ideal gas, so that we may write, for the pressure

$$p = n k T$$

where n is the number of particles per unit volume in position space, k is Boltzmann's constant, and T is the temperature. At the same time, we rewrite Eq. (4.67) so it is similar in form to (4.64); that is,

$$f_N = Ae^{-\beta E_i}$$

Use this equation to develop an expression for the pressure on a plane surface in one direction, for example, the x direction. Equate the two relations for pressure and solve for β. The result, as shown in (4.63) must be

$$\beta = \frac{1}{kT}$$

7 Discuss why the distribution function in Eq. (4.74) is a nonequilibrium function.

8 Develop Eq. (4.23).

9 Find the net accumulation of systems in an incremental volume in phase space due to motion in the p_1 momentum coordinate direction.

5 Further Aspects of Collective Phenomena: Statistics of Collisions and Fluid Behavior

The previous chapter covered a new concept: a statistical approach to collective phenomena. The result was a distribution function f_n or f for the particles in a plasma, and a "conservation" relation in phase space: Liouville's theorem for f_n and the collisionless Boltzmann equation (the Vlasov equation) for f.

The present chapter investigates three topics of further importance. First, we consider the fundamental properties of the collisionless Boltzmann equation and the nature of its solutions. Then, various methods for including collisional processes in the statistical description will be investigated. And finally, we extend our work to consider how to "integrate out" the velocity dependence of the collisionless Boltzmann equation, yielding a fluid model for a plasma.

A. ADIABATIC INVARIANTS AND CONSTANTS OF THE MOTION

We continue our investigation of the collisionless Boltzmann (Vlasov) equation by considering a collection of charged particles of one species. It may be then written as

$$\frac{df}{dt} = \frac{\partial f}{\partial t} + \frac{\partial f}{\partial x}\frac{dx}{dt} + \frac{\partial f}{\partial y}\frac{dy}{dt} + \frac{\partial f}{\partial z}\frac{dz}{dt} + \frac{\partial f}{\partial v_x}\frac{dv_x}{dt} + \frac{\partial f}{\partial v_y}\frac{dv_y}{dt} + \frac{\partial f}{\partial v_z}\frac{dv_z}{dt} = 0 \qquad (5.1)$$

Now $dx/dt = v_x, \ldots,$ and $dv_x/dt = F_x/m, \ldots,$ where v_x and F_x are the velocities and forces *per particle*. Note that this means also that we are moving *along the trajectory of a single particle.* We have also assumed that the mass is independent of velocity (nonrelativistic case) so Eq. (5.1) becomes

$$\frac{\partial f}{\partial t} + \frac{\partial f}{\partial x} v_x + \frac{\partial f}{\partial y} v_y + \frac{\partial f}{\partial z} v_z + \frac{\partial f}{\partial v_x}\frac{F_x}{m} + \frac{\partial f}{\partial v_y}\frac{F_y}{m} + \frac{\partial f}{\partial v_z}\frac{F_z}{m} = 0 \tag{5.2}$$

where again, F is the force per particle.

By means of vector notation, we can simplify Eq. (5.2) to be

$$\frac{\partial f}{\partial t} + \mathbf{v} \cdot \nabla_r f + \frac{\mathbf{F}}{m} \cdot \nabla_v f = 0 \tag{5.3}$$

Note that the operators ∇_r and ∇_v are the gradient operators in position and velocity space, respectively. That is,

$$\nabla_r = \frac{\partial}{\partial x} \hat{a}_x + \frac{\partial}{\partial y} \hat{a}_y + \frac{\partial}{\partial z} \hat{a}_z \equiv \frac{\partial}{\partial \mathbf{r}}$$

and

$$\nabla_v = \frac{\partial}{\partial v_x} \hat{a}_x + \frac{\partial}{\partial v_y} \hat{a}_y + \frac{\partial}{\partial v_z} \hat{a}_z \equiv \frac{\partial}{\partial \mathbf{v}}$$

We will now consider in greater detail the nature of the *third* term on the left-hand side of Eq. (5.3). In considering a plasma, we will look at this force per particle produced by electric, magnetic, and/or gravitational fields. The general force term would then be

$$\mathbf{F} = q(\mathbf{E} + \mathbf{v} \times \mathbf{B}) + m\mathbf{g} \tag{5.4}$$

So Eq. (5.3) becomes

$$\frac{\partial f}{\partial t} + \mathbf{v} \cdot \nabla_r f + \frac{q}{m}(\mathbf{E} + \mathbf{v} \times \mathbf{B}) \cdot \nabla_v f + \mathbf{g} \cdot \nabla_v f = 0 \tag{5.5}$$

The collisionless Boltzmann equation (Eq. (5.3)) is, of course, a partial differential equation of first order. A solution for f can, in principle, be obtained by knowing the nature of the force term and the initial and boundary conditions of the problem.

There is another way to solve this equation, however. There may be certain quantities that remain *invariant over the trajectories* of the particles. These quantities, called *constants of the motion*, result in a coupling between at least two of the μ-space coordinates utilized in the Boltzmann equation. Every new independent constant of the motion successively eliminates another coordinate until, if all $2r$ constants of the motion are known, the solution of the collisionless Boltzmann equation is obtained directly.

An example of a constant of the motion might be the total energy in a one-dimensional problem of a particle moving in a potential field. This can be expressed as

$$E_{\text{total}} = \tfrac{1}{2}mv_x^2 + \mathscr{V}(x) \tag{5.6}$$

where $\mathscr{V}(x)$ is the potential energy. If E_{total} is a constant of the motion, then there exists a relation between v_x and x. We have just simplified the problem to that of one coordinate, since once we know x, we know v_x, and vice versa. The most general solution, therefore, to the Boltzmann equation would be

$$f(q_1, \ldots, q_r; p_1, \ldots, p_r) = f(c_1, \ldots, c_{2r}) \tag{5.7}$$

Remember that the single particle has r degrees of freedom. $c_1, \ldots, c_{2r}$ are appropriate constants of the motion. We will see shortly how constants of the motion reduce the number of coordinates in plasma problems.

First, we will examine some special constants of the motion used in plasma work in more detail. Specifically, we will consider two constants: the magnetic moment μ_m, and the action J. These quantities may not always be constant but may depend upon how "slowly" certain variables change. We call these kinds of constants of the motion *adiabatic invariants*.

The previously mentioned assumed constant of the motion, the magnetic moment μ_m, allowed us to obtain a relation between the velocity perpendicular to the magnetic field and the velocity parallel to the field lines by the use of an additional constant of the motion, the energy.

However, the magnetic moment is *not* always conserved, as can be seen by examining Fig. 5.1. This is a representation of the cusped geometry presented

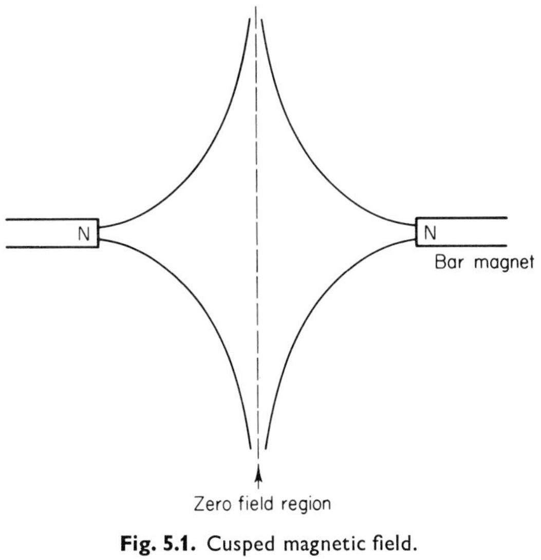

Fig. 5.1. Cusped magnetic field.

in Chapter 3. If a charged particle managed to go into the zero field region, the magnetic moment $W_\perp/B$ would become infinite unless $W_\perp$ (kinetic energy perpendicular to the field lines) became zero, which is not possible. In this case then, the magnetic moment is *not* conserved. But it is possible, under other conditions, to conserve the magnetic moment if the variation of some parameters of the motion is not too rapid. Quantities of this general type are adiabatic invariants.

We shall now develop the conditions under which the magnetic moment may be assumed constant. We previously defined the magnetic moment of a charged particle to be

$$\boldsymbol{\mu}_m = -\frac{v_\perp q r_g \mathbf{B}}{2B} = -\frac{W_\perp \mathbf{B}}{B^2} \tag{5.8}$$

We first consider the magnetic field **B** to be constant in space but varying in time. Then μ_m may also vary in time. If we differentiate Eq. (5.8) with respect to time it becomes, in component form,

$$\frac{d\mu_m}{dt} = \dot{\mu}_m = \frac{1}{B}\frac{dW_\perp}{dt} - \frac{W_\perp}{B^2}\frac{dB}{dt} \tag{5.9}$$

Note that $\dot{\mu}_m$ always acts in the same or opposite direction to **B**. That is, the *plane* of rotation of the particle does not change. If we knew $\mathbf{B}(t)$, and if we could calculate $W_\perp$, we would have $d\mu_m/dt$. This last is not easy. However, we can approximate $dW_\perp/dt$ by averaging the change in $W_\perp$ over 1 revolution of the particle and dividing by the period of revolution. The change in energy over one revolution is

$$\Delta W_\perp = \oint \mathbf{F} \cdot d\mathbf{l} = \oint q\mathbf{E} \cdot d\mathbf{l} \tag{5.10}$$

where **F** is the force on the particle. The $\mathbf{v} \times \mathbf{B}$ force does not do any work, so it does not appear in Eq. (5.10), if we let $\oint d\mathbf{l} = \oint \mathbf{v}\, dt$ around the orbit. Now, by use of Stokes' theorem and one of Maxwell's equations, $\Delta W_\perp$ becomes

$$\Delta W_\perp = q \int (\nabla \times \mathbf{E}) \cdot d\mathbf{s} = -q \int \frac{d\mathbf{B}}{dt} \cdot d\mathbf{s} \approx q \pi r_g^{\,2} \frac{dB}{dt} \tag{5.11}$$

The sign changes on the extreme right of (5.11) due to the direction of rotation of the particle opposite to the sense of the line integral in (5.10). Now since $v_\perp = r_g \omega$,

$$\Delta W_\perp = \frac{q\pi v_\perp^{\,2}}{\omega_c^{\,2}} \frac{dB}{dt} \tag{5.12}$$

and finally,

$$\frac{dW_\perp}{dt} \cong \frac{q v_\perp^{\,2} \dfrac{dB}{dt}}{2\omega_c} = \frac{W_\perp}{B}\frac{dB}{dt} \tag{5.13}$$

where we have divided (5.12) by the cyclotron period $2\pi/\omega_c$, to obtain (5.13). Placing this result in Eq. (5.9), we see that $d\mu_m/dt \approx 0$, which shows that for slowly time-varying fields $\mu_m \cong$ constant. Again, the crucial assumption here is that the surface to be used in the integral in Eq. (5.11) is a circle of radius r_g, which is perpendicular to **B**. Obviously if **B** changes rapidly over one cyclotron period, this cannot be the case. Therefore, the change in **B** during one cyclotron orbit must be much less than $|\mathbf{B}|$ for the adiabatic invariance of μ_m to hold.

For spatially varying fields, we can also show the adiabatic invariance of μ_m. Let B_z be expanded in a power series

$$B_z = B_0(1 + \alpha z + \cdots) \tag{5.14}$$

We have assumed that **B** points in the z direction. We then find that the z-component of the equation of motion is

$$m\frac{dv_z}{dt} \approx -\alpha W_\perp = -\frac{\partial B_z/\partial z}{B_0} W_\perp$$

This is so because B_z also is

$$B_0 + \left.\frac{\partial B_z}{\partial z}\right|_{z=0} z + \cdots \qquad \text{(a Taylor series)}$$

The value for α is thus

$$\alpha = \frac{\partial B_z/\partial z}{B_0}$$

Now since energy is conserved,

$$\frac{dW_\perp}{dt} = -\frac{dW_\parallel}{dt} = -mv_z\frac{dv_z}{dt} \approx \frac{W_\perp}{B_0}\frac{\partial B_z}{\partial z} v_z \tag{5.15}$$

But

$$\frac{dB}{dt} = \frac{\partial B_z}{\partial z}\frac{dz}{dt} = \frac{\partial B_z}{\partial z} v_z \tag{5.16}$$

and therefore,

$$\frac{dW_\perp}{dt} \approx \frac{W_\perp}{B_0}\frac{dB_z}{dt} \tag{5.17}$$

Thus we may obtain, for the variation of μ_m, if B varies in space,

$$\frac{d\mu_m}{dt} = \frac{1}{B}\frac{dW_\perp}{dt} - \frac{W_\perp}{B^2}\frac{dB}{dt} \cong 0$$

if (5.17) is used in (5.9).

The approximation here is that **B** varies slowly in space compared to the length $v_z \Delta t$ where Δt is the cyclotron period. This permits us to note that again $\Delta W_\perp$ changes very little in the z direction over one cyclotron period, and therefore μ_m is still constant.

Another adiabatic invariant is the longitudinal "action" of a particle contained between magnetic mirrors. If the containment of a particle is depicted in the mirror geometry of Fig. 3.10, then the containment can be made equivalent to a particle being trapped in a potential well, where the mirror reflection points are on the walls of the potential well. If the axis of the magnetic field is the z direction, then the z equation of motion for the particle can be written as

$$m \frac{d^2 z}{dt^2} = -\frac{\partial \mathscr{V}(z, t)}{\partial z} \tag{5.18}$$

where $\mathscr{V}(z, t)$ is a potential function. Figure 5.2 gives a representation of the particle being reflected in this kind of potential well. Let us also assume that

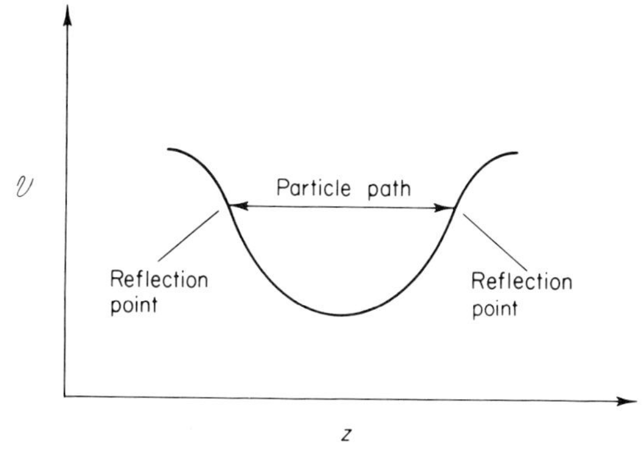

Fig. 5.2. Charged particle in a magnetic mirror "potential well."

$\mathscr{V}(z, t)$ is a function that may be changing slowly in time. The equation of motion (5.18) permits us to determine the Hamiltonian for this system, and since no loss mechanism is directly specified, the Hamiltonian may be assumed to be constant in time, except for the effects due to the variation of $\mathscr{V}$. The Hamiltonian is therefore written in the form shown in the equation below.

$$H = \frac{m}{2}\left(\frac{dz}{dt}\right)^2 + \mathscr{V}(z, t) \tag{5.19}$$

We define the *action* in the z direction to be

$$J_z = \oint m \frac{dz}{dt}\, dz \tag{5.20}$$

This integral is to be made for a complete cycle of the motion, that is, from one starting point in Fig. 5.2 through the reflections at the magnetic mirrors (2) and back again.

We shall require the time average of the Hamiltonian in the calculations. It is

$$\overline{H} = \frac{\oint [\frac{1}{2}m(dz/dt)^2 + \mathscr{V}(z,t)]\,dt}{\oint dt} \tag{5.21}$$

Now $\oint dt = \tau =$ the period of the complete oscillation between the mirrors. In terms of $\overline{H}$, J_z becomes, if $\overline{H}$ is not too different from H,

$$J_z = m \oint \left[\frac{2(\overline{H} - \mathscr{V})}{m}\right]^{1/2} dz \tag{5.22}$$

since we can solve Eq. (5.19) directly for dz/dt and thus formulate Eq. (5.20).

We now look for dJ_z/dt. If J_z is an adiabatic invariant, then dJ_z/dt can be shown to be ≈ 0. The time derivative of J_z is

$$\frac{dJ_z}{dt} = \frac{\partial J_z}{\partial \overline{H}}\frac{d\overline{H}}{dt} + \frac{\partial J_z}{\partial \mathscr{V}}\frac{d\mathscr{V}}{dt} \tag{5.23}$$

There is no explicit time dependence of J_z on t. Fom Eq. (5.22) we can obtain the following two partial derivatives.

$$\frac{\partial J_z}{\partial \overline{H}} = \oint \left[\frac{2(\overline{H} - \mathscr{V})}{m}\right]^{-1/2} dz \tag{5.24}$$

and

$$\frac{\partial J_z}{\partial \mathscr{V}} = -\oint \left[\frac{2(\overline{H} - \mathscr{V})}{m}\right]^{-1/2} dz \tag{5.25}$$

Therefore, from (5.24) and (5.25) we can see that

$$\frac{\partial J_z}{\partial \overline{H}} = -\frac{\partial J_z}{\partial \mathscr{V}} \tag{5.26}$$

Let us look again at Eq. (5.24). The integral may be obtained from Eq. (5.19) by solving it for the integrand of (5.24). This results in

$$\frac{2(\overline{H} - \mathscr{V})}{m} = \left(\frac{dz}{dt}\right)^2 \tag{5.27}$$

and from (5.27) we get

$$\left[\frac{2(\overline{H} - \mathscr{V})}{m}\right]^{-1/2} = \frac{dt}{dz} \tag{5.28}$$

Therefore we can now substitute this last equation into Eq. (5.24) and show that

$$\frac{\partial J_z}{\partial \bar{H}} = \oint \frac{dt}{\cancel{dz}} \cancel{dz} = \tau \tag{5.29}$$

Solving for dJ_z/dt in Eq. (5.23), we obtain

$$\frac{dJ_z}{dt} = \tau\left(\frac{d\bar{H}}{dt} - \frac{d\mathscr{V}}{dt}\right) \tag{5.30}$$

We must now determine $d\bar{H}/dt$ and $d\mathscr{V}/dt$. If they are approximately equal, the problem is solved.

The former may be determined by averaging over a complete mirror oscillation, the time derivative of Eq. (5.21). This is

$$\overline{\frac{dH}{dt}} = \frac{\oint \left(m \dfrac{d^2z}{dt^2}\dfrac{dz}{dt} + \dfrac{\partial \mathscr{V}}{\partial t} + \dfrac{\partial \mathscr{V}}{\partial z}\dfrac{dz}{dt}\right) dt}{\oint dt} \tag{5.31}$$

The equation of motion (5.18) permits us to cancel the first and third terms in (5.28) so all that remains is

$$\overline{\frac{dH}{dt}} = \frac{1}{\tau}\oint \frac{\partial \mathscr{V}}{\partial t}\, dt \cong \overline{\frac{d\mathscr{V}}{dt}} \tag{5.32}$$

if $d\mathscr{V}/dt$ is *nearly constant* over a complete oscillation between the mirrors. The result is then that the left-hand side of Eq. (5.30) is ≈ 0 or, in other words, that

$$\frac{dJ_z}{dt} \approx 0 \tag{5.33}$$

Under these conditions J_z may be considered an adiabatic invariant and may be used as a constant of the motion in solving the Boltzmann equation. This means that, just as the adiabatic invariance of the magnetic moment allowed us to develop a condition for reflection of particles in a magnetic mirror geometry, the constancy of J_z also produces some additional information on the behavior of charged particles in a magnetic bottle.

Figure 5.3 shows the result of having a constant value of J_z if the potential function defining the magnetic "well" in which a charged particle is reflected becomes steeper and narrower. This might be the case, for example, if the mirror ratio in a magnetic bottle were increased with time. Note that the particle travels a shorter distance in z when the sides of the well are closer together. However, since the action

$$J_z = \oint m v_z \, dz$$

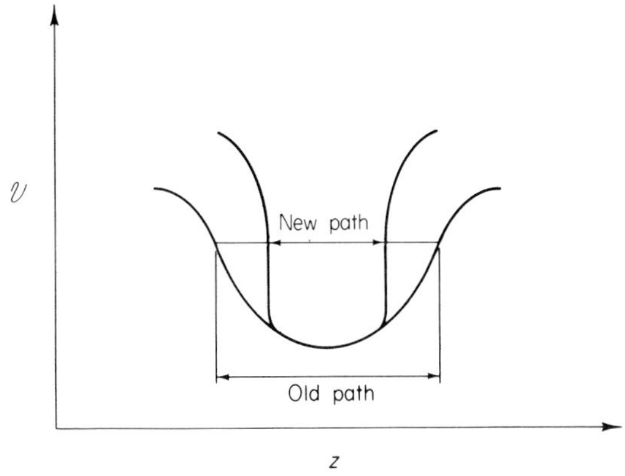

Fig. 5.3. The changing points of reflection.

must be constant (if $d\mathscr{V}/dt$ is not too large), then v_z (the velocity in the z direction) must increase, thereby increasing the energy of the particle. These *adiabatic compressions* are often used in many plasma experiments to increase the temperature of plasmas.

Let us now apply some of the foregoing work on constants of the motion to the behavior of a plasma by examining some methods of solution of the collisionless Boltzmann equation. The collisionless Boltzmann equation is a homogeneous equation. The solution of it is a function of the variables **r**, **v**, and t. The form of the solution is *not* determined directly from the solution to the equation, but from the boundary conditions for the particular problem in *time and space*.

We may eliminate some or all of the variables by use of constants of the motion. These constants are functions of **r**, **v**, and t, and remain constant *only along a particle trajectory*.

For example, suppose we have a simple velocity distribution $f(v)$ that is of the form

$$f(v) \sim \exp\left(-\frac{v^2}{\alpha^2}\right) \tag{5.34}$$

If energy is a constant of the motion, then

$$E = \tfrac{1}{2}mv^2$$

if no potential energy exists. Then we may obtain

$$f(E) = \exp\left(-\frac{2E}{m\alpha^2}\right) \tag{5.35}$$

This function is a function of the constant of the motion E, and satisfies the collisionless Boltzmann equation. That is

$$\frac{df}{dt} = \frac{\partial f(E)}{\partial E}\frac{dE}{dt} = 0 \tag{5.36}$$

since E is a constant along the trajectory of the particle. We note that Eq. (5.36) *is* the time derivative along the trajectory, because it is the collisionless Boltzmann equation.

$$\frac{df}{dt} = \frac{\partial f}{\partial t} + \mathbf{v} \cdot \nabla_r f + \frac{\mathbf{F}}{m} \cdot \nabla_v f = 0$$

Equation (5.36) has only one term, however, since we have one coordinate and one constant of the motion.

If we have two variables, perhaps in a velocity distribution in a magnetic field, f might be

$$f(v_\parallel, v_\perp) \sim \exp\left(-\frac{v_\parallel^2}{\alpha_\parallel^2}\right) \exp\left(-\frac{v_\perp^2}{\alpha_\perp^2}\right) \tag{5.37}$$

We may eliminate one variable by using the relation that

$$E = \tfrac{1}{2}m(v_\parallel^2 + v_\perp^2) = \text{constant}$$

so that (5.37) becomes

$$f(E, v_\perp) = \exp\left(-\frac{2E}{m\alpha_\parallel^2}\right) \exp\left(-\frac{v_\perp^2}{\alpha_\perp^2}\right) \exp\left(-\frac{v_\perp^2}{\alpha_\parallel^2}\right) \tag{5.38}$$

We can now use the magnetic moment

$$\mu_m = \frac{mv_\perp^2}{2B}$$

if it is an adiabatic invariant, to eliminate the variable $v_\perp$ from (5.38), producing

$$f(E, \mu_m) = \exp\left(-\frac{2E}{m\alpha_\parallel^2}\right) \exp\left(-\frac{2B\mu_m}{m\alpha_\parallel^2}\right) \exp\left(-\frac{2B\mu_m}{m\alpha_\perp^2}\right) \tag{5.39}$$

Equation (5.39) is therefore a solution to the collisionless Boltzmann equation. Note that if B changes with position, say z, then Eq. (5.37) is a function of three variables, and a third constant of the motion (perhaps the action J_z) is needed to eliminate this third coordinate.

In summary, then, the constants of the motion may be used to simplify the functional form of the solutions of the collisionless Boltzmann equation.

Any *arbitrary function* of the constants of the motion is a solution to the collisionless Boltzmann equation. The correct solution is the one that satisfies the *boundary* and/or initial conditions as well as the collisionless Boltzmann equation. Once this is done, the function $f(r, v, t)$ may be obtained directly from the relations between the constants of the motion and the coordinates.

It is important to note that this method may not be simple to use, especially if the constants of the motion are complicated functions of the coordinates. But in many cases, knowing the constants of the motion can be a great help.

B. HAMILTON–JACOBI THEORY

Another important method that we may use to solve for the distribution function is to use *Hamilton–Jacobi* theory. The method of utilization will be specified, but not the theory behind it. The reader is referred to any standard text on classical mechanics.

The method is used in the following way. First, we must transform the coordinates in Hamilton's equations to obtain a new *transformed Hamiltonian.* This transformed Hamiltonian, when placed into Hamilton's equations. must be such that the new equations will show that the new coordinates are constant. That is, their time derivatives must be identically equal to zero. We will allow the Hamiltonian to be a function of time. In order to find the transformation necessary we first return to Hamilton's equations:

$$\dot{q}_i = \frac{\partial H}{\partial p_i} \tag{5.40}$$

and

$$\dot{p}_i = -\frac{\partial H}{\partial q_i} \tag{5.41}$$

Remember that the Hamiltonian H is a function of the coordinates and momenta. However, *if* it should appear that H is *not* a function of one of the position coordinates q_i then Eq. (5.41) is identically equal to zero for that coordinate. This means that

$$\dot{p}_i \equiv 0 \qquad \text{or} \qquad p_i = \text{constant} \tag{5.42}$$

Each q_i that results in this condition is said to be *cyclic* or *ignorable.* If *all* of the position coordinates can be made to be cyclic, then, from Eq. (5.40)

$$\dot{q}_i = \frac{\partial H}{\partial p_i} = \text{constant} = f(p_1, \ldots, p_r) \tag{5.43}$$

This means that $\dot{q}_i$ is constant. The magnitude of these cyclic coordinates must be only a *linearly increasing function of time*. Then $p_1, \ldots, p_r$ are the constants of integration required for the complete solution of the problem. Note that the value of the cyclic coordinate is easily obtained directly from its corresponding constant momentum. The Hamilton–Jacobi method provides a formal procecure for seeking these coordinates. It is developed in the following discussion.

The Hamiltonian may be changed so that it is a function only of cyclic coordinates. To obtain the cyclic coordinates we may use a *canonical transformation* to obtain the new variables. The transformation is written symbolically as

$$p_i = \frac{\partial S(q_i, P_i)}{\partial q_i} = \frac{\partial S(q_i, \alpha_i)}{\partial q_i} \tag{5.44}$$

$$Q_i = \frac{\partial S(q_i, P_i)}{\partial P_i} = \frac{\partial S(q_i, \alpha_i)}{\partial \alpha_i} \tag{5.45}$$

The quantity Q_i is a cyclic coordinate. S is called the generating function and is a function of q_i and the constants (*of integration*) α_i, which are the P_i's (the momentum if Q_i is cyclic). (We have let $P_i = \alpha_i$.) The transformed Hamiltonian is called K and Hamilton's equations in terms of K and the cyclic variables become

$$\dot{Q}_i = \frac{\partial K}{\partial P_i} = 0 \tag{5.46}$$

$$\dot{P}_i = -\frac{\partial K}{\partial Q_i} = 0 \tag{5.47}$$

$\dot{Q}_i$ must be zero. Otherwise Q_i would change with time, would appear in K, and would not be an ignorable coordinate.

The generating function *may* be a function of time. If so, the relation between H and K is

$$K = H + \frac{\partial S}{\partial t} \tag{5.48}$$

Now K will be zero if S satisfies the equation

$$H(q, p, t) + \frac{\partial S}{\partial t} = 0 \tag{5.49}$$

Using the canonical transformation (5.44)

$$p_i = \frac{\partial S(q_i, \alpha_i)}{\partial q_i}$$

Eq. (5.49) becomes

$$H\left(q_1, \ldots, q_r; \frac{\partial S}{\partial q_1}, \ldots, \frac{\partial S}{\partial q_r}; t\right) + \frac{\partial S}{\partial t} = 0 \tag{5.50}$$

Equation (5.50) is called the Hamilton–Jacobi equation. The solution of Eq. (5.50) is S, the generating function, or *Hamilton's principal function.* Equation (5.50) is a first-order partial differential equation of $r + 1$ variables (r coordinates and the time). A general solution will have $r + 1$ independent constants of ingegration $\alpha_1, \ldots, \alpha_{r+1}$. Note that these are *not* necessarily also constants of the motion, but merely integration constants. A complete solution for S could be written as

$$S = S(q_1, \ldots, q_r; \alpha_1, \ldots, \alpha_r; t) \tag{5.51}$$

since the $(r + 1)$st constant can be represented as an additive constant to S.

Note again that since the momenta do not appear in Eq. (5.50) we can set

$$P_i = \alpha_i$$

We will now investigate an example to be solved by the Hamilton–Jacobi method. Consider a one-dimensional harmonic oscillator (a spring-mass problem). The Hamiltonian for such an oscillator is

$$H = \frac{p^2}{2m} + \frac{kq^2}{2} \tag{5.52}$$

k is the spring constant. Since Eq. (5.44) requires that $p = \partial S/\partial q$, then (5.50) becomes

$$\frac{1}{2m}\left(\frac{\partial S}{\partial q}\right)^2 + \frac{kq^2}{2} + \frac{\partial S}{\partial t} = 0 \tag{5.53}$$

Now since S only depends on t explicitly in the last term, the time dependence of S must be like αt. This can be seen easily. We may fix a q (cyclic variable) so that the first two terms are constant in (5.53). Therefore $\partial S/\partial t$ must be constant for all time. The result is $S \sim \alpha t$. A solution for S can then be

$$S(q, \alpha, t) = W(q, \alpha) - \alpha t \tag{5.54}$$

Equation (5.53) now becomes

$$\frac{1}{2m}\left(\frac{\partial W}{\partial q}\right)^2 + \frac{kq^2}{2} = \alpha \tag{5.55}$$

We have eliminated time from this equation by use of the integration constant α. This *may* also be a constant of the motion as well. The solution of (5.55) can be found simply and it will be

$$W = (mk)^{1/2} \int \left(\frac{2\alpha}{k} - q^2\right)^{1/2} dq \tag{5.56}$$

We then find S to be

$$S = (mk)^{1/2} \int \left(\frac{2\alpha}{k} - q^2\right)^{1/2} dq - \alpha t \tag{5.57}$$

Since $p = \partial S/\partial q$, we can find α at $t = 0$, $p = 0$, and $q = q_0$ (stationary particle) by evaluating Eq. (5.57) at these points. It is

$$\left(\frac{\partial S}{\partial q}\right)\bigg|_{t=0} = 0 = (2m)^{1/2} \left(\alpha - \frac{k{q_0}^2}{2}\right)^{1/2} \tag{5.58}$$

Therefore, $\alpha = k{q_0}^2/2$ = the initial energy stored in the spring. Now since

$$H + \frac{\partial S}{\partial t} = 0 \qquad \text{and} \qquad \frac{\partial S}{\partial t} = -\alpha \tag{5.59}$$

then

$$H = \alpha = \text{total energy of the system} \tag{5.60}$$

Note that in this case α *is* a constant of the motion, but this is not necessarily always so.

We can now solve for both p and q by use of Eqs. (5.44) and (5.45). Equation (5.45) must be inverted to obtain

$$q = q(\alpha, Q, t)$$

We proceed to obtain Q from Eq. (5.57). It is

$$Q = \frac{\partial S}{\partial \alpha} = \left(\frac{m}{k}\right)^{1/2} \int \frac{dq}{[(2\alpha/k) - q^2]^{1/2}} - t$$

Therefore

$$t + Q = \left(\frac{m}{k}\right)^{1/2} \cos^{-1}\left[q\left(\frac{k}{2\alpha}\right)^{1/2}\right]$$

and we may now solve for q

$$q = \left(\frac{2\alpha}{k}\right)^{1/2} \cos[\omega(t + Q)] \qquad \text{where} \qquad \omega = \left(\frac{k}{m}\right)^{1/2} \tag{5.61}$$

We must still find Q, but since its derivative is zero, as shown in Eq. (5.46), it is a constant and may be evaluated from the boundary conditions. Finally, we obtain the solution for p to be

$$p = \frac{\partial S}{\partial q} = (mk)^{1/2} \left(\frac{2\alpha}{k} - q^2\right)^{1/2} \tag{5.62}$$

If the system under consideration has more than *one* degree of freedom, the Hamilton–Jacobi equation becomes a *partial* differential equation in more than two variables. *If* the variables are *separable*, then this method offers an advantage. If not, there is often little to be gained by its use. The collisionless Boltzmann equation is of this type and care must be taken in the use of the Hamilton–Jacobi method.

This concludes the introduction to methods concerning the collisionless Boltzmann equation.

C. COLLISIONS IN THE BOLTZMANN EQUATION: FOKKER–PLANCK METHODS

We now will investigate the effects of collisions on the Boltzmann equation. In recapitulation, Eq. (5.3) shows how f changes with time and position in μ space due to the appearances of various slowly varying, long-range interactions (collisionless interactions) between particles and/or externally applied fields. It is presumed that the interactions can be described by the force term used on the left-hand side of the equation. If the time of interaction of these particles is of the order of, or greater than, the time of interest, then Eq. (5.3) should be sufficient to describe the problem. However, should the interaction occur over times that are so short that we cannot know the details of the interaction (collisions), we must provide some means for their inclusion. One possible scheme is to assume that f could change abruptly in an immediate way without any definitive form; we may represent this by

$$\left(\frac{\partial f}{\partial t}\right)_{\text{coll}}$$

We note that a collision essentially "throws" a particle into or out of a given incremental volume in phase space and therefore it should appear as a "forcing function" in Eq. (5.3). The forcing function may either be positive (particles get "thrown into" the incremental volume) or negative (particles get "thrown out" of the incremental volume). Equation (5.3) may then be written as

$$\frac{\partial f}{\partial t} + \mathbf{v} \cdot \nabla_r f + \frac{\mathbf{F}}{m} \cdot \nabla_v f = \left(\frac{\partial f}{\partial t}\right)_{\text{coll}} \tag{5.63}$$

Quite often, the necessity of determining which processes go into the collisional term is arbitrary, but usually, short-time interactions are placed on the right-hand side of Eq. (5.63) and long-time interactions are placed on the left-hand side. Note now that, because we have placed a term on the right-hand side of Eq. (5.3), Liouville's theorem does not apply and we also may not have an equilibrium condition. It should again be mentioned, however, that many nonequilibrium distribution functions are solutions to the *collisionless* Boltzmann equation if they can be shown to be functions only of the constants of the motion.

We must find a means for developing an expression for the collision term on the right-hand side of the Boltzmann equation, since we really have not mentioned its nature at all. One possible approach is as follows.

Let Prob($\mathbf{v}$, $\Delta\mathbf{v}$) be the probability *that* a particle with velocity $\mathbf{v}$ has velocity $\mathbf{v} + \Delta\mathbf{v}$ after a time Δt has elapsed. We shall also assume that Prob($\mathbf{v}$, $\Delta\mathbf{v}$) does not depend explicitly on the time. We may then find the velocity distribution at time t in terms of the velocity distribution at time $t - \Delta t$ by use of the following integral.

$$f(\mathbf{r}, \mathbf{v}, t) = \int f(\mathbf{r}, \mathbf{v} - \Delta\mathbf{v}, t - \Delta t)\,\text{Prob}(\mathbf{v} - \Delta\mathbf{v}, \Delta\mathbf{v})\, d\Delta\mathbf{v} \tag{5.64}$$

This integration process depicted in (5.64) is called a *Markoff process.* Note that we integrate over all velocity increments $\Delta\mathbf{v}$. That is, $\Delta\mathbf{v}$ goes from $-\infty$ to $+\infty$. It tells us that $f(\mathbf{r}, \mathbf{v}, t)$ only depends upon the velocity distribution Δt units of time previous to the time of interest, but it is an integral over *all* the velocity increments (changes) at time $t - \Delta t$. We then shall expand $f(\mathbf{r}, \mathbf{v} - \Delta\mathbf{v}, t - \Delta t)$ in a Taylor series in both time and velocity *about the point* $\mathbf{v}$, t. We must then note that *we will have to expand backward in time and velocity.* The series, neglecting terms of higher order than those written, becomes

$$f(\mathbf{r}, \mathbf{v} - \Delta\mathbf{v}, t - \Delta t) = f(\mathbf{r}, \mathbf{v}, t) - \left.\frac{\partial f}{\partial t}\right|_{t=t} \Delta t - \left.\frac{\partial f}{\partial \mathbf{v}}\right|_{\mathbf{v}=\mathbf{v}} \Delta\mathbf{v} + \left.\frac{\Delta\mathbf{v}\,\Delta\mathbf{v}}{2!}\frac{\partial^2 f}{\partial \mathbf{v}\,\partial \mathbf{v}}\right|_{\mathbf{v}=\mathbf{v}} + \cdots \tag{5.65}$$

The notation of the last term in Eq. (5.65) must be explained. The expression $\Delta\mathbf{v}\ \Delta\mathbf{v}$ is a *second-order dyadic.* We digress for a moment to discuss some general properties of these quantities.

A second-order dyadic is, essentially, a matrix of all possible products of the components of two vectors. *No* vector operation is immediately performed between these components. However, *any* vector operation may be performed on *either side* of the dyadic *with* a vector. Therefore $\mathbf{A} \cdot \mathbf{BC}$ is a vector produced by taking the scalar product of $\mathbf{A}$ with the dyadic $\mathbf{BC}$ on the left. $\mathbf{BC} \cdot \mathbf{A}$ is also a vector, and is not in general equal to $\mathbf{A} \cdot \mathbf{BC}$. We may also use a vector product operation as

$$\mathbf{A} \times \mathbf{BC} \qquad \text{or} \qquad \mathbf{BC} \times \mathbf{A}$$

The result here is not a vector but another dyadic. The components, in Cartesian coordinates, of the dyadic **BC** may be written as follows.

$$\begin{bmatrix} \hat{a}_x B_x C_x \hat{a}_x & \hat{a}_x B_x C_y \hat{a}_y & \hat{a}_x B_x C_z \hat{a}_z \\ \hat{a}_y B_y C_x \hat{a}_x & \hat{a}_y B_y C_y \hat{a}_y & \hat{a}_y B_y C_z \hat{a}_z \\ \hat{a}_z B_z C_x \hat{a}_x & \hat{a}_z B_z C_y \hat{a}_y & \hat{a}_z B_z C_z \hat{a}_z \end{bmatrix} \tag{5.66}$$

Note thay we may also operate *from the left* with the divergence and curl operators on a dyadic. We may now expand the *integrand* of Eq. (5.64) to be

$$\begin{aligned} &f(\mathbf{r}, \mathbf{v} - \Delta\mathbf{v}, t - \Delta t)\ \text{Prob}(\mathbf{v} - \Delta\mathbf{v}, \Delta\mathbf{v}) \\ &\quad = f(\mathbf{r}, \mathbf{v}, t)\ \text{Prob}(\mathbf{v}, \Delta\mathbf{v}) - \Delta t\ \text{Prob}(\mathbf{v}, \Delta\mathbf{v}) \frac{\partial f(\mathbf{r}, \mathbf{v}, t)}{\partial t} \\ &\quad - \Delta\mathbf{v} \frac{\partial}{\partial \mathbf{v}} [f(\mathbf{r}, \mathbf{v}, t)\ \text{Prob}(\mathbf{v}, \Delta\mathbf{v})] + \frac{\Delta\mathbf{v}\,\Delta\mathbf{v}}{2} \frac{\partial^2}{\partial \mathbf{v}\,\partial \mathbf{v}} [f(\mathbf{r}, \mathbf{v}, t)\ \text{Prob}(\mathbf{v}, \Delta\mathbf{v})] \end{aligned} \tag{5.67}$$

using the Taylor series expansion (5.65). Each term in the Taylor series is evaluated at $\mathbf{v}$ and t. We may now use this series directly in Eq. (5.64), producing

$$\begin{aligned} f(\mathbf{r}, \mathbf{v}, t) = &\int [f(\mathbf{r}, \mathbf{v}, t)\ \text{Prob}(\mathbf{v}, \Delta\mathbf{v})]\, d(\Delta\mathbf{v}) \\ &- \int \Delta t\ \text{Prob}(\mathbf{v}, \Delta\mathbf{v}) \frac{\partial f(\mathbf{r}, \mathbf{v}, t)}{\partial t}\, d(\Delta\mathbf{v}) \\ &- \int \Delta\mathbf{v} \frac{\partial}{\partial \mathbf{v}} [f(\mathbf{r}, \mathbf{v}, t)\ \text{Prob}(\mathbf{v}, \Delta\mathbf{v})]\, d(\Delta\mathbf{v}) \\ &+ \int \Delta\mathbf{v}\,\Delta\mathbf{v} \frac{\partial^2}{\partial \mathbf{v}\,\partial \mathbf{v}} [f(\mathbf{r}, \mathbf{v}, t)\ \text{Prob}(\mathbf{v}, \Delta\mathbf{v})]\, d(\Delta\mathbf{v}) \end{aligned} \tag{5.68}$$

We shall now define several integrals. First, let

$$\int \text{Prob}(\mathbf{v}, \Delta\mathbf{v})\, d(\Delta\mathbf{v}) = 1 \tag{5.69}$$

This means that a particle is certain to make *some* kind of a collision. Next, we define

$$\langle \Delta\mathbf{v} \rangle = \int \text{Prob}(\mathbf{v}, \Delta\mathbf{v})\, \Delta\mathbf{v}\, d(\Delta\mathbf{v}) \tag{5.70}$$

Equation (5.70) is the expression for the average velocity change. And finally, we write

$$\langle \Delta\mathbf{v}\,\Delta\mathbf{v} \rangle = \int \text{Prob}(\mathbf{v}, \Delta\mathbf{v})\, \Delta\mathbf{v}\,\Delta\mathbf{v}\, d(\Delta\mathbf{v}) \tag{5.71}$$

which is a dyadic, as was $\Delta\mathbf{v}\,\Delta\mathbf{v}$. These definitions may be used in Eq. (5.68). The first term on the right of (5.68) is then just $f(\mathbf{r}, \mathbf{v}, t)$, since the integration is over $\Delta\mathbf{v}$ only. Therefore $f(\mathbf{r}, \mathbf{v}, t)$ may be removed from that integral and Eq. (5.69) may be applied directly. The second term in (5.68) is

$$-\Delta t \frac{\partial f(\mathbf{r}, \mathbf{v}, t)}{\partial t}$$

by the same reasoning. The third term becomes

$$-\frac{\partial}{\partial \mathbf{v}} [f(\mathbf{r}, \mathbf{v}, t)\langle\Delta\mathbf{v}\rangle]$$

while the fourth term is

$$\frac{1}{2}\frac{\partial^2}{\partial \mathbf{v}\,\partial \mathbf{v}} [f(\mathbf{r}, \mathbf{v}, t)\langle\Delta\mathbf{v}\,\Delta\mathbf{v}\rangle]$$

Placing these terms in Eq. (5.68) produces

$$f(\mathbf{r}, \mathbf{v}, t) = f(\mathbf{r}, \mathbf{v}, t) - \Delta t \frac{\partial f(\mathbf{r}, \mathbf{v}, t)}{\partial t} - \frac{\partial}{\partial \mathbf{v}} [f(\mathbf{r}, \mathbf{v}, t)\langle\Delta\mathbf{v}\rangle] + \frac{1}{2}\frac{\partial^2}{\partial \mathbf{v}\,\partial \mathbf{v}} [f(\mathbf{r}, \mathbf{v}, t)\langle\Delta\mathbf{v}\,\Delta\mathbf{v}\rangle] \tag{5.72}$$

Equation (5.72) now permits us to solve for $\Delta t\, \partial f/\partial t$. It is

$$\Delta t \frac{\partial f}{\partial t} = -\frac{\partial}{\partial \mathbf{v}} (f\langle\Delta\mathbf{v}\rangle) + \frac{1}{2}\frac{\partial^2}{\partial \mathbf{v}\,\partial \mathbf{v}} (f\langle\Delta\mathbf{v}\,\Delta\mathbf{v}\rangle) \tag{5.73}$$

since the left-hand side of (5.72) cancels the first term on the right-hand side of (5.72). This last equation, in fact, determines the net change in f over a collision time Δt, which is the required quantity for the right-hand side of the Boltzmann equation. Equation (5.73) is called the *Fokker–Planck equation.* Dividing through by Δt, the collision time, gives the exact formulation necessary for inclusion in the Boltzmann equation.

Note that the Fokker–Planck equation has two terms of opposite sign, which could result in zero net change in f over a collision. $\langle\Delta\mathbf{v}\rangle$ is the average net change in velocity. $m\langle\Delta\mathbf{v}\rangle/\Delta t$ has the dimension of force and tends to slow down or speed up particles until they reach the average velocity $[\langle\Delta\mathbf{v}\rangle = 0]$. This process is called *dynamical friction.*

The $\langle\Delta\mathbf{v}\,\Delta\mathbf{v}\rangle$ term in Eq. (5.73) represents a diffusion in velocity space. If $\langle\Delta\mathbf{v}\,\Delta\mathbf{v}\rangle$ is not zero, it implies a "spreading" of points in velocity space. In equilibrium, diffusion in velocity space is exactly balanced by dynamical friction. A Maxwell–Boltzmann distribution, for example, exists in equilibrium, by having particles diffuse away from the average velocity in both directions, while dynamical friction brings them back. In fact, for a Maxwell–Boltzmann distribution, the Fokker–Planck equation (5.73) has the value zero. Figure 5.4 shows this pictorially. The problem, of course, is to be able

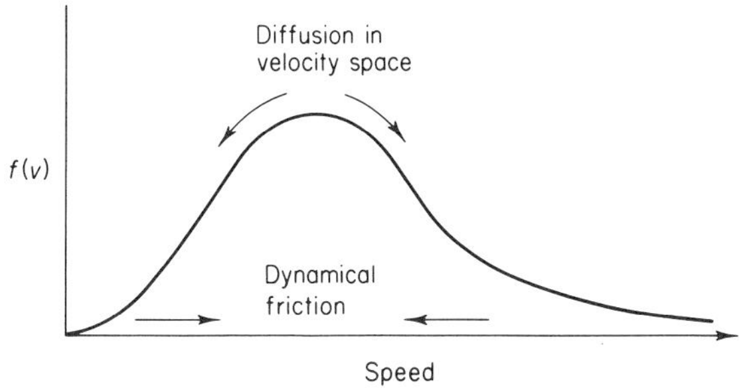

Fig. 5.4. Dynamical friction and velocity space diffusion.

to calculate $\langle \Delta \mathbf{v} \rangle$ and $\langle \Delta \mathbf{v}\, \Delta \mathbf{v} \rangle$ for each particular case in question. Further work with this equation will appear in Chapter 9.

D. DIFFUSION AND MOBILITY FROM THE BOLTZMANN EQUATION

We shall now develop two expressions from the Boltzmann equation to present the notions of diffusion and mobility, which have been introduced previously. A perturbation-type approach is useful in obtaining these terms.

Let us assume that the distribution function f can be broken up into two parts, an "equilibrium" term and a perturbation part as shown in Eq. (5.74):

$$f(\mathbf{r}, \mathbf{v}, t) = f_0(\mathbf{r}, \mathbf{v}, t) + f_1(\mathbf{r}, \mathbf{v}, t) \tag{5.74}$$

If we integrate f over all velocity space, we should obtain the density of particles in position space as follows.

$$n(\mathbf{r}, t) = N \int f(\mathbf{r}, \mathbf{v}, t)\, dv_x\, dv_y\, dv_z \tag{5.75}$$

where N is the total number of particles. Term-by-term integration of Eq. (5.75) using the perturbation (5.74) then yields

$$n(\mathbf{r}, t) = N \int f_0(\mathbf{r}, \mathbf{v}, t)\, dv_x\, dv_y\, dv_z + N \int f_1(\mathbf{r}, \mathbf{v}, t)\, dv_x\, dv_y\, dv_z \tag{5.76}$$

Unless specified otherwise, the limits on the integrals over velocity space in the following work will be from $-\infty$ to $+\infty$ in the three coordinate directions.

The equilibrium distribution f_0 will now be defined to produce $n(\mathbf{r}, t)$ alone. Therefore it must be true that

$$N \int f_1(\mathbf{r}, \mathbf{v}, t)\, dv_x\, dv_y\, dv_z \equiv 0 \tag{5.77}$$

since f_1 cannot produce any net number of particles.

Equation (5.77) implies that unless f_1 is zero everywhere, it must be positive and negative in different regions of *velocity* space. We shall also note that

$$f_1 \ll f_0$$

so that f itself is *always* positive. We shall assume a steady state (no explicit time dependence) condition. Equation (5.5), neglecting magnetic and gravitational fields, becomes, using this perturbation expansion

$$\mathbf{v} \cdot \nabla_r(f_0 + f_1) + \frac{q\mathbf{E}}{m} \cdot \nabla_v(f_0 + f_1) = \left(\frac{\partial f_0}{\partial t}\right)_{\text{coll}} + \left(\frac{\partial f_1}{\partial t}\right)_{\text{coll}} \tag{5.78}$$

$\mathbf{E}$ is the electric field produced by charge separation. We shall neglect the terms $\nabla_r f_1$ and $\nabla_v f_1$ on the left-hand side of Eq. (5.78). We may do this because the gradients of f_0 will dominate on the left-hand side. Also

$$\left(\frac{\partial f_0}{\partial t}\right)_{\text{coll}}$$

must be identically equal to zero, since by definition, f_0 is an equilibrium solution, which means that there is no net transport or motion due to f_0.

f_1 is generated by the "forcing" terms in Eq. (5.78). These are not the collisions, which are normally damping terms, but are rather the spatial gradients ($\nabla_r f_0$) and the electric field forces $\mathbf{E}/m$. The collisional term should therefore have a negative sign, to show the damping. We shall assume that it is of the form

$$-\nu_m f_1$$

where $-\nu_m$ is the momentum transfer collision frequency. Equation (5.78) then becomes

$$\mathbf{v} \cdot \nabla_r f_0 + \frac{q\mathbf{E}}{m} \cdot \nabla_v f_0 = -\nu_m f_1 \tag{5.79}$$

The particle current $\mathbf{\Gamma}$ can now be determined, with the help of the distribution function, as the product of the average velocity (which is defined to be zero in equilibrium) times the density of the particles in position space. The particle current is

$$\begin{aligned}\boldsymbol{\Gamma} = n\langle \mathbf{v}\rangle &= N \int \mathbf{v} f \, dv_x \, dv_y \, dv_z \\ &= N \int \mathbf{v}(f_0 + f_1) \, dv_x \, dv_y \, dv_z \\ &= N \int \mathbf{v} f_1 \, dv_x \, dv_y \, dv_z\end{aligned}$$

Note that since f_0 is an equilibrium term, it cannot contribute to the particle current. We can now solve Eq. (5.79) for f_1 as

$$f_1 = -\frac{\mathbf{v}}{v_m} \cdot \nabla_r f_0 - \frac{q\mathbf{E}}{m v_m} \cdot \nabla_r f_0 \tag{5.80}$$

The particle current then becomes

$$\boldsymbol{\Gamma} = -N \int \mathbf{v}\left[\frac{\mathbf{v}}{v_m} \cdot \nabla_r f_0 + \frac{q\mathbf{E}}{m v_m} \cdot \nabla_v f_0\right] dv_x \, dv_y \, dv_z \tag{5.81}$$

The first term on the right of Eq. (5.81) is the *diffusion* flux, and the second term is the *mobility* flux in the particle current.

Let us investigate the first term. If it is written in the components of the three coordinate directions, either of two integrals will appear. They are, for example,

$$N \int \frac{v_x^2}{v_m} \frac{\partial f_0}{\partial x} \, dv_x \, dv_y \, dv_z \qquad \text{or} \qquad N \int \frac{v_x v_y}{v_m} \frac{\partial f_0}{\partial y} \, dv_x \, dv_y \, dv_z \tag{5.82}$$

The first term of expression (5.82) may be rewritten as

$$N \frac{\partial}{\partial x} \int \frac{v_x^2}{v_m} f_0 \, dv_x \, dv_y \, dv_z$$

since the coordinates are independent. The second term can be written as

$$N \frac{\partial}{\partial y} \int \frac{v_x v_y}{v_m} f_0 \, dv_x \, dv_y \, dv_z$$

From this, it can be seen that the second terms in expression (5.82) have a value of zero since the function f_0 (the equilibrium distribution) is an even function. This is so, since otherwise particles would be lost from the plasma.

v_x and v_y are both odd, uncorrelated functions. From each of the three coordinate directions, then, we obtain for the diffusion particle current

$$\begin{aligned}\boldsymbol{\Gamma}_{\text{diff}} = &-N\frac{\partial}{\partial x}\int\frac{v_x^2 f_0}{v_m}\,dv_x\,dv_y\,dv_z\,\hat{a}_x\\ &-N\frac{\partial}{\partial y}\int\frac{v_y^2 f_0}{v_m}\,dv_x\,dv_y\,dv_z\,\hat{a}_y\\ &-N\frac{\partial}{\partial z}\int\frac{v_z^2 f_0}{v_m}\,dv_x\,dv_y\,dv_z\,\hat{a}_z\end{aligned}\tag{5.83}$$

We may then note that this is essentially a gradient operation on the integrand if we set $v_x^2 + v_y^2 + v_z^2 = v^2$. Then $v_x^2 \approx v^2/3$, and so on. We may then change Eq. (5.83) to

$$\begin{aligned}\boldsymbol{\Gamma}_{\text{diff}} &= -N\frac{\partial}{\partial x}\int\frac{v^2 f_0}{3v_m}\,dv_x\,dv_y\,dv_z\,\hat{a}_x - N\frac{\partial}{\partial y}\int\frac{v^2 f_0}{3v_m}\,dv_x\,dv_y\,dv_z\,\hat{a}_y\\ &\quad -N\frac{\partial}{\partial z}\int\frac{v^2 f_0}{3v_m}\,dv_x\,dv_y\,dv_z\,\hat{a}_z\\ &= -N\nabla_r\int\frac{v^2 f_0}{3v_m}\,dv_x\,dv_y\,dv_z = -\left\langle\frac{v^2}{3v_m}\right\rangle\nabla_r n(\mathbf{r}, t)\\ &= -D\nabla_r n(\mathbf{r}, t)\end{aligned}\tag{5.84}$$

where

$$D = \left\langle\frac{v^2}{3v_m}\right\rangle$$

as in Eq. (2.43), assuming that $v_m = v/l$, where l is the mean free path.

We have now shown that the free diffusion coefficient may be obtained in *two* ways. We first used a single-particle model in Chapter 2. Then we used a statistical model in this chapter and obtained the same result. Hence, the limits of both models approach the same values, which is quite convincing of the validity of the diffusion coefficient.

The next step will be to look at the second term in Eq. (5.81), the mobility. To do so, we first take the partial derivative of $v_x^2 + v_y^2 + v_z^2 = v^2$ with respect to v_x. This is

$$2v_x\frac{\partial v_x}{\partial v_x} + 2v_y\frac{\partial v_y}{\partial v_x} + 2v_z\frac{\partial v_z}{\partial v_x} = 2v\frac{\partial v}{\partial v_x}\tag{5.85}$$

Since the coordinates are independent, Eq. (5.85) yields

$$v_x = v\,\frac{\partial v}{\partial v_x}$$

and therefore

$$\frac{\partial v_x}{\partial v} = \frac{v}{v_x} \tag{5.86}$$

The integrals for the mobility term of Eq. (5.81) are of the form

$$-N \int \left(\frac{v_x\, q E_x}{m v_m}\right) \frac{v_x}{v}\,\frac{\partial f_0}{\partial v}\, dv_x\, dv_y\, dv_z\, \hat{a}_x \tag{5.87}$$

or the form

$$-N \int \left(\frac{v_x\, q E_x}{m v_m}\right) \frac{v_y}{v}\,\frac{\partial f_0}{\partial v}\, dv_x\, dv_y\, dv_z\, \hat{a}_x \tag{5.88}$$

Integral (5.87) can be then changed so that its integrand becomes only a function of v as shown below.

$$-N \int \left(\frac{v q E_x}{3 m v_m}\right) \frac{\partial f_0}{\partial v}\, dv_x\, dv_y\, dv_z\, \hat{a}_x \tag{5.89}$$

since $v_x^{\,2} \simeq v^2/3$.

The integral is now independent of direction and as a result may be changed to a single integral in the v direction by multiplying it by $4\pi v^2$. We have therefore converted to a spherical coordinate representation for v. Equation (5.89) then becomes

$$-N \int_0^\infty \frac{4\pi v^3 q E_x}{3 m v_m}\,\frac{\partial f_0}{\partial v}\, dv\, \hat{a}_x \tag{5.90}$$

We may integrate (5.90) by parts to give

$$-\left.\frac{N 4\pi q E_x f_0\, v^3}{3 m v_m}\right|_0^\infty \hat{a}_x + N \int_0^\infty f_0\, \frac{\partial}{\partial v}\left(\frac{4\pi q v^3}{3 m v_m}\right) E_x\, dv\, \hat{a}_x \tag{5.91}$$

If now $f_0\, v^3 = 0$ at both 0 and ∞, then (5.91) is simply

$$N \int_0^\infty f_0\, \frac{\partial}{\partial v}\left(\frac{4\pi q v^3}{3 m v_m}\right) E_x\, dv\, \hat{a}_x \tag{5.92}$$

If $v_m =$ constant, then integral (5.92) is

$$\frac{qn}{m v_m}\, E_x\, \hat{a}_x \tag{5.93}$$

This is the value of integral (5.87).

We must now turn our attention to the integral (5.88). This integral will be zero because of the assumption that v_x and v_y are uncorrelated.

Similar results for the y and z components of the integrals show that the total particle current due to the electric field is

$$\mathbf{\Gamma}_{\text{mob}} = \frac{qn}{m\nu_m}(E_x \hat{a}_x + E_y \hat{a}_y + E_z \hat{a}_z) \tag{5.94}$$

We note that $qn/m\nu_m$ is the *same* as the mobility that we have defined previously for a dc electric field with collisions. The net particle current is thus the sum of both diffusion and mobility currents, that is

$$\mathbf{\Gamma} = -D\nabla_r(n) + \mu \mathbf{E} n \qquad \text{where} \qquad \mu = \frac{qn}{m\nu_m} \tag{5.95}$$

Again, the result is the same as that obtained from the single-particle model. We have now extended the statistical model to a limit where we have integrated "out" over all velocities in the Boltzmann equation, assuming that collisions have a perturbing effect on the equilibrium distribution function f_0, and we have obtained the identical single-particle motion results for diffusion and mobility from the statistical model.

E. THE MOMENTS OF THE BOLTZMANN EQUATION

The next matter to be considered is the integration over the velocity in the Boltzmann equation itself *without* perturbations, to ascertain its effects. To obtain the most information, we will multiply the Boltzmann equation by a function of velocity $Q(v)$. Then we will integrate all terms of Eq. (5.63) over all velocity space in order to obtain a set of equations that are then in terms of the *ensemble average* velocities and densities. These equations are significant on a more macroscopic level than the Boltzmann equation. This process is called *taking the moments* of the Boltzmann equation.

Assume that the function is of the form $Q(v) \sim v^p$, where p has the values $0, 1, 2, \ldots$. If $p = 0$, then the first term on the left-hand side of the Boltzmann equation when integrated is

$$N \int \frac{\partial f}{\partial t}\, dv_x\, dv_y\, dv_z = \frac{\partial}{\partial t} n(\mathbf{r}, t) \tag{5.96}$$

since $Q(v) = 1$. The integrated second term in the Boltzmann equation is

$$N \int \mathbf{v} \cdot \nabla_r f \, dv_x \, dv_y \, dv_z = N\nabla_r \cdot \int f\mathbf{v} \, dv_x \, dv_y \, dv_z$$
$$= \nabla_r \cdot n\langle \mathbf{v} \rangle \qquad (5.97)$$

since

$$\nabla_r \cdot (f\mathbf{v}) = f\nabla_r \cdot \mathbf{v} + \mathbf{v} \cdot \nabla_r f \qquad \text{and} \qquad \nabla_r \cdot \mathbf{v} = 0$$

The third term becomes

$$N \int \mathbf{F} \cdot \nabla_v f \, dv_x \, dv_y \, dv_z = -N \int f\nabla_v \cdot \mathbf{F} \, dv_x \, dv_y \, dv_y \equiv 0 \qquad (5.98)$$

Equation (5.98) is zero, as we can show if we integrate by parts and pass to the limit over a large volume, and let the force term be only $\mathbf{F} = m\mathbf{g} + q\mathbf{E} + q(\mathbf{v} \times \mathbf{B})$.

The right-hand side of the Boltzmann equation, when integrated, becomes

$$N \int \left(\frac{\partial f}{\partial t}\right)_{\text{coll}} dv_x \, dv_y \, dv_z \equiv 0 \qquad (5.99)$$

since the collisional processes do not introduce any net average flux of particles in space, *unless* they result in ionization or recombination. Then, the right-hand side may not be zero.

Combining all the terms gives the 0th moment of the Boltzmann equation, namely,

$$\frac{\partial n(\mathbf{r}, t)}{\partial t} + \nabla_r \cdot [n(\mathbf{r}, t)\langle \mathbf{v} \rangle] = 0 \qquad (5.100)$$

This is the equation of continuity, which is identical in form to the equation for the conservation of charge.

If collisions *are* included, Eq. (5.100) becomes

$$\frac{\partial n(\mathbf{r}, t)}{\partial t} + \nabla_r \cdot [n(\mathbf{r}, t)\langle \mathbf{v} \rangle] = N \int \left(\frac{\partial f}{\partial t}\right)_{\text{coll}} dv_x \, dv_y \, dv_z \qquad (5.101)$$

Since we do not explicitly know the form of $(\partial f/\partial t)_{\text{coll}}$, we must leave the collision term in integral form.

The first moment is obtained with $p = 1$. We multiply the Boltzmann equation through by $m\mathbf{v}$ and integrate over velocity space as before. The first term in the equation becomes

$$Nm \frac{\partial}{\partial t} \int f\mathbf{v} \, dv_x \, dv_y \, dv_z = \frac{\partial}{\partial t} [mn(\mathbf{r}, t)\langle \mathbf{v} \rangle] \qquad (5.102)$$

The second term in the Boltzmann equation is

$$Nm \int \mathbf{v}(\mathbf{v} \cdot \nabla_r f)\, dv_x\, dv_y\, dv_z \tag{5.103}$$

We can separate expression (5.103) into components, and find that there are two types of terms. The first type, and its integral, is of the form

$$\hat{a}_x Nm \int v_x\left(v_x \frac{\partial f}{\partial x}\right) dv_x\, dv_y\, dv_z = \frac{\partial}{\partial x}[n(\mathbf{r}, t)m\langle v_x^2\rangle]\, \hat{a}_x \tag{5.104}$$

The second type is

$$\hat{a}_y Nm \int v_x\left(v_y \frac{\partial f}{\partial x}\right) dv_x\, dv_y\, dv_z = \frac{\partial}{\partial x}[n(\mathbf{r}, t)m\langle v_x v_y\rangle]\, \hat{a}_y \tag{5.105}$$

The representation $n(\mathbf{r}, t)m\langle v_x v_y\rangle$ exists *if* there is a correlation between v_x and v_y that appears through the distribution function. The complete integral of Eq. (5.103) is then

$$\begin{aligned} Nm \int \mathbf{v}(\mathbf{v} \cdot \nabla_r f)\, dv_x\, dv_y\, dv_z \\ = \nabla_r \cdot [&\hat{a}_x n(\mathbf{r}, t)m\langle v_x^2\rangle\hat{a}_x + \hat{a}_y n(\mathbf{r}, t)m\langle v_y v_x\rangle\hat{a}_x + \hat{a}_z n(\mathbf{r}, t)m\langle v_z v_x\rangle\hat{a}_x \\ &+ \hat{a}_x n(\mathbf{r}, t)m\langle v_x v_y\rangle\hat{a}_y + \hat{a}_y n(\mathbf{r}, t)m\langle v_y^2\rangle\hat{a}_y + \hat{a}_z n(\mathbf{r}, t)m\langle v_z v_y\rangle\hat{a}_y \\ &+ \hat{a}_x n(\mathbf{r}, t)m\langle v_x v_z\rangle\hat{a}_z + \hat{a}_y n(\mathbf{r}, t)m\langle v_y v_z\rangle\hat{a}_z + \hat{a}_z n(\mathbf{r}, t)m\langle v_z^2\rangle\hat{a}_z] \end{aligned} \tag{5.106}$$

A shorthand notation for the right-hand side of Eq. (5.106) would be

$$\nabla_r \cdot [n(\mathbf{r}, t)m\langle \mathbf{vv}\rangle]$$

where $\langle \mathbf{vv}\rangle$ is a second-order dyadic as was defined previously. It is a complete representation of Eq. (5.106).

The third term on the left-hand side of the first moment of the Boltzmann equation is

$$N \int \mathbf{v}(\mathbf{F} \cdot \nabla_v f)\, dv_x\, dv_y\, dv_z \tag{5.107a}$$

and this can be broken into components of the form

$$N \int v_x F_x \frac{\partial f}{\partial v_x}\, dv_x\, dv_y\, dv_z\, \hat{a}_x$$

or

$$N \int v_x F_y \frac{\partial f}{\partial v_y}\, dv_x\, dv_y\, dv_z\, \hat{a}_x$$

The first term above may be integrated by parts to give

$$-N \int f F_x \, dv_x \, dv_y \, dv_z = -n(\mathbf{r}, t)\langle F_x \rangle$$

while the second term has the value zero, since

$$\frac{\partial v_x}{\partial v_y} \equiv 0 \qquad \text{and} \qquad \frac{\partial v_x}{\partial v_x} \equiv 1$$

The complete third term (5.107a) is then merely

$$N \int \mathbf{v}(\mathbf{F} \cdot \nabla_v f) \, dv_x \, dv_y \, dv_z = -n(\mathbf{r}, t)\langle \mathbf{F} \rangle \tag{5.107b}$$

The integral of the collision term is again zero if there is no net momentum transferred by collisions. Otherwise the form of the integral is

$$N \int m\mathbf{v} \left(\frac{\partial f}{\partial t} \right)_{\text{coll}} dv_x \, dv_y \, dv_z$$

The results above will be used in writing the complete first moment equation subsequently.

We now can find the second moment of the Boltzmann equation by multiplying Eq. (5.63) by $\frac{1}{2}mv^2$ and then integrating as before. The first term on the left-hand side is then

$$N \int \frac{1}{2} mv^2 \frac{\partial f}{\partial t} \, dv_x \, dv_y \, dv_z = N \frac{\partial}{\partial t} \int \frac{1}{2} mv^2 f \, dv_x \, dv_y \, dv_z$$

$$= \frac{\partial}{\partial t} \left[\frac{1}{2} mn(\mathbf{r}, t)\langle v^2 \rangle \right] \tag{5.108}$$

This is equivalent to the time rate of change of the average kinetic energy stored in the systems. Note it is a scalar.

The second term on the left-hand side of the Boltzmann equation becomes, when taking the second moment,

$$N \int \tfrac{1}{2} mv^2 (\mathbf{v} \cdot \nabla_r f) \, dv_x \, dv_y \, dv_z \tag{5.109}$$

Writing this out in components, a typical term is

$$N \int \frac{1}{2} mv^2 v_x \frac{\partial f}{\partial x} \, dv_x \, dv_y \, dv_z$$

This is equal to

$$N \frac{\partial}{\partial x} \int \frac{1}{2} mv^2 v_x f \, dv_x \, dv_y \, dv_z = \frac{\partial}{\partial x} \left[n(\mathbf{r}, t) \frac{1}{2} m \langle v^2 v_x \rangle \right] \tag{5.110}$$

The net result summed over all components is then

$$\nabla_r \cdot [\tfrac{1}{2}n(\mathbf{r}, t)m\langle v^2\mathbf{v}\rangle] \tag{5.111}$$

which is equivalent to a transport of energy flux. The third term on the left-hand side of the second moment of the Boltzmann equation is

$$N\int \frac{1}{2} mv^2\left(\frac{\mathbf{F}}{m}\cdot\nabla_v f\right) dv_x\, dv_y\, dv_z \tag{5.112}$$

A typical term in (5.112) is

$$N\int \frac{1}{2} mv^2 \frac{F_x}{m}\frac{\partial f}{\partial v_x} dv_x\, dv_y\, dv_z \tag{5.113a}$$

Integrating this term by parts, we get

$$N\frac{1}{2} v^2 F_x f\Big]_{-\infty}^{+\infty} - N\int \frac{f}{2}\frac{\partial}{\partial v_x}(v^2 F_x)\, dv_x\, dv_y\, dv_z$$

The first term goes to zero at $\pm\infty$. Noting the nature of the force (since F_x is not dependent on v_x, but may be dependent on v_y or v_z, etc.), we obtain for (5.113a)

$$-\int \frac{f}{2} F_x \frac{\partial}{\partial v_x} v^2\, dv_x\, dv_y\, dv_z \tag{5.113b}$$

Since $v^2 = v_x{}^2 + v_y{}^2 + v_z{}^2$ and

$$\frac{\partial(v^2)}{\partial v_x} = 2v_x$$

We can then obtain, for the expression (5.113b)

$$-\langle F_x v_x\rangle n(\mathbf{r}, t)$$

The net result for the third term [Eq. (5.112)] summed over components is

$$-\langle \mathbf{F}\cdot\mathbf{v}\rangle n(\mathbf{r}, t) \tag{5.114}$$

The collision integral in the second moment of (5.63),

$$N\int \frac{1}{2} mv^2\left(\frac{\partial f}{\partial t}\right)_{\text{coll}} dv_x\, dv_y\, dv_z \tag{5.115}$$

has a value of zero if there is no net energy change transmitted by collisions. Otherwise it is written in its complete form as shown above in (5.115).

We can now summarize and write together the zeroth, first, and second moments of the Boltzmann equation. They are written in both collisionless and collisional forms.

0th moment

$$\frac{\partial n(\mathbf{r}, t)}{\partial t} + \nabla_r \cdot [n(\mathbf{r}, t)\langle \mathbf{v} \rangle] = \begin{cases} 0 \\ \text{or} \\ N \int \left(\dfrac{\partial f}{\partial t}\right)_{\text{coll}} dv_x\, dv_y\, dv_z \end{cases} \tag{5.116}$$

1st moment

$$\frac{\partial}{\partial t} n(\mathbf{r}, t) m \langle \mathbf{v} \rangle + \nabla_r \cdot [n(\mathbf{r}, t) m \langle \mathbf{v}\mathbf{v} \rangle] - n(\mathbf{r}, t)\langle \mathbf{F} \rangle = \begin{cases} 0 \\ \text{or} \\ N \int m\mathbf{v} \left(\dfrac{\partial f}{\partial t}\right)_{\text{coll}} dv_x\, dv_y\, dv_z \end{cases} \tag{5.117}$$

2nd moment

$$\frac{\partial}{\partial t} \frac{1}{2} n(\mathbf{r}, t) m \langle v^2 \rangle + \nabla_r \cdot \left[\frac{1}{2} n(\mathbf{r}, t) m \langle v^2 \mathbf{v} \rangle\right] - \langle \mathbf{F} \cdot \mathbf{v} \rangle n(\mathbf{r}, t) = \begin{cases} 0 \\ \text{or} \\ N \int \dfrac{1}{2} m v^2 \left(\dfrac{\partial f}{\partial t}\right)_{\text{coll}} dv_x\, dv_y\, dv_z \end{cases} \tag{5.118}$$

The right-hand sides of Eqs. (5.116)–(5.118) are equal to zero if there is no net mass transfer, momentum transfer, or energy transfer, respectively. These three equations are essentially the conservation relations for particles, momentum, and energy, respectively. They are simpler in form than the Boltzmann equation from which they have been derived, but some information, namely, the velocity dependence of the particles and the nature of the distribution function, has been removed. These are essentially fluid-type equations and there are many plasma problems for which their solution is perfectly adequate. However, experimental verification of such solutions may be unobtainable under the conditions where the plasma velocity distribution must be considered. Under these conditions, the Boltzmann equation approach is often used.

It is theoretically possible to continue taking moments of the Boltzmann equation to as high an order as desired. Such an operation is, in fact, necessary for a complete solution to a problem using the moment approach, since portions of each moment equation often contain information that is contained in the next higher order moment equation. This is especially noticeable in the terms involving the collisions. Often, however, and as will be shown subsequently, the collision terms may be simplified and the series of moment equations truncated. This will be discussed in Chapter 9.

SUGGESTED READING

Cowling, T. G., "Magnetohydrodynamics," Wiley (Interscience), New York, 1959.
Goldstein, S., "Classical Mechanics," Addison-Wesley, Reading, Massachusetts, 1956. This book includes a good development of the Hamilton–Jacobi method.
Rose, D. J., and Clark, M., "Plasmas and Controlled Fusion," The M.I.T. Press, Cambridge, Massachusetts, 1961.
Schmidt, G., "Physics of High Temperature Plasmas—An Introduction," Academic Press, New York, 1966.
Spitzer, L., "Physics of Fully Ionized Gases," 2nd ed., Wiley (Interscience), New York, 1962.
Thompson, W. B., "An Introduction to Plasma Physics," 2nd ed., Pergamon Press, New York, 1964.
Uman, M., "Introduction to Plasma Physics," McGraw-Hill, New York, 1964.

Problems

1 In Problem 2 of Chapter 4 we determined that the angular momentum of a charged particle orbiting in a magnetic field was

$$mr^2\dot{\theta} + qrA_\theta$$

(a) Show that this is a constant of the motion.
(b) Then show that we may form a distribution function $f(c_1, c_2)$ which is a solution to the collisionless Boltzmann equation, where c_1 is the energy and c_2 is the angular momentum. What coordinates could be eliminated with these two constants?
(c) Can we also use the magnetic moment as an additional constant of the motion? If so, what additional coordinates can be eliminated?

2 Consider a charged particle that is moving with velocity v that is perpendicular to a dc magnetic field B. If the dc field is quickly reversed in direction, what happens to the particle's path and its energy? After transients have died away, what is the velocity and radius of gyration of the particle. Is the magnetic moment constant? Hint: In this case, energy may not be conserved, because the time-changing magnetic field has an associated electric field that can do work.

3 Use the Hamilton–Jacobi theory to solve the problem of a charged particle orbiting about a uniform dc magnetic field line with axial symmetry. That is, find the appropriate momenta P_r, P_θ, and P_z needed to ignore their associated Q's. Then obtain the Q's and the q's.

4 Calculate the energy as a function of time of a *neutral* particle, bouncing perfectly elastically between two walls that move toward each other with relative velocity v. Assume that the particle is moving with velocity v toward one of the walls initially. Compute the action for the particle. Is it a constant? If not, why not?

5 Assume that we have determined Prob(**v**, Δ**v**) for a plasma to be the following function.

$$\text{Prob}(\mathbf{v}, \Delta\mathbf{v}) = \exp\left(-\frac{|\Delta\mathbf{v}|^2}{2m\ell T}\right)$$

Calculate the two terms in the Fokker–Planck equation. Which dominates? Hint: Use components of the vectors and/or dyadics where necessary.

6 Take the third moment of the Boltzmann equation with collisions. What can you say about the result in view of the momentum and energy conservation expressed in the first and secomd moments?

7 Determine if any of the momenta P_r, P_θ, and P_z found in Problem 3 are also constants of the motion.

8 Another form of velocity distribution that is often used as that for plasmas confined in magnetic bottles is the "loss-cone" distribution. It is essentially an anisotropic Maxwellian distribution like that shown in Eq. (4.74) but all of those particles whose velocity vectors lie in the escape cone are missing. That is, the function is

$$f \sim \exp\left(-\frac{v_\parallel^{\,2}}{\alpha_\parallel^{\,2}} - \frac{v_\perp^{\,2}}{\alpha_\perp^{\,2}}\right) \qquad \text{for} \qquad \frac{v_\parallel^{\,2}}{v_\perp^{\,2}} > \cot^2\theta_{\max} = R - 1$$

and

$$f \equiv 0 \qquad \text{for} \qquad \frac{v_\parallel^{\,2}}{v_\perp^{\,2}} < \cot^2\theta_{\max} = R - 1$$

We have noted in previous work that the constants $\alpha_\parallel^{\,2}$ and $\alpha_\perp^{\,2}$ are related to the temperature and average energy.

(a) Show, for an anisotropic Maxwellian without a loss-cone distribution (i.e., the full distribution), that

$$\frac{1}{2}m\langle v_\parallel^{\,2}\rangle = \frac{2\alpha_\parallel^{\,2}}{m} \qquad \text{and} \qquad \frac{1}{2}m\langle v_\perp^{\,2}\rangle = \frac{2\alpha_\perp^{\,2}}{m}$$

(b) Find $\frac{1}{2}m\langle v_{\parallel}^2\rangle$ and $\frac{1}{2}m\langle v_{\perp}^2\rangle$ when considering the loss-cone distribution described above. Compare these results to those of part (a).

9 Find the expression for the mirror reflection potential as shown in Eq. (5.18) if

$$\mathbf{B} = B_0(1 + \alpha|z|)\hat{a}_z$$

6 Simple Applications of the Fluid and Statistical Models of a Plasma

The purpose of this chapter is to draw together much of the preceding work. We have had some acquaintance with particle motion, collisions, and statistical treatments, and now some specific applications to plasma behavior will be made.

The present chapter first introduces the plasma fluid hydromagnetic (or magnetohydrodynamic) equations. We will obtain these by utilizing the moments of the Boltzmann equation, which were found by "integrating out" the velocity dependence and by considering only average values for the velocity terms.

The results are useful when the plasma is treated as a "continuum," that is, a "mathematical substance" assumed to have no particulate structure. This continuum or fluid model is very useful in many circumstances (usually for very low frequencies) and several examples of this will be presented from the point of view of wave motion and diagnostics as well as from that of magnetic confinement.

We will also introduce some parameters commonly used to describe plasmas. We will see how they appear when considering plasma behavior and how they are useful. Already, certain "figures of merit" have been mentioned: density, magnetic field strength, volume, temperature, and so on. The quantities developed here are also of this nature, but perhaps more basic to a description of the plasma than the ones previously mentioned. Finally, a brief glimpse into the vast area of the interaction of radiation with a plasma is presented.

We first wish to apply some of the previously developed statistical equations directly to plasmas. In particular, Eqs. (5.116), (5.117), and (5.118) may be applied to plasma behavior whenever the energy distribution of the plasma is *not* to be taken into account *other* than by integration over velocity space.

A. HYDROMAGNETICS (MAGNETOHYDRODYNAMICS)

We may now deduce the hydromagnetic equations for a plasma, that is, when we consider the plasma to be a conducting fluid.

Equation (5.117) the momentum equation, is reproduced here.

$$\frac{\partial}{\partial t} n(\mathbf{r}, t)\langle m\mathbf{v}\rangle + \nabla_r \cdot [n(\mathbf{r}, t)m\langle \mathbf{vv}\rangle] - n(\mathbf{r}, t)\langle \mathbf{F}\rangle$$
$$= N \int m\mathbf{v}\left(\frac{\partial f}{\partial t}\right)_{\text{coll}} dv_x\, dv_y\, dv_z \tag{6.1}$$

In particular, the $\langle \mathbf{vv}\rangle$ term, which is the average of a dyadic, must be investigated and expanded. Let us assume that

$$\mathbf{vv} = (\mathbf{v}_0 + \mathbf{v}_r)(\mathbf{v}_0 + \mathbf{v}_r) \tag{6.2}$$

where $\mathbf{v}_r$ is the "random" velocity and $\mathbf{v}_0$ is the average velocity. Then

$$\langle \mathbf{vv}\rangle = \langle \mathbf{v}_0 \mathbf{v}_0 + \mathbf{v}_r \mathbf{v}_0 + \mathbf{v}_0 \mathbf{v}_r + \mathbf{v}_r \mathbf{v}_r\rangle \tag{6.3}$$

$\mathbf{v}_r$ may be thought of as a local *spatial* departure or fluctuation from equilibrium. It is not a perturbation of the distribution function. The middle two terms on the right-hand side of Eq. (6.3) are zero since the average of a random velocity must be zero. Therefore

$$\langle \mathbf{vv}\rangle = \langle \mathbf{v}_0\rangle\langle \mathbf{v}_0\rangle + \langle \mathbf{v}_r \mathbf{v}_r\rangle \tag{6.4}$$

since $\langle \mathbf{v}_0 \mathbf{v}_0\rangle = \langle \mathbf{v}_0\rangle\langle \mathbf{v}_0\rangle$ by definition.

The second term on the *left* of Eq. (6.1) may now be written as the sum of two terms, namely,

$$\nabla_r \cdot [n(\mathbf{r}, t)m\langle \mathbf{v}_0\rangle\langle \mathbf{v}_0\rangle] + \nabla_r \cdot [n(\mathbf{r}, t)m\langle \mathbf{v}_r \mathbf{v}_r\rangle] \tag{6.5}$$

The first term in expression (6.5) may be expanded by assuming that it is a product of two functions

$$\mathbf{n}(\mathbf{r}, t)m\langle \mathbf{v}_0\rangle \qquad \text{and} \qquad \langle \mathbf{v}_0\rangle$$

The term then becomes

$$\nabla_r \cdot [n(\mathbf{r}, t)m\langle \mathbf{v}_0\rangle\langle \mathbf{v}_0\rangle] = n(\mathbf{r}, t)m\langle \mathbf{v}_0\rangle\nabla_r \cdot \langle \mathbf{v}_0\rangle$$
$$+ \langle \mathbf{v}_0\rangle\nabla_r \cdot [n(\mathbf{r}, t)m\langle \mathbf{v}_0\rangle] \tag{6.6}$$

We can now write the following for the second term on the left-hand side of (6.1), using the expansion (6.4).

$$\begin{aligned}\nabla_r \cdot [n(\mathbf{r}, t)m\langle \mathbf{v}\mathbf{v}\rangle] &= \nabla_r \cdot [n(\mathbf{r}, t)m\langle \mathbf{v}_0 \mathbf{v}_0\rangle] \\ &\quad + \nabla_r \cdot [n(\mathbf{r}, t)m\langle \mathbf{v}_r \mathbf{v}_r\rangle] \\ &= n(\mathbf{r}, t)m\langle \mathbf{v}_0\rangle(\nabla_r \cdot \langle \mathbf{v}_0\rangle) \\ &\quad + \langle \mathbf{v}_0\rangle\nabla_r \cdot [n(\mathbf{r}, t)m\langle \mathbf{v}_0\rangle] \\ &\quad + \nabla_r \cdot (n(\mathbf{r}, t)m\langle \mathbf{v}_r \mathbf{v}_r\rangle) \end{aligned} \tag{6.7}$$

We now return to the first term in Eq. (6.1). The *first* term on the left-hand side of Eq. (6.1) may be expanded as

$$\frac{\partial}{\partial t}[n(\mathbf{r},t)\, m\langle \mathbf{v}\rangle] = n(\mathbf{r},t)m\frac{\partial\langle \mathbf{v}\rangle}{\partial t} + m\langle \mathbf{v}\rangle\frac{\partial n(\mathbf{r},t)}{\partial t} \tag{6.8}$$

We may further expand this term by using the zeroth moment of the Boltzmann equation with collisions. It is reproduced below.

$$\frac{\partial}{\partial t}n(\mathbf{r}, t) + \nabla_r \cdot (n(\mathbf{r}, t)\langle \mathbf{v}\rangle) = N\int\left(\frac{\partial f}{\partial t}\right)_{\text{coll}} dv_x\, dv_y\, dv_z \tag{6.9}$$

Equation (6.9) is often called the continuity equation. Now, if Eq. (6.9) is multiplied through by $m\langle \mathbf{v}\rangle$ we get

$$m\langle \mathbf{v}\rangle\frac{\partial n(\mathbf{r}, t)}{\partial t} + m\langle \mathbf{v}\rangle\nabla_r \cdot (n(\mathbf{r}, t)\langle \mathbf{v}\rangle) = m\langle \mathbf{v}\rangle N\int\left(\frac{\partial f}{\partial t}\right)_{\text{coll}} dv_x\, dv_y\, dv_z \tag{6.10}$$

Now, it is true that

$$m\langle \mathbf{v}\rangle\frac{\partial n(\mathbf{r}, t)}{\partial t} = \frac{\partial}{\partial t}[n(\mathbf{r}, t)m\langle \mathbf{v}\rangle] - n(\mathbf{r}, t)m\frac{\partial\langle \mathbf{v}\rangle}{\partial t} \tag{6.11}$$

This can be seen if $\partial[n(\mathbf{r}, t)m\langle \mathbf{v}\rangle]/\partial t$ is expanded as the derivative of a product.

We are now able to find an equivalence between the first term on the left-hand side of Eq. (6.1) and the work we have just done, by using (6.11) in conjunction with (6.10).

This relation is

$$\frac{\partial}{\partial t}[n\,(\mathbf{r},t)m\langle \mathbf{v}\rangle] = m\langle \mathbf{v}\rangle\frac{\partial}{\partial t}n(\mathbf{r}, t) + n(\mathbf{r}, t)m\frac{\partial\langle \mathbf{v}\rangle}{\partial t} \tag{6.12}$$

$$\begin{aligned} &= m\langle \mathbf{v}\rangle N\int\left(\frac{\partial f}{\partial t}\right)_{\text{coll}} dv_x\, dv_y\, dv_z \\ &\quad - m\langle \mathbf{v}\rangle\nabla_r \cdot [n(\mathbf{r}, t)\langle \mathbf{v}\rangle] + n(\mathbf{r}, t)m\frac{\partial\langle \mathbf{v}\rangle}{\partial t} \end{aligned} \tag{6.13}$$

Equation (6.13) comes from the substitution of Eq. (6.10) into Eq. (6.12).

We can combine all of the foregoing results into the momentum equation (6.1) to arrive at

$$mn(\mathbf{r}, t)\frac{\partial\langle\mathbf{v}\rangle}{\partial t} + mn(\mathbf{r}, t)\langle\mathbf{v}\rangle(\nabla_r \cdot \langle\mathbf{v}\rangle)$$

$$= n(\mathbf{r}, t)\langle\mathbf{F}\rangle - \nabla_r \cdot \mathbf{P} + N\int m\mathbf{v}\left(\frac{\partial f}{\partial t}\right)_{\text{coll}} dv_x\, dv_y\, dv_z$$

$$- m\langle\mathbf{v}\rangle N\int\left(\frac{\partial f}{\partial t}\right)_{\text{coll}} dv_x\, dv_y\, dv_z \tag{6.14}$$

That is, we have substituted in Eq. (6.13) for the first term on the left-hand side of (6.1) and Eq. (6.7) for the second term.

The quantity $\mathbf{P}$ is $mn(\mathbf{r}, t)\langle\mathbf{v}_r\,\mathbf{v}_r\rangle$. It is called the *kinetic stress* or pressure *tensor*. (It is still a dyadic, just the same.)

Equation (6.14) is still the first moment of the Boltzmann equation with the velocity effects integrated out. Only averages of velocity terms remain *except* in the collision integrals, where knowledge of $(\partial f/\partial t)_{\text{coll}}$ is required before the velocity dependence can be removed. For each species of particles in a plasma, a *separate* momentum equation (6.14) may be written.

We will now define some quantities based on average values which are useful for plasma studies. If we consider a mixture of electrons and ions of masses m_e and m_i, and charges q_e and q_i, respectively, we may then define the following.

Mass Density

$$\rho_m = n_e m_e + n_i m_i \tag{6.15}$$

Particle Current

$$\boldsymbol{\Gamma} = n_e\langle\mathbf{v}_e\rangle + n_i\langle\mathbf{v}_i\rangle \tag{6.16}$$

Mass Current

$$\mathbf{J}_m = n_e m_e\langle\mathbf{v}_e\rangle + n_i m_i\langle\mathbf{v}_i\rangle \tag{6.17}$$

Electric Charge Density

$$\rho_E = n_e q_e + n_i q_i \tag{6.18}$$

Electric Current Density

$$\mathbf{J}_E = n_e q_e\langle\mathbf{v}_e\rangle + n_i q_i\langle\mathbf{v}_i\rangle \tag{6.19}$$

If we neglect creation and annihilation of particles, as well as collisions, the continuity equations (see Eq. (6.9)) for each species of particle become

$$\frac{\partial}{\partial t}(n_e m_e) + \nabla_r \cdot (n_e m_e \langle \mathbf{v}_e \rangle) = 0 \tag{6.20}$$

for the electrons, and

$$\frac{\partial}{\partial t}(n_i m_i) + \nabla_r \cdot (n_i m_i \langle \mathbf{v}_i \rangle) = 0 \tag{6.21}$$

for the ions. We may add Eqs. (6.20) and (6.21) to obtain a continuity equation for the plasma as a whole, which is

$$\frac{\partial \rho_m}{\partial t} + \nabla_r \cdot (\rho_m \mathbf{v}) = 0 \tag{6.22}$$

$\mathbf{v}$ is defined as $\mathbf{J}_m/\rho_m$ and is the net average mass velocity. We can also utilize the equation of conservation of charge for the plasma

$$\frac{\partial \rho_E}{\partial t} + \nabla_r \cdot \mathbf{J}_E = 0 \tag{6.23}$$

Let us now rewrite Eq. (6.14) by using some of the foregoing results. Recall that there are separate momentum equations for the electrons and the ions; that the momentum equation is the momentum balance on an incremental volume of plasma; and that the quantities that appear in it must have dimensions per unit volume. For the electrons and ions the net average forces $\langle \mathbf{F} \rangle$ per unit volume will be

$$\langle \mathbf{F}_e \rangle = n_e q_e \mathbf{E} + n_e q_e (\langle \mathbf{v}_e \rangle \times \mathbf{B}) - n_e m_e \mathbf{g} \tag{6.24}$$

and

$$\langle \mathbf{F}_i \rangle = n_i q_i \mathbf{E} + n_i q_i (\langle \mathbf{v}_i \rangle \times \mathbf{B}) - n_i m_i \mathbf{g} \tag{6.25}$$

Note that only electric, magnetic and, gravitational forces are included in the foregoing equations.

Equation (6.14) considers only *net* momentum transfer and therefore only electron–ion collisions can contribute; *not* collisions between particles of the same species, since no net momentum change is obtained. The revised momentum equations now become

$$\begin{aligned}\frac{\partial}{\partial t}(n_e m_e \langle \mathbf{v}_e \rangle) &+ \nabla_r \cdot (n_e m_e \langle \mathbf{v}_e \rangle \langle \mathbf{v}_e \rangle) \\ &= n_e q_e [\mathbf{E} + \langle \mathbf{v}_e \rangle \times \mathbf{B}] - n_e m_e \mathbf{g} - \nabla_r \cdot \mathbf{P}_e \\ &\quad + N_e \int m_e \mathbf{v}_e \left(\frac{\partial f_e}{\partial t} \right)_{\text{e–i coll}} dv_x \, dv_y \, dv_z \end{aligned} \tag{6.26}$$

for the electrons, and for the ions,

$$\frac{\partial}{\partial t}(n_i m_i\langle \mathbf{v}_i\rangle) + \nabla_r \cdot (n_i m_i \langle \mathbf{v}_i\rangle\langle \mathbf{v}_i\rangle)$$
$$= n_i q_i[\mathbf{E} + \langle \mathbf{v}_i\rangle \times \mathbf{B}] - n_i m_i \mathbf{g} - \nabla_r \cdot \mathbf{P}_i$$
$$+ N_i \int m_i \mathbf{v}_i \left(\frac{\partial f}{\partial t}\right)_{\text{i-e coll}} dv_x\, dv_y\, dv_z \tag{6.27}$$

Note that in Eq. (6.27) the distribution function in the collision integral must be that for the ions. We have also used Eq. (6.13) to remove the second collision integral from (6.26) and (6.27).

Obviously, if the net momentum gained by the electrons results in the same net momentum loss by the ions in their collisional processes, the remaining collision terms in Eqs. (6.26) and (6.27) must be *equal in magnitude and opposite in sign.* If (6.26) and (6.27) are added, there results a total momentum conservation equation for the plasma, with the collision terms canceling out. This momentum equation is

$$\rho_m \frac{\partial \mathbf{v}}{\partial t} + \mathbf{v}\frac{\partial \rho_m}{\partial t} + \nabla_r \cdot (n_e m_e \langle \mathbf{v}_e\rangle\langle \mathbf{v}_e\rangle + n_i m_i \langle \mathbf{v}_i\rangle\langle \mathbf{v}_i\rangle)$$
$$= \rho_E \mathbf{E} + \mathbf{J}_E \times \mathbf{B} - \nabla_r \cdot \mathbf{P}_e - \nabla_r \cdot \mathbf{P}_i - \rho_m \mathbf{g} \tag{6.28}$$

If the plasma can be considered incompressible, the pressure dyadics isotropic, and drifts, shear, and so on, are neglected, then (6.28) becomes

$$\rho_m \frac{\partial \mathbf{v}}{\partial t} = \rho_E \mathbf{E} + \mathbf{J}_E \times \mathbf{B} - \nabla_r p - \rho_m \mathbf{g} \tag{6.29}$$

The term $\nabla_r p$ is then the gradient of a scalar pressure. That is, $\mathbf{P}$ has become a scalar since we have assumed isotropy of the plasma.

B. HYDROMAGNETIC EQUILIBRIUM

Equation (6.29) may now be written for a *steady state* time-invariant *equilibrium*, neglecting gravitational and space change effects. The result is

$$\nabla_r p = \mathbf{J}_E \times \mathbf{B} \tag{6.30}$$

This equilibrium equation is quite interesting because it gives us the relation between magnetic field, current density, and pressure gradient. It does not, however, show that the velocity distribution of the plasma when it is in *hydromagnetic* equilibrium may be out of *thermal* equilibrium and hence destroy this type of equilibrium. We may examine the nature of this equation with the help of Maxwell's equations in the same steady state:

$$\nabla_r \times \mathbf{B} = \mu_0 \mathbf{J}_E \tag{6.31}$$

$$\nabla_r \times \mathbf{E} = 0 \tag{6.32}$$

$$\nabla_r \cdot \mathbf{B} = 0 \tag{6.33}$$

$$\nabla_r \cdot \mathbf{D} = 0 \tag{6.34}$$

Taking the divergence of Eq. (6.31) results in

$$\nabla_r \cdot \mathbf{J}_E = 0 \tag{6.35}$$

since

$$\nabla_r \cdot (\nabla_r \times \mathbf{B}) \equiv 0$$

Now Eq. (6.30) implies that $\mathbf{J}_E$ and $\mathbf{B}$ are both perpendicular to the scalar pressure gradient. This may be expressed mathematically as

$$\mathbf{B} \cdot \nabla_r p = 0 \tag{6.36}$$

$$\mathbf{J}_E \cdot \nabla_r p = 0 \tag{6.37}$$

The pressure, since it is assumed to be a scalar, can have isobaric surfaces ($p =$ constant) defined. These surfaces must be normal to $\nabla_r p$. Equation (6.30) states that $\mathbf{B}$ and $\mathbf{J}_E$ must be perpendicular to $\nabla_r p$. Therefore, $\mathbf{B}$ and $\mathbf{J}_E$ must lie on a surface of constant pressure.

The magnetic field shape ought to have some bearing on the shape of the plasma *equilibrium* surfaces. By suitably designing the magnetic field, a set of nested surfaces composed of magnetic lines of force may be constructed where, presumably, a plasma could be contained if momentum equilibrium can be achieved. It should be added that, as discussed previously, even if these equilibrium surfaces could be made to exist, a plasma may not remain in energy equilibrium, and therefore the equilibrium surfaces are often only used as a beginning point for plasma production. The magnetic mirror and cusp fields have surfaces of containment, which may be developed from the expressions for their magnetic field lines and the following discussion. In principle, the nested set of surfaces should correspond to a set of pressure isobars, as shown in Fig. 6.1. These nested surfaces *may* produce a system in which a plasma can be contained, that is, it may remain in this configuration for a very long time, much longer than would be the case without this field configuration. If we arrange $\mathbf{J}_E$ (these are the currents flowing in the plasma) and $\mathbf{B}$ to be such that $\mathbf{J}_E \times \mathbf{B}$ points everywhere inward, then, for momentum equilibrium to exist, the kinetic pressure must increase as we

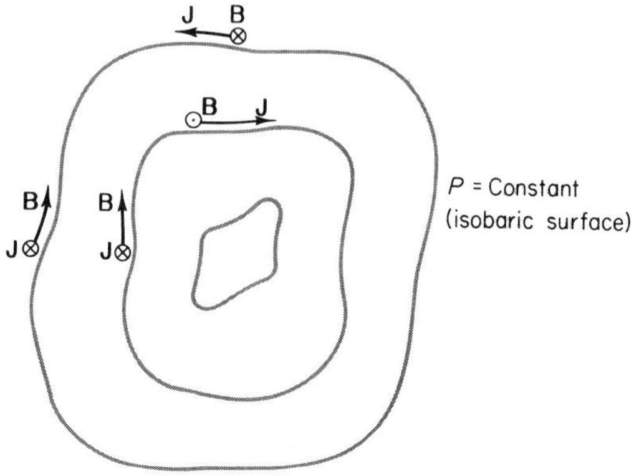

Fig. 6.1. Plasma magnetohydrodynamic confinement surfaces.

penetrate the plasma from the outside. Now since $\mathbf{J}_E \times \mathbf{B}$ has the dimensions of force per unit volume, it is equivalent (in equilibrium) to the pressure gradient term, which has the same units. We may insert Eq. (6.31) into (6.30) to eliminate $\mathbf{J}_E$, yielding

$$\nabla_r p = \frac{(\nabla_r \times \mathbf{B}) \times \mathbf{B}}{\mu_0} = -\frac{\nabla_r(B^2)}{2\mu_0} + \frac{(\mathbf{B} \cdot \nabla_r)\mathbf{B}}{\mu_0} \tag{6.38}$$

We have used the vector identity

$$\nabla_r(\mathbf{A} \cdot \mathbf{B}) = (\mathbf{A} \cdot \nabla_r)\mathbf{B} + (\mathbf{B} \cdot \nabla_r)\mathbf{A} + \mathbf{A} \times (\nabla_r \times \mathbf{B}) + \mathbf{B} \times (\nabla_r \times \mathbf{A}) \tag{6.39}$$

to obtain Eq. (6.38). If the magnetic field lines are straight and parallel, then

$$(\mathbf{B} \cdot \nabla_r)\mathbf{B} = 0$$

so Eq. (6.38) becomes

$$\nabla_r p = -\frac{\nabla_r(B^2)}{2\mu_0} \tag{6.40}$$

or

$$\nabla_r\left[p + \frac{B^2}{2\mu_0}\right] = 0 \tag{6.41}$$

An integration of (6.41) gives

$$p + \frac{B^2}{2\mu_0} = \text{constant} \tag{6.42}$$

This last expression states that the term $B^2/2\mu_0$ is equivalent to a pressure that is produced by the magnetic field. If the magnetic field is B_0 when there is no plasma present ($p = 0$), then the constant term in Eq. (6.42) is $B_0^2/2\mu_0$. The quantity β, for a plasma,

$$\beta = \frac{p}{B_0^2/2\mu_0} \tag{6.43}$$

is defined as the ratio of kinetic pressure to magnetic pressure. In general, the lower β is, the more likely a plasma is to be confined, if β is computed by taking the kinetic pressure over the magnetic pressure when no plasma is present. β should never be greater than 1 for this type of confinement since otherwise Eq. (6.42) will no longer be satisfied, and confinement will not be possible. Typically, in the laboratory, β may have values of less than 0.001 for cold tenuous plasmas, up to values of the order of unity for hot dense pulsed plasmas.

Equation (6.42) may be written in terms of the field B_0 as

$$p + \frac{B^2}{2\mu_0} = \frac{B_0^2}{2\mu_0} \tag{6.44}$$

C. PLASMA DIAMAGNETISM

The last equation showed that, if plasma were present ($p \neq 0$), then the magnetic field inside the plasma would be less than what it would be if plasma were not present. This means that the plasma appears to be "diamagnetic," that is, it excludes magnetic field lines from itself. If the kinetic pressure is assumed to be of the form

$$p = nkT \tag{6.45}$$

then measurement of the net *decrease* in magnetic field should result in knowledge of the product of temperature and density of the plasma (energy density). The ability to measure the diamagnetism of a plasma depends on its magnitude (if the measurement is made in a static case) or its rate of change (if it is a dynamic situation). Two separate methods are often used for these two cases. A Hall effect probe can be used in the steady state case and an integrated voltage from a pickup coil may be used for the dynamic case. Figure 6.2 shows a sketch of a diamagnetic pickup loop.

As the plasma builds up or collapses, it forces out B lines, which cut the loop, inducing a voltage that is proportional to dB/dt. The loop may be placed around, just outside of, or inside the plasma. In each case, care must be taken to assure that the correct measurement of the change of magnetic field in the plasma is made. In particular, the loop should be shielded to avoid responding

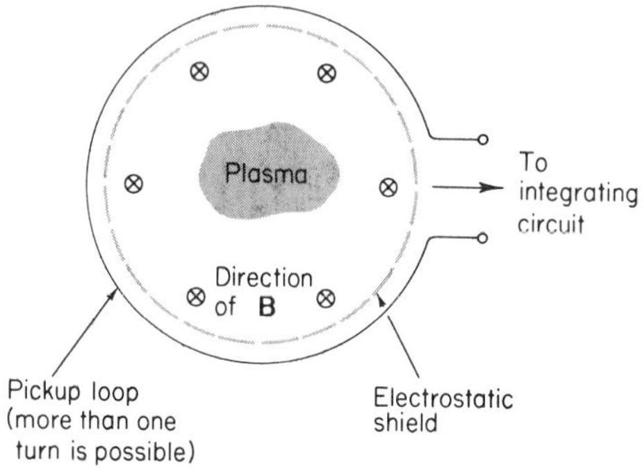

Fig. 6.2. Diamagnetic loop.

to plasma particles directly, and compensated to remove the effects of dc magnetic field ripple. Integration of the output voltage of the loop with respect to time measures the net diamagnetic field. This is depicted in Fig. 6.3.

In no case should the integrated diamagnetic signal fall below the 0 axis. However, certain experiments sometimes show that very effect occurring. This means that the plasma has a higher magnetic field inside it than the field where no plasma is present, that is, it is *paramagnetic*.

How is it possible for a plasma to exhibit paramagnetism? A single-particle approach in a static uniform magnetic field shows the plasma to be diamagnetic. Consider Fig. 6.4. Here, electrons and ions spiral as shown about

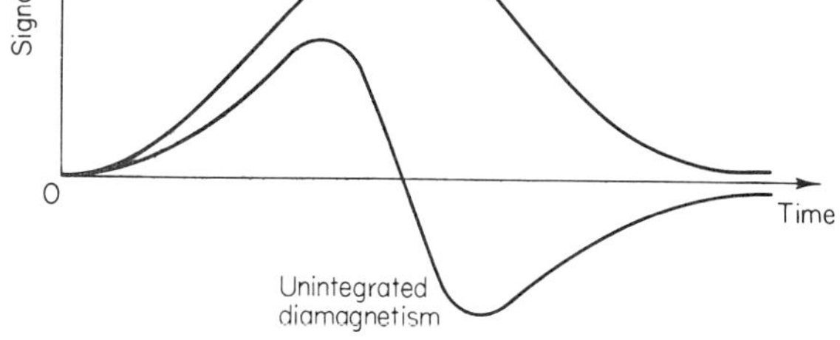

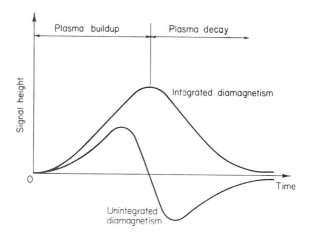

Fig. 6.3. Typical diamagnetic signal.

the external dc magnetic field lines. Their motion is such as to produce a magnetic field in opposition to the external field. Hence, the net field decreases. Notice both particles show this diamagnetic behavior.

However, should external currents, waves, or motion of the plasma result in drift currents that produce magnetic fields that *aid* the external field, then the plasma can behave as a paramagnetic body. Therefore, knowledge of such motions is essential before full faith can be placed in the measurements of the diamagnetic properties of a plasma.

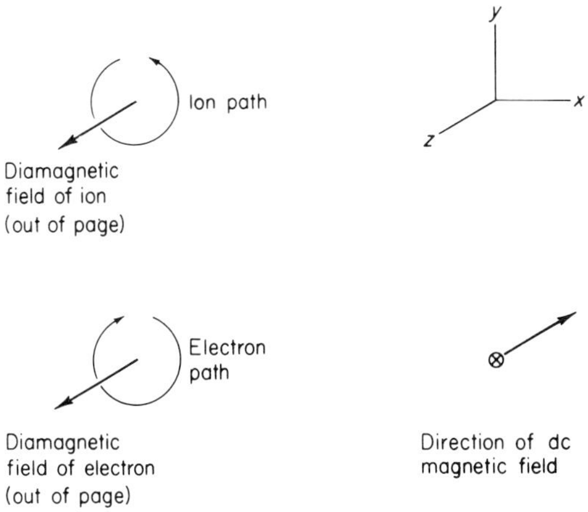

Fig. 6.4. Diamagnetic self-fields of ions and electrons.

D. MAGNETIC CONFINEMENT

A further brief discussion of confining surfaces is now in order. We have shown that a series of nested surfaces of confinement may be produced when a plasma is placed in a region where the magnetic field increases outward from the place where the plasma is found. The cusp was such an example.

The use of magnetic mirror coils should inhibit end losses. However, there are always some particles that will escape through the loss cone and eventually the confinement is lost. An obvious scheme to eliminate completely such losses is to use a toroidal or closed system rather than such an open-ended system. A toroidal magnetic field will certainly eliminate end losses, but other difficulties will now develop due to the curvature of the field.

First, the magnetic field is no longer uniform across a section of the torus. This means that the surfaces of constant magnetic pressure, although nested, are not concentric. This in itself is not bad, but the nonuniform magnetic field results in a $\nabla_r B$ drift of the electrons and ions.

Examining the cross section of the magnetic field in a toroid shows that, by comparing the field directions with those of Fig. 3.7, the electrons will drift to the top of the toroid and the ions will drift to the bottom, producing an undesirable charge separation.

However, various forms of compensating fields can be added to the basic toroidal field configuration to produce compensating drifts and improve confinement characteristics. One possibility is to use a "sheared" magnetic field, in which the field lines twist as they go around the torus. Figure 6.5 shows such a toroidal system. The direction of charge drift is shown in cross section.

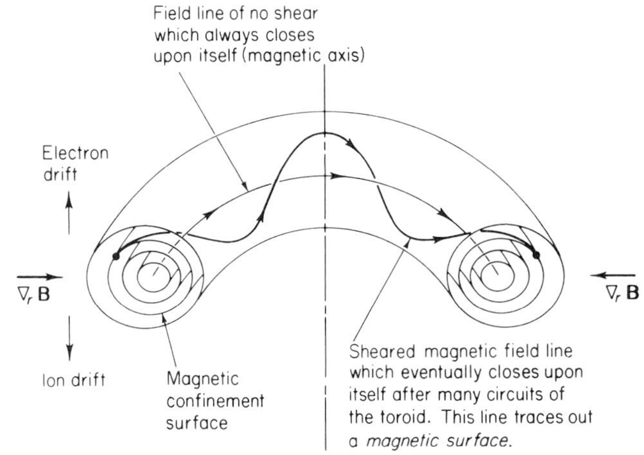

Fig. 6.5. Toroidal confinement geometry.

We now proceed to a discussion of a plasma effect that exhibits slow motion. We shall find that, even if we consider the plasma to be a fluid, we may observe movement of the plasma *through* the magnetic field and/or motion of the plasma *with* the magnetic field.

It should now become clear that further examination of Eq. (6.30) may also show a basic property of conducting plasmas: they may be accelerated. That is, if an external current is passed through a plasma in a magnetic field such that it is perpendicular to **B**, a body force is set up that can *accelerate* the plasma (or conducting fluid).

The reverse situation is also true. That is, if some external force *pushes* the plasma through a magnetic field, an electromotive force may be set up that can produce electrical energy. This electromotive force can be used to measure the velocity of the plasma or to produce a significant amount of electric power for external consumption.

These two schemes have been investigated for a long period of time in the hope of utilizing plasma accelerators (rocket engines), magnetohydrodynamic power generators, and electromagnetic flowmeters. Figure 6.6 shows such an accelerator–generator device.

If electric current is forced through the plasma in the direction shown, the $\mathbf{J}_E \times \mathbf{B}$ body force can produce acceleration in the direction indicated in the figure. The pressure gradient term $\nabla_r p$ must then be in the $\mathbf{J}_E \times \mathbf{B}$ direction. If the device is to be operated as a generator, then forcing the plasma (or conducting fluid) through the channel in the direction shown will produce current through a load. Here the pressure gradient must be set up by an external pressure source.

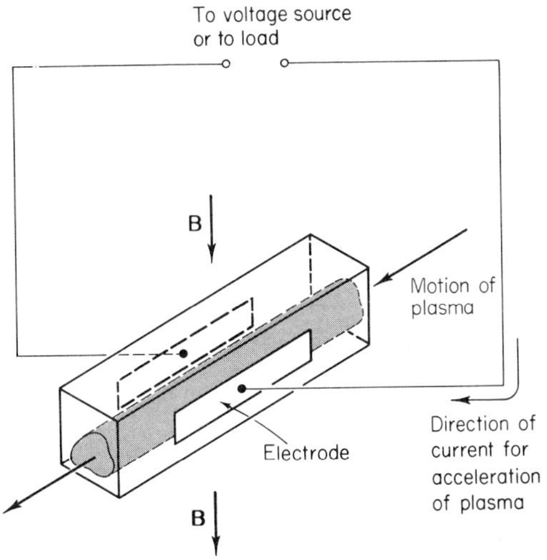

Fig. 6.6. Hydromagnetic channel (accelerator–generator device).

To further expand upon these effects and others, we also now make use of another fundamental equation, Ohm's Law:

$$\mathbf{J}_E = \sigma \mathbf{E}_L \tag{6.46}$$

where σ is the plasma conductivity. This may be the conductivity derived in Chapter 3, the values obtained from other theoretical calculations, or from experimental measurements.

The electric field $\mathbf{E}_L$ used in Eq. (6.46) must be that observed by the charged particles themselves as they move. If they are moving through a magnetic field with a velocity $\mathbf{v}$, the *net* electric field that the plasma "feels" at each point is

$$\mathbf{E}_L = \mathbf{E} + \mathbf{v} \times \mathbf{B} \tag{6.47}$$

where **E** is the field measured by a *fixed* observer at the point in question and **v** is the velocity of the particle at that point, relative to the observer. So, for a moving fluid in a magnetic field, Ohm's Law becomes

$$\mathbf{J}_E = \sigma(\mathbf{E} + \mathbf{v} \times \mathbf{B}) \tag{6.48}$$

properly written in the laboratory frame of reference. We will call a set of coordinates that are fixed in the laboratory *Eulerian* coordinates. If we use coordinates that move with the fluid (or the individual particles), they will be called *Lagrangian* coordinates. Equation (6.46) is the correct formulation of Ohm's Law in Lagrangian coordinates, and (6.48) is the correct formulation in Eulerian coordinates.

The following development uses Eq. (6.48) as the desired formulation. Note first that if $\sigma \to \infty$, then Eq. (6.48) becomes

$$\mathbf{E} + \mathbf{v} \times \mathbf{B} = 0$$

The solution for **v** is then the $(\mathbf{E} \times \mathbf{B})/B^2$ drift velocity previously developed.

Taking the curl of Eq. (6.48) results in

$$\nabla_r \times \mathbf{J}_E = \sigma(\nabla_r \times \mathbf{E}) + \sigma \nabla_r \times (\mathbf{v} \times \mathbf{B}) \tag{6.49}$$

If we use Maxwell's equations to replace $\nabla_r \times \mathbf{E}$ and $\mathbf{J}_E$, we get for Eq. (6.49)

$$\nabla_r \times \frac{(\nabla_r \times \mathbf{B})}{\mu_0} = -\sigma \frac{\partial \mathbf{B}}{\partial t} + \sigma \nabla_r \times (\mathbf{v} \times \mathbf{B}) \tag{6.50}$$

Note that we have neglected the displacement current with respect to conduction current in obtaining Eq. (6.50). This means that we have made a low-frequency approximation to Maxwell's equations $[\partial \mathbf{D}/\partial t \ll \mathbf{J}_E]$. Now

$$\nabla_r \times (\nabla_r \times \mathbf{B}) = \nabla_r(\nabla_r \cdot \mathbf{B}) - \nabla_r^2 \mathbf{B} = -\nabla_r^2 \mathbf{B} \tag{6.51}$$

since $\nabla_r \cdot \mathbf{B} = 0$.

Then we may use Eq. (6.51) in Eq. (6.50) to obtain

$$\frac{\partial \mathbf{B}}{\partial t} = \nabla_r \times (\mathbf{v} \times \mathbf{B}) + \frac{1}{\sigma \mu_0} \nabla_r^2 \mathbf{B} \tag{6.52}$$

E. HYDROMAGNETIC WAVE MOTION AND DIFFUSION

The two terms on the right-hand side of Eq. (6.52) correspond to *flow* and *diffusion* properties of the plasma, respectively. If $1/\sigma\mu_0$ is not too small (σ is not too large), then we may be able to neglect the first term on the right-hand side of Eq. (6.52). This means that σ must be large enough (for the frequency under consideration) to neglect the displacement current, but small enough to allow us to neglect the flow term. The result is that (6.52) may be written as

$$\frac{\partial \mathbf{B}}{\partial t} = \frac{1}{\sigma \mu_0} \nabla_r^2 \mathbf{B} \tag{6.53}$$

This equation has the form of the time-dependent *diffusion* equation and implies that magnetic field lines can diffuse through the plasma. However, if the conductivity of the plasma is "too high," then the diffusion term must be neglected and instead we have the following form of Eq. (6.52).

$$\frac{\partial \mathbf{B}}{\partial t} = \nabla_r \times (\mathbf{v} \times \mathbf{B}) \tag{6.54}$$

This equation is a "flow-type" equation and implies that the magnetic field does not move unless $\mathbf{v}$ is nonzero. This shows that the magnetic field lines are "frozen" into the plasma. We can determine that not only do these field lines remain fixed in the plasma, but they also behave as though a constant mass of plasma were associated with each of them. The first consequence of this is the possibility of acceleration or power generation described previously. The second result, as we shall see, is a wave type of equation, in which the magnetic lines of force behave as stretched strings and vibrate with their associated plasma mass.

Let us assume that the plasma does not move too far away from a steady state condition, so that we can write the magnetic field as a perturbation expression as shown in Eq. (6.55).

$$\mathbf{B} = \mathbf{B}_0 + \mathbf{b}(\mathbf{r}, t) \tag{6.55}$$

$\mathbf{B}_0$ is a constant and is much larger than $\mathbf{b}$. All the time and spatial variations are assumed to be contained in $\mathbf{b}$. The momentum equation (6.29) now becomes, if gravity is neglected

$$\rho_m \dot{\mathbf{v}} = \mathbf{J}_E \times \mathbf{B} - \nabla_r p \tag{6.56}$$

This is a departure from steady state, since if there is no plasma motion, the momentum equation is simply Eq. (6.30), which is Eq. (6.56) with the left-hand side set equal to zero.

Maxwell's equation

$$\nabla_r \times \mathbf{B} = \mu_0 \mathbf{J}_E \tag{6.57}$$

which is again written in the low-frequency approximation, may be used to eliminate $\mathbf{J}_E$ from Eq. (6.56) so that we can write

$$\rho_m \dot{\mathbf{v}} = \frac{(\nabla_r \times \mathbf{B}) \times \mathbf{B}}{\mu_0} - \nabla_r p \tag{6.58}$$

Note that the first term on the right-hand side of Eq. (6.58) is the same as the entire right side of Eq. (6.38). We may therefore use the same expansion for this term as previously. This allows us to write, for (6.58)

$$\rho_m \dot{\mathbf{v}} = -\frac{\nabla_r B^2}{2\mu_0} + \frac{(\mathbf{B} \cdot \nabla_r)\mathbf{B}}{\mu_0} - \nabla_r p \tag{6.59}$$

or finally,

$$\rho_m \dot{\mathbf{v}} = -\nabla_r \left[\frac{B^2}{2\mu_0} + p\right] + \frac{(\mathbf{B} \cdot \nabla_r)\mathbf{B}}{\mu_0} \tag{6.60}$$

The quantity in the square brackets is a constant by the same reasoning used in the previous development leading to Eq. (6.42). Its gradient is zero in equilibrium. Therefore, Eq. (6.60) becomes

$$\rho_m \dot{\mathbf{v}} = \frac{(\mathbf{B} \cdot \nabla_r)\mathbf{B}}{\mu_0} = \frac{(\mathbf{B}_0 \cdot \nabla_r)}{\mu_0}\mathbf{b} \tag{6.61}$$

if we apply the perturbation definition of $\mathbf{B}$ from Eq. (6.55). We have neglected $\mathbf{b}$ in terms where we are not taking the gradient and $\mathbf{B}_0$ does not appear in those terms in which we *do* take the gradient. The left-hand side and the second term on the right-hand side of (6.60) are perturbations. We are thus justified in removing the first term on the right from (6.60) (which has the value zero in equilibrium) in going to the perturbed equation (6.61).

Equation (6.54) may now be used with the perturbation expansion for $\mathbf{B}$ and a vector identity to expand the cross product. It becomes

$$\frac{\partial \mathbf{B}}{\partial t} = \mathbf{v}(\nabla_r \cdot \mathbf{B}) + (\mathbf{B} \cdot \nabla_r)\mathbf{v} - \mathbf{B}(\nabla_r \cdot \mathbf{v}) - (\mathbf{v} \cdot \nabla_r)\mathbf{B}$$

or, in other terms

$$\dot{\mathbf{b}} = \frac{\partial \mathbf{b}}{\partial t} = (\mathbf{B}_0 \cdot \nabla_r)\mathbf{v} \tag{6.62}$$

since $\nabla_r \cdot \mathbf{v}$ and $\nabla_r \cdot \mathbf{B}$ are both zero. Also, since

$$(\mathbf{v} \cdot \nabla_r)\mathbf{B} = (\mathbf{v} \cdot \nabla_r)\mathbf{b}$$

and is therefore the product of two perturbed quantities ($\mathbf{b}$ and $\mathbf{v}$), we may neglect this term with respect to $(\mathbf{B}_0 \cdot \nabla_r)\mathbf{v}$, which only involves one perturbation term, because it is of higher order.

If we differentiate (6.62) with respect to time, then (6.61) may be substituted for $\dot{\mathbf{v}}$, giving

$$\ddot{\mathbf{b}} = \frac{(\mathbf{B}_0 \cdot \nabla_r)(\mathbf{B}_0 \cdot \nabla_r)}{\rho_m \mu_0}\mathbf{b} \tag{6.63}$$

This is a wave-type equation. If, for example, we will consider plasma waves that vary only in the z direction, then $\partial/\partial y = \partial/\partial x = 0$. Equation (6.63) then becomes

$$\frac{\partial^2 \mathbf{b}}{\partial t^2} = \frac{B_0{}^2}{\rho_m \mu_0} \frac{\partial^2 \mathbf{b}}{\partial z^2} \tag{6.64}$$

This is the wave equation defining *Alfvén waves*, which are effectively vibrations of the magnetic field lines that are "frozen" in the plasma. To examine their nature, we first note that the wave equation for a vibrating string is

$$\frac{\partial^2 y}{\partial t^2} = \frac{T}{\rho_m} \frac{\partial^2 y}{\partial z^2} \tag{6.65}$$

where T is the tension, ρ_m is the mass density, and y is the lateral displacement of the string that lies on the z axis. By comparing Eqs. (6.65) and (6.64) we see that $B_0{}^2/\mu_0$ is equivalent to the tension of the string and ρ_m is the equivalent mass associated with it.

In order to obtain the same type of equation for Alfvén waves, we have had to make a low-frequency and high-conductivity approximation. This implies that these waves are usually the lowest-frequency wave phenomena appearing in a plasma. The *phase velocity* of the Alfvén wave is

$$v_{\text{phase}} = \frac{B}{(\rho_m \mu_0)^{1/2}} \tag{6.66}$$

in units of meters per second. Alfvén waves may appear in both laboratory and astrophysical plasmas, where, in the latter, the frequencies may be extremely low, of the order of 1 cycle per hour or even less. The Alfvén wave frequencies are lower than the plasma and cyclotron frequencies of both the ions and electrons.

F. PLASMA PARAMETERS

It now begins to be clear that we must consider some quantities that offer a measure of various plasma properties. Generally, the particle density, magnetic field, and temperature (or the distribution function) are all that are needed to fully specify a plasma. However, it is often more convenient to consider characteristic frequencies (or times) and lengths, which may be derived from the density, magnetic field, and temperature of a plasma. These formulations help to determine, for example, what frequencies and wavelengths of waves and/or instabilities may appear in a particular plasma. We will now introduce two of these new terms that are of much importance in plasma work.

Previous expressions defining the statistical velocity distribution for electrons or ions show the existence of functions of the form

$$f(\mathbf{r}, \mathbf{v}, t) \sim \exp\left(-\frac{m_e v^2}{2kT_e} - \frac{q_e\Phi(\mathbf{r}, t)}{kT_e}\right) \tag{6.67}$$

if the electrons experience a potential field defined by a scalar potential function Φ, which may vary in space and time. Should the potential function be independent of velocity, then Eq. (6.67) for the electrons may be integrated over velocity space to obtain

$$n_e(\mathbf{r}, t) = n_{0_e}(t) \exp\left(-\frac{q_e\Phi(\mathbf{r}, t)}{kT_e}\right) \tag{6.68}$$

For small values of $\Phi(\mathbf{r}, t)$ we may expand the potential in Eq. (6.68) in a power series and get, by neglecting higher-order terms,

$$n_e(\mathbf{r}, t) \simeq n_{0_e}(t)\left[1 - \frac{q_e\Phi(\mathbf{r}, t)}{kT_e}\right] \tag{6.69}$$

Equation (6.69) states that as the potential increases, the density of electrons also increases (since q_e is negative). The net charge density in a plasma must be the difference between the "equilibrium" and "perturbed" charge densities. (Perturbation in this case means the existence of the potential function $\Phi(\mathbf{r}, t)$, which changes the density. If Φ were zero, a pure Maxwellian velocity distribution would result.) The equilibrium electron density n_{0e} is balanced by an equivalent equilibrium ion density n_{0_i} to preserve charge neutrality. Therefore, Poisson's equation becomes

$$\nabla_r{}^2\Phi(\mathbf{r}, t) = \frac{q_e(n_e(\mathbf{r}, t) - n_{0e})}{\varepsilon_0}$$

or, using the expansion from Eq. (6.68)

$$\nabla_r{}^2\Phi(\mathbf{r}, t) = \frac{q_e}{\varepsilon_0}\left[n_{0e}\frac{q_e\Phi(\mathbf{r}, t)}{kT_e}\right] \tag{6.70}$$

where n_{0e} is the electron equilibrium density and n_e is the density including the departure from equilibrium (now assumed to be only due to electrons). Equation (6.70) may be rewritten in the form

$$\nabla_r{}^2\Phi - \frac{\Phi}{\lambda_{D_e}^2} = 0 \tag{6.71}$$

where

$$\lambda_{D_e} = \left(\frac{\varepsilon_0 kT_e}{n_{0e}q_e{}^2}\right)^{1/2} \tag{6.72}$$

This quantity is called the *Debye length* for electrons. The solution to Eq. (6.71) may be obtained by a separation of variables.

If we only examine spherically symmetric solutions for Φ, then the radial coordinate is the only variable that needs to be considered. Equation (6.71), in spherical coordinaties but only considering the radial dependence, becomes

$$\frac{1}{r}\frac{d^2}{dr^2}\Phi(r) - \frac{\Phi(r)}{\lambda_{D_e}^2} = 0 \tag{6.73}$$

This ordinary differential equation can be solved for Φ. The solution is of the form

$$\Phi(r) = \frac{C}{r}\exp(-r/\lambda_{D_e}) \tag{6.74}$$

where C is a constant to be determined from the boundary conditions. The important thing to note here is that the potential decays faster than the potential from a charge that does not have any other charges surrounding it. The function in (6.74) effectively results in a Debye "shielded" potential from every charge in the plasma. Figure 6.7 shows a plot of these two potentials. The longer the Debye length, the less the shielding effect over a given

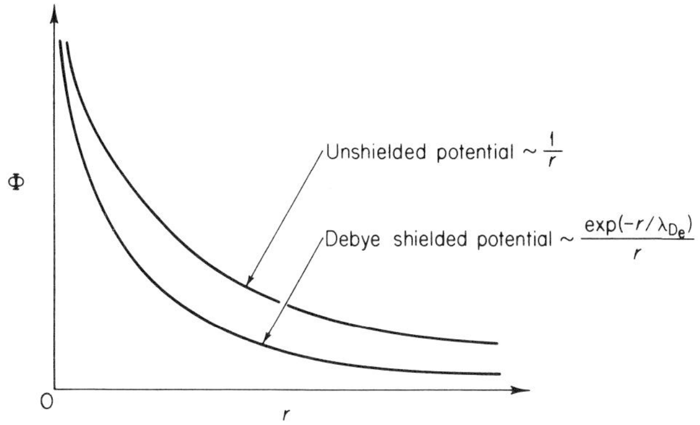

Fig. 6.7. Comparison of Debye shielded and unshielded electrostatic potentials from a point charge.

distance. If the plasma's dimensions are less than λ_{D_e} it is likely that charge of one polarity or another may collect at the walls of the vessel holding the plasma and form *sheaths* extending throughout the plasma volume. These sheaths tend to destroy the charge neutrality of a plasma and may change the behavior of the plasma greatly.

For a density of 10^{12} particles/cm^3 and a temperature of 1000 eV, λ_{D_e} is approximately 1 mm. Often, if a plasma is larger than a Debye length, the outer boundaries will be covered with a sheath of thickness $\approx \lambda_{D_e}$. It should be again noted that if the potential energy of a charge in the electric field in a plasma is of the order of the particle's kinetic energy, the electric field may alter the energy distribution. This electric field may be produced by neighboring charges *or* by externally applied sources.

A Debye length exists for ions as well as for electrons. If the plasma is isothermal and electrically neutral, then it is such that $\lambda_{D_i} = \lambda_{D_e}$.

The concept of the Debye length may also be useful in considering the effects of various probes or other diagnostic elements that are immersed in a plasma. Potentials may develop or be applied to these devices. The conditions under which they perturb the plasma can be obtained by considering that sheaths of the order of a Debye length will develop if the probe is floating, but may not develop if its bias potential, if any, is of the order of the local potentials present in the plasma. Bias potentials greater than the local potentials often result in thickening of the sheaths and a consequent increase in the perturbing effects.

We may compute the number of particles inside a sphere of radius λ_{D_e} to be

$$n_{\lambda D_e} = \frac{4}{3}\pi\lambda_{D_e}^3 n_{0_e} = \frac{4}{3}\pi\left(\frac{\varepsilon_0 k T_e}{n_{0_e} q_e^2}\right)^{3/2} n_{0_e} \tag{6.75}$$

It is required that this number not be too small, or the shielding effects predicted by the theory will not take place. A simplified condition for good shielding is that the average interparticle spacing be considerably less than the Debye length. This condition may be expressed as

$$\left(\frac{1}{n_{0_e}}\right)^{1/3} \ll \lambda_{D_e} \tag{6.76}$$

The conditions for the existence of a "true" plasma can be summarized as the following inequality.

$$\left(\frac{1}{n_{0_e}}\right)^{1/3} \ll \lambda_{D_e} \ll L \tag{6.77}$$

where L is some characteristic dimension of the plasma. This means that if inequality (6.77) is satisfied, the plasma will behave much like the models we have been considering. In addition, the previous work concerning collision cross sections resulted in a situation where the total cross section for Coulomb collisions became infinite when large impact parameters were included. We now have some justification in truncating the integration over all possible impact parameters to *exclude all impact parameters greater than* λ_{D_e}. This results in a finite Coulomb collision cross section, and we may now feel relatively confident in doing this truncation, since we have predicted a shielding effect for distances greater than λ_{D_e}. In the following development we will determine the value of the cross section when the truncation at the Debye length is made.

Equation (2.36) resulted in an expression for a differential scattering cross section that may be integrated over all possible scattering angles. We now rewrite it as

$$n_0 \frac{d\sigma}{d\Omega} = \frac{n_0 K^2}{4g^4 \sin^4(\chi/2)} \tag{2.36}$$

where

$$K = \frac{q_e q_e}{4\pi\varepsilon_0 m_r}$$

assuming electron–electron collisions. Note that the solid angle Ω is $2\pi(1 - \cos\chi)$ and that $d\Omega = 2\pi \sin\chi\, d\chi$, since we have assumed that the interaction is independent of the azimuthal scattering angle ε. If Eq. (2.36) is integrated over all possible angles χ, a divergence of the integral will obviously result at $\chi = 0$. However, since

$$\tan\left(\frac{\chi}{2}\right) = \frac{K}{g^2 b} \tag{6.78}$$

we can let

$$\tan\left(\frac{\chi_{\min}}{2}\right) = \frac{K}{g^2 \lambda_{D_e}} \tag{6.79}$$

and remove the singularity. $\chi_{\min}$ is the smallest scattering angle and hence is produced at the largest impact parameter, which in this case has been set to λ_{D_e}. The smallest possible impact parameter will exist for 180° scattering. The momentum transfer cross section is then obtained by integrating Eq. (2.36) and making the foregoing substitution. This cross section is

$$\sigma_m = \frac{1}{4\pi} \int_{\sin(\chi_{\min}/2)}^{1} \frac{K^2 d(\sin(\chi/2))}{4g^2 \sin^4(\chi/2)} \tag{6.80}$$

The result of integrating Eq. (6.80) with the lower limit defined by Eq. (6.79) is

$$\sigma_m = \frac{4\pi K^2}{g^4} \ln\left[\frac{g^2}{K} \lambda_{D_e}\right] \tag{6.81}$$

which is now the expression for momentum transfer collision cross section.

Note that the cross section is *not* infinite. We may let

$$\Lambda = \frac{g^2}{K} \lambda_{D_e}$$

so that

$$\sigma_m = \frac{4\pi K^2}{g^4} \ln \Lambda \tag{6.82}$$

In practice, $\ln \Lambda \approx 20$.

We now extend the foregoing development to consider some characteristic *times* for a plasma. The *angular relaxation time*, or time required for particles to diffuse in velocity space due to small angle collisions, must be of the order of the Coulomb collision time. We define this time as

$$\tau_\theta = \frac{1}{n_e \sigma_m g} = \frac{g^3}{4\pi n_e K^2 \ln \Lambda} \tag{6.83}$$

The value of n_e is the density of scattering centers, and if other types of scattering occur, separate angular relaxation times can also be defined for these processes. For example, in a plasma, electrons are colliding with other electrons, ions, and neutrals, while the ions are colliding with electrons, each other, and neutrals. How does one determine the dominant process? Each type of collision has its cross section, and its particular angular relaxation time may be calculated. Typically, the ordering of the angular relaxation times is

$$\tau_{ee} < \tau_{ei} < \tau_{ii}$$

where the subscripts *ee*, *ei*, and *ii* refer to electron–electron, electron–ion (or ion–electron), and ion–ion scattering processes. These times have approximately the ratio

$$1 : \left(\frac{m_i}{m_e}\right)^{1/2} : \frac{m_i}{m_e} \tag{6.84}$$

Normally, then, the electron–electron scattering time is the fastest, by a considerable margin. However, ion–neutral and electron–neutral collisions may be dominant if the precentage of ionization is low.

We now reintroduce another of the characteristic terms dealing with plasmas. It will be introduced in two ways, so as to show that (1) it is a "natural" property of a plasma and (2) it may be used to determine a measurement of plasma density. This quantity is the plasma frequency, first mentioned in connection with the plasma permittivity in Chapter 3.

Consider a "slab" of plasma composed of equal numbers of electrons and singly charged ions as shown in Fig. 6.8. If the ions are considered to be

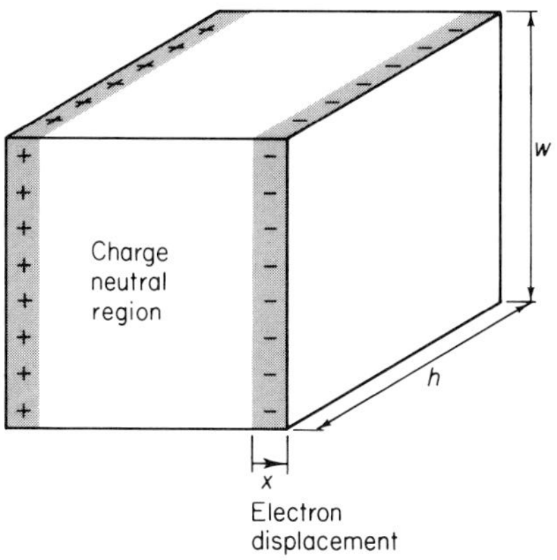

Fig. 6.8. Plasma slab, with electrons displaced from the ions.

fixed, due to their large mass, we can investigate the behavior of the system if all the electrons are initially displaced a distance x to the right. The immediate result of this displacement is the creation of an electric field, directed so as to restore charge neutrality. Note that the main body of the plasma is still electrically neutral; but two sheaths of net negative and positive charge are produced by the electron displacement as shown on the two sides of the "slab" of plasma.

If we now consider the right-hand side of Fig. 6.8 (the negative charge region) we may use Gauss's law to determine the electric field. It is

$$\oint \mathbf{E} \cdot d\mathbf{s} = \int_{\text{plasma}} \frac{\rho_E \, dV}{\varepsilon_0} \tag{6.85}$$

If we assume that $\mathbf{E}$ has only a component in the negative x direction and only appears along the face of the slab where it meets the neutral region, then Eq. (6.85) may be written as

$$Ehw = \frac{q_e n_e hwx}{\varepsilon_0}$$

neglecting fringing fields. The slab is of height w and depth h. In vector notation, the electric field is

$$\mathbf{E} = \frac{q_e n_e}{\varepsilon_0} x \hat{a}_x \tag{6.86}$$

The net force on any individual electron due to E must be in the x direction, and obeys the following scalar equation

$$F_e = q_e E = -\frac{q_e^2 n_e}{\varepsilon_0} x = m_e \frac{d^2 x}{dt^2} \tag{6.87}$$

since the force must act in the $-x$ direction initially, for the displacement shown in Fig. 6.8. Equation (6.87), therefore, can be rewritten as the following second-order linear ordinary differential equation.

$$\frac{d^2x}{dt^2} + \left(\frac{q_e^2 n_e}{m_e \varepsilon_0}\right) x = 0 \tag{6.88}$$

This homogeneous equation may be solved by assuming a solution for x as

$$x = Ae^{pt}$$

Substitution into (6.88) results in

$$p^2 + \left(\frac{q_e^2 n_e}{m_e \varepsilon_0}\right) = 0$$

and therefore

$$p = \pm j\left(\frac{q_e^2 n_e}{m_e \varepsilon_0}\right)^{1/2} = \pm j\omega_{p_e} \tag{6.89}$$

The two solutions for x are therefore oscillatory in form. Their natural frequency ω_{p_e} is the *electron plasma frequency*, and is a normal mode of such a system.

The frequency ω_{p_e} is proportional to known fixed quantities except for the value of n_e. Therefore, if a method existed for measuring ω_{p_e}, then n_e could be obtained. If n_e is expressed in particles per cubic meter, then

$$\omega_{p_e} = 56.4(n_e)^{1/2} \qquad \text{radians/second} \tag{6.90}$$

We may also redefine the *ion plasma frequency* ω_{p_i} to be

$$\omega_{p_i} = \left(\frac{q_i^2 n_i}{m_i \varepsilon_0}\right)^{1/2} \tag{6.91}$$

This frequency is much lower than ω_{p_e} since the ions have a much greater mass.

The second method of obtaining ω_{p_e} will result in some additional insight into how it may be measured. If we assume a time-varying electric field applied to a single-particle model, with no collisions and no magnetic field, we may use the results of Chapter 3 to obtain the plasma conductivity as

$$\sigma = \frac{jn_e q_e^2}{m_e \omega} \qquad \text{from Eq. (3.45)}$$

assuming $e^{-j\omega t}$ time dependence. We may then write Eq. (3.52)

$$\nabla_r \times \mathbf{H} = -j\omega\varepsilon_0\left[1 - \frac{\sigma}{j\omega\varepsilon_0}\right]\mathbf{E} \tag{3.52}$$

without tensor notation since σ is a scalar. Equation (3.52) becomes, using the conductivity defined above,

$$\nabla_r \times \mathbf{H} = -j\omega\varepsilon_0\left[1 - \frac{n_e q_e^{\,2}}{m_e\omega^2\varepsilon_0}\right]\mathbf{E} = -j\omega\varepsilon_0\left[1 - \frac{\omega_{p_e}^2}{\omega^2}\right]\mathbf{E} \qquad (6.92)$$

The quantity in the brackets is dimensionless and may be considered to be the dielectric constant of the plasma namely,

$$\frac{\varepsilon}{\varepsilon_0} = 1 - \frac{\omega_{p_e}^2}{\omega^2} \qquad (6.93)$$

G. WAVE PROPAGATION IN PLASMAS—INTRODUCTORY REMARKS

If we should attempt to propagate a plane monochromatic wave through the plasma medium, the solutions to the wave equation will be of the form

$$e^{j(kz-\omega t)} \qquad (6.94)$$

where k is the wave number

$$\frac{2\pi}{\lambda} = \frac{2\pi f}{v_p} = \frac{\omega}{v_p} = k \qquad (6.95)$$

v_p is the velocity of propagation in the plasma defined as

$$v_p = \frac{1}{(\mu_0\,\varepsilon)^{1/2}} \qquad (6.96)$$

As long as $\varepsilon = \varepsilon_0(1 - \omega_{p_e}^2/\omega^2)$ is positive, v_p will be real and plane electromagnetic waves can propagate.

Whenever ε is negative, then v_p and k are imaginary and propagating solutions do not exist. Since $\omega_{p_e}^2$ is proportional to plasma density, the result is that, as plasma density increases, a point will be reached where $\omega_{p_e}^2/\omega^2 > 1$ and then propagation is cut off. Experimentally, this point is often observed to appear with great swiftness and the plasma may very sharply cut off the transmission of radiation through it as its density builds up.

At this point we can learn something about the plasma. Since the frequency of a transmitted signal is presumed to be known, the plasma density is measurable at the instant of cutoff. If enough different frequencies can be propagated through the plasma, a wide range of densities can be measured, often dynamically as the plasma changes its properties with time.

Figure 6.9 shows schematically a typical wave propagation experiment. Here the radiation can traverse the plasma and is detected by the reception arm of the microwave system. When the density is such that $\omega \leq \omega_{p_e}$, the signal at the receiver should drop in amplitude. It will probably not be completely cut off due to scattering from other parts of the experimental apparatus, but in practice this base level can usually be reduced to a very low value. A more advanced technique is to provide a second path, *not* through the plasma, for the microwaves. Comparison of phase shifts then produces microwave interference fringes, which can be used to measure a wide range of plasma densities at a single frequency. Usually, the ion plasma frequency is neglected in these measurements.

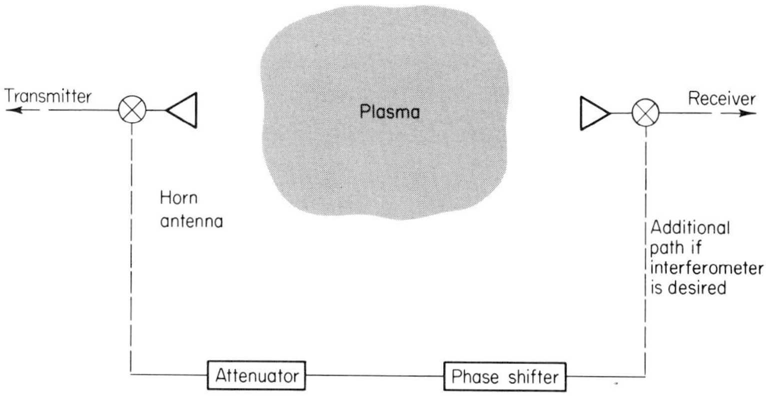

Fig. 6.9. Plasma–microwave system.

There are two difficulties with such a measuring process. First, the wavelength of the radiation ought to be fairly small compared to the dimensions of the plasma, since otherwise the plane wave assumption breaks down. This tends to limit the available frequencies to microwave bands (and higher, up to laser frequencies) for laboratory plasmas. As a result, there is a lower bound on the density that can be measured. Second, the plasma density may vary spatially and such propagation experiments do not measure the density profile very well. They will exhibit cutoff if *some* wavelength-sized portion of the plasma has enough electron density to produce the cutoff condition.

However, even with these limitations, plasma frequency cutoff or interferometric measurements are quite useful and provide a reasonably good and simple diagnostic technique.

The propagation vector (wave number) $\mathbf{k}$ is evidently a function of the properties of the plasma, as well as the frequency. In general

$$k = k\,(\text{plasma properties}, \omega) \tag{6.97}$$

For example, in our previous development

$$k = \omega(\varepsilon_0 \mu_0)^{1/2}\left[1 - \frac{\omega_{p_e}^2}{\omega^2}\right]^{1/2} \tag{6.98}$$

and thus k is a function of the plasma properties, since ω_{p_e} depends upon the density. Should the dielectric constant be a tensor, then by the definition of the propagation vector, k will be a tensor as well.

The wave equation for plane monochromatic waves in a plasma may be obtained in the usual way from Maxwell's equations. It is

$$\nabla_r \times (\nabla_r \times \mathbf{E}) + \mu_0 \varepsilon \frac{\partial^2 \mathbf{E}}{\partial t^2} = 0 \tag{6.99}$$

providing ε does not vary in space or time. If ε does vary, and it usually will, then Eq. (6.99) is only an approximation to reality. If the approximation is not a good one, that is, ε may vary significantly over a wavelength or wave period, then the wave equation must include these variations as well. The correct wave equation may be obtained from Maxwell's equations by noting that ε now has derivatives with respect to position and/or time. This will be discussed subsequently in Chapter 8. Equation (6.99) is a *homogeneous* equation, and has wave solutions of the form

$$\mathbf{E}_0 \exp[j(\mathbf{k} \cdot \mathbf{r} - \omega t)]$$

However, this does not mean that such solutions correspond to physically realizable conditions, and the initial and boundary conditions of the problem must be used to determine the validity of the solutions.

Substitution of the exponential-type solution into Eq. (6.99) produces

$$-[\mathbf{k} \times (\mathbf{k} \times \mathbf{E}_0) + \mu_0 \varepsilon\omega^2 \mathbf{E}_0] \exp[j(\mathbf{k} \cdot \mathbf{r} - \omega t)] = 0 \tag{6.100}$$

This last equation results in a relationship between $\mathbf{k}$ and ω that must be satisfied if solutions are to exist. The relation is expressed as the following equation.

$$-[\mathbf{k} \times (\mathbf{k} \times \quad) + \mu_0 \varepsilon\omega^2]\mathbf{E}_0 = 0 \tag{6.101}$$

The quantity in the brackets is called a *dispersion relation*. If we write out the components of (6.101), assuming $\mathbf{k} = k_x \hat{a}_x + k_y \hat{a}_y + k_z \hat{a}_z$, we obtain

$$\begin{bmatrix} (\omega^2\mu_0\varepsilon - k_y{}^2 - k_z{}^2) & k_x k_y & k_x k_z \\ k_y k_x & (\omega^2\mu_0\varepsilon - k_x{}^2 - k_z{}^2) & k_y k_z \\ k_z k_x & k_z k_y & (\omega^2\mu_0\varepsilon - k_x{}^2 - k_y{}^2) \end{bmatrix} \times \begin{bmatrix} E_x \\ E_y \\ E_z \end{bmatrix} = 0 \tag{6.102}$$

in matrix notation.

The determinant of the coefficients of E is set equal to zero, giving a set of *normal modes* for the system. These normal modes are the only possible *resonant frequencies* for the system. They do not in themselves guarantee that only these frequencies will be excited. In fact, if a forcing function is used, such as an ac source for a wave, then the propagation if it exists, will occur at the frequency of the forcing function after *transients* have died away.

An example that should clarify this point is that of a series RLC circuit. The normal modes for such a circuit are two in number:

$$p_{1,2} = -\frac{R}{2L} \pm \left[\left(\frac{R}{2L}\right)^2 - \frac{1}{LC}\right]^{1/2} \tag{6.103}$$

Note that it is possible for the system to oscillate *by itself* if the condition

$$\frac{1}{LC} > \left(\frac{R}{2L}\right)^2$$

is satisfied. However, if an ac source is applied to the RLC circuit, a signal will appear at the frequency of the source in the output even if the condition for oscillation is *not* satisfied.

Equation (6.102) is the dispersion relation if ε is a not a tensor. If ε is a tensor, and can be written as

$$\boldsymbol{\varepsilon} = \begin{vmatrix} \varepsilon_{xx} & \varepsilon_{xy} & \varepsilon_{xz} \\ \varepsilon_{yx} & \varepsilon_{yy} & \varepsilon_{yz} \\ \varepsilon_{zx} & \varepsilon_{zy} & \varepsilon_{zz} \end{vmatrix}$$

then the wave equation, and Eq. (6.101), will have more terms in its components. The result is that (6.102) becomes

$$\begin{bmatrix} (\omega^2\mu_0\varepsilon_{xx} - k_y^2 - k_z^2) & (k_xk_y + \omega^2\mu_0\varepsilon_{xy}) & (k_xk_z + \omega^2\mu_0\varepsilon_{xz}) \\ (k_yk_x + \omega^2\mu_0\varepsilon_{yx}) & (\omega^2\mu_0\varepsilon_{yy} - k_x^2 - k_z^2) & (k_yk_z + \omega^2\mu_0\varepsilon_{yz}) \\ (k_zk_x + \omega^2\mu_0\varepsilon_{zx}) & (k_zk_y + \omega^2\mu_0\varepsilon_{zy}) & (\omega^2\mu_0\varepsilon_{zz} - k_x^2 - k_y^2) \end{bmatrix} \times \begin{bmatrix} E_x \\ E_y \\ E_z \end{bmatrix} = 0 \tag{6.104}$$

Note that the *only* place where the tensor notation for ε appears is in the time derivative part of the wave equation.

If we consider a plasma in a magnetic field that is directed along one axis (say the z direction), then the expression for $\boldsymbol{\varepsilon}$ becomes

$$\boldsymbol{\varepsilon} = \begin{vmatrix} \varepsilon_{xx} & \varepsilon_{xy} & 0 \\ \varepsilon_{yx} & \varepsilon_{yy} & 0 \\ 0 & 0 & \varepsilon_{zz} \end{vmatrix} \tag{6.105}$$

which, of course, greatly simplifies Eq. (6.104). Similarly, if $\varepsilon_{xx} = \varepsilon_{yy}$ and $\varepsilon_{xy} = -\varepsilon_{yx}$, further simplifications result. Recall that for the cold plasma permittivity, these latter conditions are satisfied if the magnetic field is uniform and constant.

This discussion of dispersion relations and waves in plasmas is continued in Chapter 7.

We note that the development above has only used the cold plasma model to arrive at a dispersion relation. The following material shows briefly how a statistical "hot" plasma model may be used to find a dispersion relation.

We return to the collisionless Boltzmann equation to obtain an expression for oscillations involving the nature of the distribution function. Again, we will use a perturbation analysis of the distribution function. We will assume it to be

$$f(\mathbf{r}, \mathbf{v}, t) = f_0(\mathbf{v}) + f_1(\mathbf{r}, \mathbf{v}, t) \tag{6.106}$$

which means that the *time* and *spatial* variations are contained in f_1. This is a slightly different expansion than that used previously. Before, it was assumed that f_0 also varied in space and time. If we consider only electrostatic forces, the collisionless Boltzmann equation for electrons is

$$\frac{\partial f}{\partial t} + \mathbf{v} \cdot \nabla_r f + \frac{q_e}{m_e} \mathbf{E} \cdot \nabla_v f = 0 \tag{6.107}$$

Equation (6.107) becomes, after using the perturbation (6.106)

$$\frac{\partial f_1}{\partial t} + \mathbf{v} \cdot \nabla_r f_1 + \frac{q_e}{m_e} \mathbf{E} \cdot \nabla_v f_0 = 0 \tag{6.108}$$

Note that f_0 is still a function only of velocity. Therefore f_1 does not appear in the third term of (6.108) because of this, and $\partial f_0/\partial \mathbf{v}$ dominates. We can obtain $\mathbf{E}$ from f_1 if we use Poisson's equation, since f_1 is the change in charge from equilibrium. The equilibrium condition f_0 results in zero net charge and zero net electric field. Poisson's equation becomes

$$\nabla_r \cdot \mathbf{E} = \frac{\rho_E}{\varepsilon_0} = \frac{N q_e \int f_1 \, dv_x \, dv_y \, dv_z}{\varepsilon_0} \tag{6.109}$$

if we assume the plasma ions are stationary and the electron motion produces the net electrostatic charge. Equations (6.108) and (6.109) then form a complete set and may be solved simultaneously.

Now should electrostatic waves develop, they must be related both to a disturbance in f_1 as well as to the resulting disturbance in $\mathbf{E}$ because they are produced by space charges. Therefore, a "wave" of f_1 may result.

If we assume a wave-type soution for both E and f_1 of the form

$$\exp[j(\mathbf{k} \cdot \mathbf{r} - \omega t)],$$

then Eq. (6.108) becomes

$$\left[-j\omega f_{10} + j\mathbf{v}\cdot\mathbf{k}f_{10} + \frac{q_e}{m_e}\mathbf{E}_0\cdot\nabla_v f_0\right]\exp[j(\mathbf{k}\cdot\mathbf{r}-\omega t)] = 0 \tag{6.110}$$

where $f_1 = f_{10}\exp[j(\mathbf{k}\cdot\mathbf{r}-\omega t)]$ and $\mathbf{E} = \mathbf{E}_0\exp[j(\mathbf{k}\cdot\mathbf{r}-\omega t)]$. We can then solve (6.110) for f_{10} in terms of $\mathbf{E}_0$, getting

$$f_{10} = \frac{j}{\mathbf{v}\cdot\mathbf{k}-\omega}\frac{q_e}{m_e}\mathbf{E}_0\cdot\nabla_v f_0 \tag{6.111}$$

Poisson's equation (6.109) becomes, under the same conditions,

$$j\mathbf{k}\cdot\mathbf{E}_0 = \frac{Nq_e}{\varepsilon_0}\int f_{10}\,dv_x\,dv_y\,dv_z \tag{6.112}$$

We can now obtain the relationship between $\mathbf{k}$ and ω from (6.111) and (6.112) which are two equations in two unknowns. Integrating (6.111) over velocity space results in

$$N\int f_{10}\,dv_x\,dv_y\,dv_z = \frac{jNq_e}{m_e}\mathbf{E}_0\cdot\int\frac{1}{\mathbf{k}\cdot\mathbf{v}-\omega}\nabla_v f_0\,dv_x\,dv_y\,dv_z \tag{6.113}$$

The left-hand side of (6.113) can now be found in terms of $\mathbf{E}_0$ by using (6.112). We get

$$N\int f_{10}\,dv_x\,dv_y\,dv_z = \frac{\varepsilon_0}{q_e}j\mathbf{k}\cdot\mathbf{E}_0 = \frac{jNq_e}{m_e}\mathbf{E}_0\cdot\int\frac{1}{\mathbf{k}\cdot\mathbf{v}-\omega}\nabla_v f_0\,dv_x\,dv_y\,dv_z \tag{6.114}$$

We will assume electrostatic longitudinal oscillations. Then $\mathbf{E}$ must point in the direction of propagation,† so that

$$\mathbf{E} = |\mathbf{E}|\frac{\mathbf{k}}{|\mathbf{k}|} \tag{6.115}$$

This permits us to make (6.114) become

$$\frac{\varepsilon_0}{q_e}j|\mathbf{E}_0|\frac{k^2}{|\mathbf{k}|} = \frac{jNq_e}{m_e}|\mathbf{E}_0|\frac{\mathbf{k}}{|\mathbf{k}|}\cdot\int\frac{\nabla_v f_0}{(\mathbf{k}\cdot\mathbf{v}-\omega)}\,dv_x\,dv_y\,dv_z \tag{6.116}$$

The j's and E_0's cancel on both sides. Multiplying both sides of (6.116) by $\varepsilon_0|\mathbf{k}|/q_e k^2$ results in the desired equation, that is,

$$1 = \frac{N\varepsilon_0 q_e^2}{m_e}\frac{\mathbf{k}}{k^2}\cdot\int\frac{\nabla_v f_0}{(\mathbf{k}\cdot\mathbf{v}-\omega)}\,dv_x\,dv_y\,dv_z \tag{6.117}$$

† The validity of this electrostatic approximation will be discussed in Chapter 8.

Note that (6.117) is a dispersion relation, because it shows the relation between $\mathbf{k}$ and ω. This tells us what values of $\mathbf{k}$ are possible for given values of ω and vice versa. Note that in this case the *form* of f_0 is needed, and hence the shape of the velocity distribution function determines whether certain values of $\mathbf{k}$ or ω exist. For a given value of ω, there will be some value of $\mathbf{v}$ such that the denominator of (6.117) "blows up." We will see subsequently that this condition results in a novel behavior of the dispersion relation. We have not used collisions in arriving at Eq. (6.117). As a result, we should not expect to see any absorption of energy of a wave as it propagates through a plasma using this model. We shall see that the presence of this singularity, however, has as its somewhat astonishing result a damping or growth (negative damping) of a wave as it propagates through a plasma.

This chapter has now concluded its introduction to various plasma properties. It has only scratched the surface and its only object has been to show some of the things that may be done with the fundamentals covered in the previous chapters.

Subsequently, a greater depth of many of these topics will be considered with most of the subjects covered in this chapter, along with several other important topics.

Suggested Reading

Alfvén, H., and Fälthammer, G. G., "Cosmical Electrodynamics," 2nd ed., Oxford Univ. Press, London and New York, 1963. This book is a basic work on magnetohydrodynamic waves and astrophysical plasmas.

Allis, W. P., Buchbaum, S. J., and Bers, A., "Waves in Anisotropic Plasmas," M.I.T. Press, Cambridge, Massachusetts, 1963.

Bennett, W. H., Magnetic Self-Focusing Streams, *Phys. Rev.* **45**, 890 (1934). This article describes the original linear pinch development.

Cowling, T. G., "Magnetohydrodynamics," Wiley (Interscience), New York, 1959.

Heald, M. A., and Wharton, C. B., "Plasma Diagnostics with Microwaves," Wiley, New York, 1965. This book offers good applications of much of the material presented here.

Huddlestone, R. H., and Leonard, S. L., eds., "Plasma Diagnostic Techniques," Academic Press, New York, 1965. This work discusses many different types of plasma diagnostics.

Rose, D. J., and Clark, M., "Plasmas and Controlled Fusion," M.I.T. Press, Cambridge, Massachusetts, 1961.

Schmidt, G., "Physics of High Temperature Plasmas—An Introduction," Academic Press, New York, 1966.

Stix, T. H., "The Theory of Plasma Waves," McGraw-Hill, New York, 1963.

Tanenbaum, B. S., "Plasma Physics," McGraw-Hill, New York, 1967.

Thompson, W. B., "An Introduction to Plasma Physics," 2nd ed., Pergamon Press, New York, 1964.

Problems

1. Prove that no confinement surface using currents and magnetic fields can exist in a completely closed spherical geometry. If this is true, then "ball lightning" must be confined by another process—perhaps only atmospheric pressure. Calculate the value of nkT (energy density) that could be confined by atmospheric pressure alone at the earth's surface (760 Torr). For a 1 eV plasma, what is the density of particles in the "ball"?

2. Calculate the energy density as a function of time in a plasma as observed by a diamagnetic loop, given the following data: dc magnetic field in the absence of plasma, 10^4 G; number of turns on the loop, 75; diameter of the loop, 15 cm; diameter of the plasma, 10 cm. The unintegrated signal coming from the loop is displayed below. If the temperature is 1 keV, what is the density as a function of time?

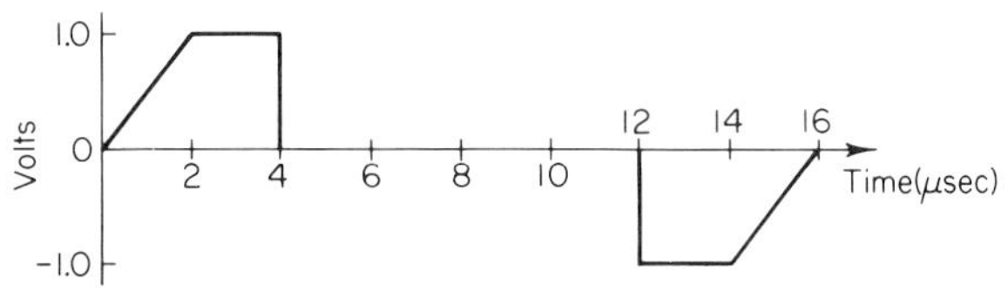

3. The work of this chapter showed that many quantities can be the variables in the plasma equations. For example, if we let $p = nkT$, n and/or T may vary. If only one quantity changes, the other term *may* be assumed constant. Now suppose that in a slab of plasma, the electrons of temperature T drift relative to the ions with a uniform velocity v, which is to be across the slab as shown in the accompanying figure. Obviously, a space-charge field is set up. What is the electron density, assuming that the ions are fixed? We may use the fluid-mechanical approximation. The width of the slab is L. Use Eq. (6.30) and Ohm's Law, assuming that (6.31) can be used to solve for B in terms of J_E.

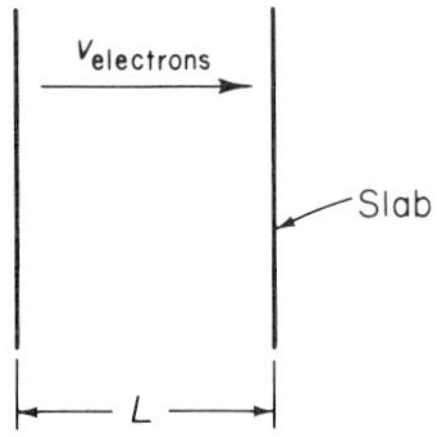

4. When the conductivity, in the magnetohydrodynamic approximation, is not infinite, then magnetic field lines can diffuse through the material. Estimate the time constant for a magnetic field to penetrate a unit length of

(a) mercury,
(b) aluminum,
(c) stainless steel, and
(d) gaseous plasma with a density of 10^{12} electrons per cubic centimeter and a collision frequency of 10^7 Hz.

5. Derive the wave equation for Alfvén waves, but include the effects of conductivity. This means that *both* terms in Eq. (6.52) must be included.

6. How fast would mercury have to flow across a 50-kG magnetic field to produce an electric field of 10 V/cm? If the width of a magnetohydrodynamic channel is 10 cm and we assume that the average velocity of mercury through the channel can be used, how much pressure on the mercury is needed to supply 1000 W to a 100-ohm load resistor, not counting the power needed to supply the magnets or frictional losses in the channel?

7. Compute the approximate phase velocity of Alfvén waves in the ionosphere.

8. Examine Fig. 1.1. Plot curves of constant $\lambda_{\mathrm{D_e}}$ of values 100, 1, 0.1, 0.01, and 0.0001 meters. Also plot curves of constant $N_{\lambda_{\mathrm{D_e}}}$ of values 10^{14}, 10^{12}, 10^{10}, 10^8, 10^6, and 10^4 particles. What should the approximate minimum dimensions of a fusion reactor be if a true plasma is to be utilized? Sketch the curves on the figure.

9. Plot the phase shift per meter of an electromagnetic wave of frequency ω progagating through a cold hydrogen plasma of uniform density as this density is changed from zero to cutoff. Over what range would the measurement of phase shift produce the most accurate measurement of density?

10. Several confining schemes have been proposed involving external current flowing around the plasma. Another scheme is to force current through the plasma, to generate a self-confining magnetic field. Suppose a plasma column parallel to the z axis has a uniform current density flowing through it in the z direction. Calculate the steady state magnetic field and the radial density profile that will be confined by this *linear pinch*. Do you think that this system will be stable? Why? Use the equilibrium momentum equations and assume that only electrons move and have a temperature T_{e}.

11. Another confining arrangement is the θ pinch. In this scheme a single-turn coil is pulsed with extremely high currents. Calculate the pressure inside the loop, making a long solenoid approximation for a current of

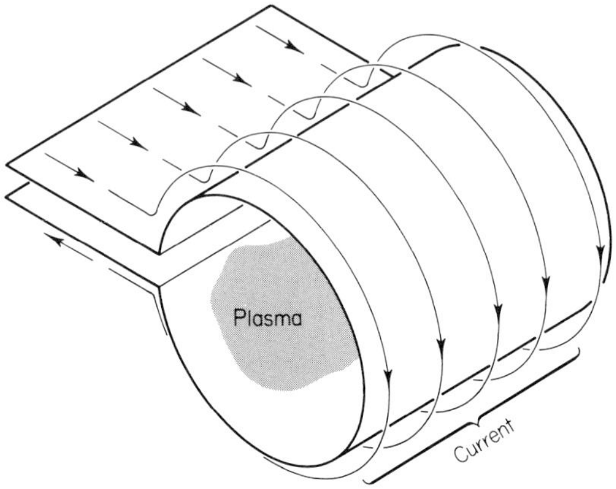

10^6 A flowing around the loop. What value of nkT will be confined? If $n = 10^{12}$ particles per cubic centimeter what temperature will be achieved? Is thermonuclear fusion possible with this device? Is the θ pinch more stable than the linear z pinch of Problem 6? Why?

12. If a conducting wire is placed inside a plasma (Langmuir probe), a potential may be applied between this probe and "ground." A plot of the current drawn by the probe from the plasma as a function of probe voltage *may* look like that in the accompanying figure. Why is the current higher when the voltage is highly positive than when it is highly negative? Would the Debye length relative to the probe diameter affect the characteristic of this probe, and if so, how?

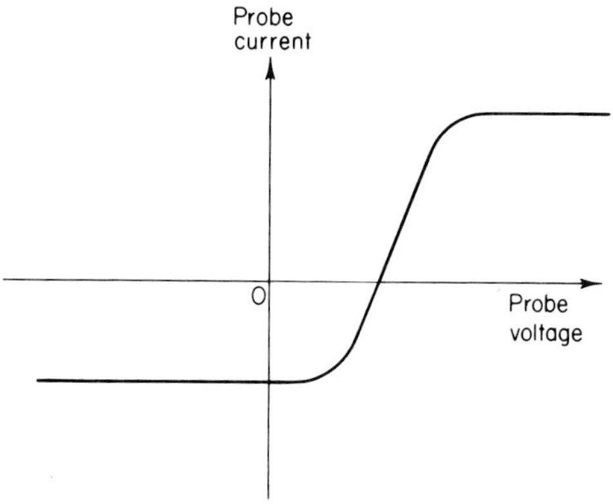

7 Waves in Cold Plasmas

The previous chapter touched on a very important body of plasma knowledge, and perhaps one of the most interesting and tantalizing areas of interest: waves in plasmas.

The presence in a plasma of many varieties of natural frequencies and lengths makes matters far more complicated than in free space, but at the same time produces some very interesting calculations for different types of wave behavior, many of which have been verified in the laboratory.

When temperature effects are considered, the possibility of wave growth or damping appears, and much effort is directed toward understanding this problem.

A. WAVES AND THE WAVE EQUATION

To begin a more specific discussion of the nature of waves in plasmas, we must first define what we mean by a wave. In general, a wave is a phenomenon that satisfies a wave-type equation. Perhaps this is begging the question, but if we can proceed with the development, the reasons for this statement will appear in due course.

For simplicity, we first note that we can easily derive a wave-type equation directly from Maxwell's equations, which are restated here.

$$\nabla_r \times \mathbf{H} = \mathbf{J} + \frac{\partial \mathbf{D}}{\partial t} \tag{7.1}$$

$$\nabla_r \times \mathbf{E} = -\frac{\partial \mathbf{B}}{\partial t} \tag{7.2}$$

$$\nabla_r \cdot \mathbf{B} = 0 \tag{7.3}$$

$$\nabla_r \cdot \mathbf{D} = \rho_E \tag{7.4}$$

We shall also need the constitutive relations

$$\mathbf{J} = \sigma \mathbf{E} \tag{7.5}$$

$$\mathbf{D} = \varepsilon \mathbf{E} \tag{7.6}$$

and

$$\mathbf{B} = \mu \mathbf{H} \tag{7.7}$$

σ, ε, and μ refer to the nature of the medium; in this case, of course, we will be using the values appropriate to a plasma. These may be tensors, especially if the plasma is immersed in a dc magnetic field.

Let us now take the curl of Eq. (7.2), which is

$$\nabla_r \times (\nabla_r \times \mathbf{E}) = -\frac{\partial}{\partial t}(\nabla_r \times \mathbf{B}) \tag{7.8}$$

assuming that we can interchange time and spatial derivatives.

Then using Eq. (7.7) in (7.8) we obtain

$$\nabla_r \times (\nabla_r \times \mathbf{E}) = -\mu \frac{\partial}{\partial t}(\nabla_r \times \mathbf{H}) \tag{7.9}$$

We can now substitute Eq. (7.1) into Eq. (7.9), which yields

$$\nabla_r \times (\nabla_r \times \mathbf{E}) = -\mu \frac{\partial}{\partial t}\left(\mathbf{J} + \frac{\partial \mathbf{D}}{\partial t}\right) \tag{7.10}$$

Using (7.6) and (7.5) to eliminate **J** and **D**, we get

$$\nabla_r \times (\nabla_r \times \mathbf{E}) = -\mu\sigma \frac{\partial \mathbf{E}}{\partial t} - \varepsilon\mu \frac{\partial^2 \mathbf{E}}{\partial t^2} \tag{7.11}$$

This partial differential equation is a general wave-type equation. By this we mean that whenever any equation of this type appears, we can always get some kind of wave-type solutions to the equation. The solutions may appear to be nonpropagating waves, but we shall consider these disturbances as "wave type" in nature. Equation (7.11) assumes that ε, μ, and σ do not vary in space or time.

Often, a simplification in Eq. (7.11) is made in using the following identity.

$$\nabla_r \times (\nabla_r \times \mathbf{E}) = \nabla_r(\nabla_r \cdot \mathbf{E}) - \nabla_r^2 \mathbf{E} \tag{7.12}$$

It is not, however, always advisable to use this identity because of the difficulty in defining $\nabla_r^2 \mathbf{E}$. And if $\nabla_r \cdot \mathbf{E}$ is not zero everywhere, there is little to be gained in using (7.12) to replace the $\nabla_r \times (\nabla_r \times \mathbf{E})$ term in Eq. (7.8).

The first term on the right-hand side of Eq. (7.11) may result in growth or decay of the waves, depending on the nature of σ and μ. Also, if σ, μ, or ε

do change with position or time, the wave equation must be rederived to include their spatial and temporal dependences.

We noted several examples where wave-type equations appeared in Chapter 6. In this chapter we are going to attempt to make some order out of the complexity of waves that result, especially when a plasma is immersed in a magnetic field. The first problem to be considered is one where we assume that the plasma is infinite in extent and has no thermal motion of its ions or electrons. Any motion will be due to the electric and magnetic fields themselves. No "heating" will be assumed by this motion (cold plasma).

We are initially only interested in classifying the types of waves that are possible, and *not* the methods by which they are excited. We are thus looking for the natural resonant frequencies (eigenvalues or modes) of a wave–plasma system.

We will consider the wave equation (7.11) and assume that the permeability μ is that of free space μ_0, and also that we can combine the conductivity into an effective permittivity. By the very nature of this latter assumption, we are assuming that we can break up the wave into sinusoidally varying components $e^{-j\omega t}$. This should not really be a great cause for worry since most sources for plasma waves can usually be decomposed into Fourier components. By this we mean that we can take the Fourier (or Laplace) transform of the dependent variable.

B. EIGENVALUES OF THE *RLC* CIRCUIT (ORDINARY DIFFERENTIAL EQUATIONS)

Let us again go over how we obtain *dispersion relations* similar to those in Chapter 6. The dispersion relation will be extremely useful in determining the nature of the plasma eigenvalues. To show this, we first examine a simple example. Suppose we wish to find the natural modes of an *RLC* circuit as shown in Fig. 7.1.

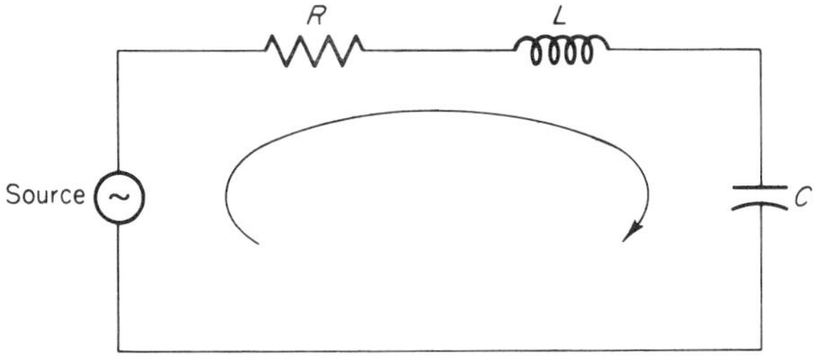

Fig. 7.1. The series *RLC* circuit.

We can find the natural frequencies of this circuit without worrying about the nature of the source, except to assume that it can have a Fourier representation.

The differential equation for the current in the system is

$$\mathscr{V}(t) = L\frac{di}{dt} + Ri + \frac{1}{C}\int i\, dt \tag{7.13}$$

If we differentiate (7.13) once with respect to time to remove the integration sign, we get

$$\frac{d\mathscr{V}(t)}{dt} = L\frac{d^2 i}{dt^2} + R\frac{di}{dt} + \frac{i}{C} \tag{7.14}$$

The eigenfrequencies of this system are simply its natural modes, which we find by solving the *homogeneous* equation

$$L\frac{d^2 i}{dt^2} + R\frac{di}{dt} + \frac{i}{C} = 0 \tag{7.15}$$

that is, the equation with all sources off. The natural modes are then found, if $i \sim e^{-j\omega t}$, to be

$$\omega_1 = j\frac{R}{2L} + \left[\frac{1}{LC} - \left(\frac{R}{2L}\right)^2\right]^{1/2}$$

$$\omega_2 = j\frac{R}{2L} - \left[\frac{1}{LC} - \left(\frac{R}{2L}\right)^2\right]^{1/2}$$

ω_1 and ω_2 may be real, complex, or imaginary, depending upon the values of R, L, and C.

In the same way, even if we had a more complicated circuit with coupled equations, the method of solving for the eigenvalues of homogeneous linear differential equations would be the same. It is, simply stated:

Write down a set of homogeneous equations describing the system. Set up the determinant of the coefficients of the dependent variables by assuming that all dependent variables vary as $e^{-j\omega t}$. Solve the determinant for the eigenvalues by setting it equal to zero.

C. EIGENVALUES OF THE WAVE EQUATION (PARTIAL DIFFERENTIAL EQUATIONS)

For systems described by *partial* differential equations the situation is much the same, *except* that we must also have some way to specify the variations in the other coordinate directions.

For the wave equation, the remaining coordinate is space. We shall consider a spatial variation of the form

$$\exp[j(\mathbf{k} \cdot \mathbf{r})]$$

where **k** is the propagation vector and **r** is the position vector. This is equivalent to saying that all spatial variations may also be broken up into Fourier components in space. The method used to find the eigenvalues is basically the same as that for ordinary differential equations. The eigenvalues are obtained by writing a transformed set of homogeneous partial differential wave equations and setting the determinant of the coefficients of the dependent variables equal to zero. This produces an eigenvalue equation as before except for one difference: The eigenvalues now show a functional relation between the temporal and spatial Fourier frequencies, that is, ω and **k**. The relation can differ for each eigenvalue if the determinant has multiple roots. This means that the values of ω, for example, are determined by what values of **k** are assumed as well as by the nature of the medium for which the differential equation was written. Note also then that the quantity $\exp[j(\mathbf{k} \cdot \mathbf{r} - \omega t)]$ defines a propagating disturbance, and hence a wave, if **k** and ω have real parts.

Let us do a simple example with Eq. (7.11). If we assume that $\sigma = 0$, or that σ may be combined with ε, (7.11) becomes

$$\nabla_r \times (\nabla_r \times \mathbf{E}) = -\varepsilon\mu \frac{\partial^2 \mathbf{E}}{\partial t^2} \tag{7.16}$$

Assuming a temporal variation $\exp(-j\omega t)$ and a spatial variation $\exp[j(\mathbf{k} \cdot \mathbf{r})]$, we obtain

$$-\mathbf{k} \times (\mathbf{k} \times \mathbf{E}_0) = \omega^2 \mu\varepsilon \mathbf{E}_0 \tag{7.17}$$

from Eq. (7.16). We may change the form to that of an operator, namely

$$[\mathbf{k} \times (\mathbf{k} \times \quad) + \omega^2 \mu\varepsilon]\mathbf{E}_0 = 0 \tag{7.18}$$

where $\mathbf{E}_0$ is the Fourier amplitude of **E**.

We may then write Eq. (7.18) in components, since it is a vector equation. Assuming Cartesian coordinates, ε to be a scalar, and **k** to be in the z direction only, we get, for the three components of (7.18), the following.

x Component

$$-k_z{}^2 E_{0x} + \omega^2 \mu\varepsilon E_{0x} = 0$$

y Component

$$-k_z{}^2 E_{0y} + \omega^2 \mu\varepsilon E_{0y} = 0$$

z Component

$$\omega^2 \mu \varepsilon E_{0z} = 0$$

The components above can be written in the matrix form below so that the determinant of the coefficients of $\mathbf{E}_0$ appears directly.

$$\begin{bmatrix} -k_z^2 + \omega^2 \mu \varepsilon & 0 & 0 \\ 0 & -k_z^2 + \omega^2 \mu \varepsilon & 0 \\ 0 & 0 & \omega^2 \mu \varepsilon \end{bmatrix} \times \begin{bmatrix} E_{0x} \\ E_{0y} \\ E_{0z} \end{bmatrix} = 0 \tag{7.19}$$

The eigenvalue equation, which is the determinant multiplied out and set equal to zero, is

$$(\omega^2 \mu \varepsilon)(-k_z^2 + \omega^2 \mu \varepsilon)(-k_z^2 + \omega^2 \mu \varepsilon) = 0 \tag{7.20}$$

There are three possible solutions, hence three possible modes, to the eigenvalue equation (7.20). They are, for the first mode

$$k_z^2 = \omega^2 \mu \varepsilon \tag{7.21a}$$

Similarly, for the second mode,

$$k_z^2 = \omega^2 \mu \varepsilon \tag{7.21b}$$

and for the third mode,

$$\omega^2 \mu \varepsilon = 0 \tag{7.21c}$$

Thus there are three eigenmodes (two of which are identical) for this problem. The first two are propagating disturbances, while the third mode is a time-invariant nonpropagating mode. It can be seen that the three relations in Eq. (7.21) are just those which respectively make the coefficients of E_{0x}, E_{0y} and E_{0z} in Eq. (7.18) vanish. Therefore the first mode (coefficients of $E_x = 0$) corresponds to a mode that has the electric field pointing in the x direction, the second mode has an electric field pointing in the y direction, and the third mode has E pointing along the z axis. This latter mode is the *non-propagating* one. Equation (7.21) also shows this, since in continua we know that we cannot allow a component of $\mathbf{E}$ only in the direction of propagation. Happily, the dispersion relation has borne this out.

In practice, the complete response of the system may be a sum of the normal mode responses, plus an impressed solution, corresponding to a steady state "forced" response.

To summarize, Eq. (7.20) is a *dispersion relation* between ω and $\mathbf{k}$ which tells us something about the eigenvalues of the system. If the system has natural modes, then Eq. (7.20) tells us what they are. However, both ω and $\mathbf{k}$ appear in (7.20), so the eigenvalues give us a relation between these variables. This is the definition of the dispersion relation.

D. EIGENVALUES OF WAVES IN COLD PLASMAS—THE DISPERSION RELATION

Equation (6.104) in the previous chapter can now be thought of as a set of eigenvalue equations used to describe a plasma system. The determinant of the coefficients of $\mathbf{E}$ set equal to zero is the dispersion relation. We recall that Eq. (6.104) is the set of three component equations of the wave equation, assuming a tensor dielectric constant. It has, however, not yet been specified for a plasma.

We shall include in $\boldsymbol{\varepsilon}$, the tensor permittivity, the magnetic field in the z direction as well as the effects of both ions and electrons. Since the conductivities of the ions and electrons add as conductivities in parallel, we can use Eq. (3.48) with no collisions written separately for the combination of ions and electrons, and added together to produce the complete plasma *conductivity* in Eq. (7.22), p. 174.

We will now change this to a permittivity. First, we note that the sign of the charge of the electrons and ions is carried along in their respective values of ω_c. Remember that the electrons and ions move in "parallel," so their respective conductivities will add directly.

To obtain the permittivity, a development similar to that used to derive Eq. (3.54) is used, but we now use Eq. (7.22) in the development. The total permittivity tensor is shown in Eq. (7.23), p. 174.

We make the following substitutions for the plasma frequencies.

$$\omega_{p_i}^2 = \frac{n_i q_i^2}{m_i \varepsilon_0} \quad \text{and} \quad \omega_{p_e}^2 = \frac{n_e q_e^2}{m_e \varepsilon_0}$$

where ω_{p_i} is the ion plasma frequency and ω_{p_e} is the electron plasma frequency. Normally $\omega_{p_i} \ll \omega_{p_e}$, and in many cases it can be neglected (set ≈ 0). These substitutions permit us to write (7.23) in the simplified form given in Eq. (7.24), p. 175. We further modify the dielectric tensor by making the following substitutions:

$$\Omega_e = |\omega_{c_e}| \quad \text{and} \quad \Omega_i = |\omega_{c_i}|$$

so that Eq. (7.24) takes the form shown in Eq. (7.25), p. 175.

$$\sigma_{\text{total}} = \begin{bmatrix} j\omega\left[\dfrac{n_i q_i^2}{m_i}\dfrac{1}{\omega^2-\omega_{c_i}^2}+\dfrac{n_e q_e^2}{m_e}\dfrac{1}{\omega^2-\omega_{c_e}^2}\right] & -\left[\dfrac{\omega_{c_i} n_i q_i^2}{m_i(\omega^2-\omega_{c_i}^2)}+\dfrac{\omega_{c_e} n_e q_e^2}{m_e(\omega^2-\omega_{c_e}^2)}\right] & 0 \\ \left[\dfrac{\omega_{c_i} n_i q_i^2}{m_i(\omega^2-\omega_{c_i}^2)}+\dfrac{\omega_{c_e} n_e q_e^2}{m_e(\omega^2-\omega_{c_e}^2)}\right] & j\omega\left[\dfrac{n_i q_i^2}{m_i}\dfrac{1}{\omega^2-\omega_{c_i}^2}+\dfrac{n_e q_e^2}{m_e}\dfrac{1}{\omega^2-\omega_{c_e}^2}\right] & 0 \\ 0 & 0 & \dfrac{j}{\omega}\left[\dfrac{n_i q_i^2}{m_i}+\dfrac{n_e q_e^2}{m_e}\right] \end{bmatrix} \tag{7.22}$$

$$\varepsilon = \varepsilon_0 \begin{bmatrix} 1-\left[\dfrac{n_i q_i^2}{m_i\varepsilon_0(\omega^2-\omega_{c_i}^2)}+\dfrac{n_e q_e^2}{m_e\varepsilon_0(\omega^2-\omega_{c_e}^2)}\right] & -\left[\dfrac{j\omega_{c_i} n_i q_i^2}{m_i\varepsilon_0\,\omega(\omega^2-\omega_{c_i}^2)}+\dfrac{j\omega_{c_e} n_e q_e^2}{m_e\varepsilon_0\,\omega(\omega^2-\omega_{c_e}^2)}\right] & 0 \\ \left[\dfrac{j\omega_{c_i} n_i q_i^2}{m_i\varepsilon_0\,\omega(\omega^2-\omega_{c_i}^2)}+\dfrac{j\omega_{c_e} n_e q_e^2}{m_e\varepsilon_0\,\omega(\omega^2-\omega_{c_e}^2)}\right] & 1-\left[\dfrac{n_i q_i^2}{m_i\varepsilon_0(\omega^2-\omega_{c_i}^2)}+\dfrac{n_e q_e^2}{m_e\varepsilon_0(\omega^2-\omega_{c_e}^2)}\right] & 0 \\ 0 & 0 & 1-\left[\dfrac{n_i q_i^2}{m_i\varepsilon_0\omega^2}+\dfrac{n_e q_e^2}{m_e\varepsilon_0\,\omega^2}\right] \end{bmatrix} \tag{7.23}$$

$$\varepsilon = \varepsilon_0 \begin{bmatrix} 1 - \left[\dfrac{\omega_{\mathrm{pi}}^2}{(\omega^2 - \omega_{\mathrm{ci}}^2)} + \dfrac{\omega_{\mathrm{pe}}^2}{(\omega^2 - \omega_{\mathrm{ce}}^2)}\right] & -j\left[\dfrac{\omega_{\mathrm{ci}}\,\omega_{\mathrm{pi}}^2}{\omega(\omega^2 - \omega_{\mathrm{ci}}^2)} + \dfrac{\omega_{\mathrm{ce}}\,\omega_{\mathrm{pe}}^2}{\omega(\omega^2 - \omega_{\mathrm{ce}}^2)}\right] & 0 \\ j\left[\dfrac{\omega_{\mathrm{ci}}\,\omega_{\mathrm{pi}}^2}{\omega(\omega^2 - \omega_{\mathrm{ci}}^2)} + \dfrac{\omega_{\mathrm{ce}}\,\omega_{\mathrm{pe}}^2}{\omega(\omega^2 - \omega_{\mathrm{ce}}^2)}\right] & 1 - \left[\dfrac{\omega_{\mathrm{pi}}^2}{(\omega^2 - \omega_{\mathrm{ci}}^2)} + \dfrac{\omega_{\mathrm{pe}}^2}{(\omega^2 - \omega_{\mathrm{ce}}^2)}\right] & 0 \\ 0 & 0 & 1 - \left[\dfrac{\omega_{\mathrm{pi}}^2}{\omega^2} + \dfrac{\omega_{\mathrm{pe}}^2}{\omega^2}\right] \end{bmatrix} \tag{7.24}$$

$$\varepsilon = \varepsilon_0 \begin{bmatrix} 1 - \left[\dfrac{\omega_{\mathrm{pi}}^2}{(\omega^2 - \Omega_{\mathrm{i}}^2)} + \dfrac{\omega_{\mathrm{pe}}^2}{(\omega^2 - \Omega_{\mathrm{e}}^2)}\right] & -j\left[\dfrac{\Omega_{\mathrm{i}}\,\omega_{\mathrm{pi}}^2}{\omega(\omega^2 - \Omega_{\mathrm{i}}^2)} - \dfrac{\Omega_{\mathrm{e}}\,\omega_{\mathrm{pe}}^2}{\omega(\omega^2 - \Omega_{\mathrm{e}}^2)}\right] & 0 \\ j\left[\dfrac{\Omega_{\mathrm{i}}\,\omega_{\mathrm{pi}}^2}{\omega(\omega^2 - \Omega_{\mathrm{i}}^2)} - \dfrac{\Omega_{\mathrm{e}}\,\omega_{\mathrm{pe}}^2}{\omega(\omega^2 - \Omega_{\mathrm{e}}^2)}\right] & 1 - \left[\dfrac{\omega_{\mathrm{pi}}^2}{(\omega^2 - \Omega_{\mathrm{i}}^2)} + \dfrac{\omega_{\mathrm{pe}}^2}{(\omega^2 - \Omega_{\mathrm{e}}^2)}\right] & 0 \\ 0 & 0 & 1 - \dfrac{\omega_{\mathrm{pi}}^2 + \omega_{\mathrm{pe}}^2}{\omega^2} \end{bmatrix} \tag{7.25}$$

Note that the off-diagonal terms in (7.25) are the negatives of each other, and two of the main diagonal terms are equal to each other. We can further manipulate the permittivity tensor as follows. First, let

$$R = 1 - \left[\frac{\omega_{\mathrm{pi}}^2 \omega}{\omega^2(\omega + \Omega_{\mathrm{i}})} + \frac{\omega_{\mathrm{pe}}^2 \omega}{\omega^2(\omega - \Omega_{\mathrm{e}})}\right] \tag{7.26}$$

and

$$L = 1 - \left[\frac{\omega_{\mathrm{pi}}^2 \omega}{\omega^2(\omega - \Omega_{\mathrm{i}})} + \frac{\omega_{\mathrm{pe}}^2 \omega}{\omega^2(\omega + \Omega_{\mathrm{e}})}\right] \tag{7.27}$$

and

$$P = 1 - \left[\frac{\omega_{\mathrm{pi}}^2 + \omega_{\mathrm{pe}}^2}{\omega^2}\right] \tag{7.28}$$

The next step is to define the quantities

$$S = \tfrac{1}{2}(R + L) \qquad \text{and} \qquad D = \tfrac{1}{2}(R - L)$$

The simple final form of the permittivity tensor made by substituting the foregoing quantities in (7.25) is

$$\boldsymbol{\varepsilon} = \varepsilon_0 \begin{vmatrix} S & -jD & 0 \\ jD & S & 0 \\ 0 & 0 & P \end{vmatrix} \tag{7.29}$$

This is the form we shall use in determining the dispersion relation for a cold plasma in a magnetic field. We return to Eq. (6.104) and apply this form of the permittivity tensor [Eq. (7.29)] to it, arriving at the dispersion relation.

However, before proceeding with this development, which still looks rather formidable, we shall first make some additional simplifying assumptions, corresponding to the peculiarities of the problem of a plasma in a magnetic field.

First, let us assume that the plasma is symmetric about the magnetic field axis, or that it is at least possible to rotate the x and y axes about the magnetic field axis so that the propagation vector $\mathbf{k}$ lies in the xz plane only.

This is shown schematically in Fig. 7.2. A result of this rotation is that we may separate completely the components of $\mathbf{k}$ in terms of the magnitude of $\mathbf{k}$ and the angle θ. We will not normally lose any information about the problem in fixing the direction of $\mathbf{k}$ in this way, since the plasma should be symmetric about the z axis. The components of $\mathbf{k}$ are

$$k_z = k_{\parallel} = k \cos\theta$$

$$k_x = k_{\perp} = k \sin\theta$$

and

$$k_y = 0$$

as seen from Fig. 7.2.

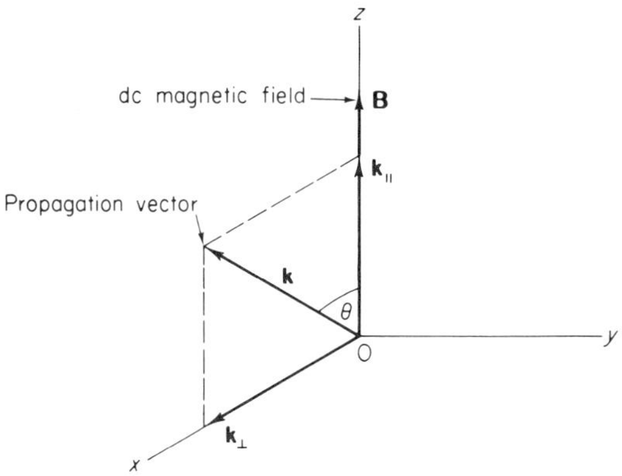

Fig. 7.2. Orientation of the propagation vector relative to a dc magnetic field and a Cartesian coordinate system.

Finally, we obtain the plasma dispersion relation from Eq. (6.104):

$$\begin{vmatrix} \omega^2\mu_0\varepsilon_0 S - k^2\cos^2\theta & -j\omega\mu_0\varepsilon_0 D & k^2\sin\theta\cos\theta \\ j\omega^2\mu_0\varepsilon_0 D & \omega^2\mu_0\varepsilon_0 S - k^2\sin^2\theta - k^2\cos^2\theta & 0 \\ k^2\sin\theta\cos\theta & 0 & \omega^2\mu_0\varepsilon_0 P - k^2\sin^2\theta \end{vmatrix} = 0 \tag{7.30}$$

Note that each term in the permittivity tensor (7.29) has been multiplied by ε_0.

We proceed to place this dispersion relation into a form designed to obtain as much information as possible about the eigenvalues. We now let $k_0{}^2 = \omega^2\mu_0\varepsilon_0$, so that (7.30) becomes

$$\begin{vmatrix} k_0{}^2S - k^2\cos^2\theta & -jk_0{}^2D & k^2\sin\theta\cos\theta \\ jk_0{}^2D & k_0{}^2S - k^2 & 0 \\ k^2\sin\theta\cos\theta & 0 & k_0{}^2P - k^2\sin^2\theta \end{vmatrix} = 0 \tag{7.31}$$

We multiply out the determinant of (7.31) to obtain the following:

$$k_0{}^6[S^2P - D^2P] - k_0{}^4[k^2SP + k^2SP\cos^2\theta + k^2S^2\cos^2\theta - k^2D^2\sin^2\theta] + k_0{}^2[k^4P\cos^2\theta + k^4S\sin^2\theta] = 0 \tag{7.32}$$

Dividing Eq. (7.32) through by $k_0{}^6$ and rearranging it in powers of (k/k_0), we obtain

$$\left(\frac{k}{k_0}\right)^4[S\sin^2\theta + P\cos^2\theta] - \left(\frac{k}{k_0}\right)^2[SP + S^2\sin^2\theta + SP\cos^2\theta - D^2\sin^2\theta] + [S^2P - D^2P] = 0 \tag{7.33}$$

Note that $S^2 - D^2 = RL$, so that (7.33) becomes

$$\left(\frac{k}{k_0}\right)^4 [S \sin^2 \theta + P \cos^2 \theta] - \left(\frac{k}{k_0}\right)^2 [RL \sin^2 \theta + SP(1 + \cos^2 \theta)] + PRL = 0 \tag{7.34}$$

This is still the plasma dispersion relation. It remains a function of ω and k and the plasma parameters. Equation (7.34) is quadratic in $(k/k_0)^2$ [biquadratic in (k/k_0)], so there must be two roots for $(k/k_0)^2$. The general form of Eq. (7.34) is

$$A\left(\frac{k}{k_0}\right)^4 - B\left(\frac{k}{k_0}\right)^2 + C = 0 \tag{7.35}$$

Equation (7.35) is a concise form of Eq. (7.34). It can also be seen from its coefficients that this dispersion relation is a function of θ, which is the angle between the propagation vector and the z axis. The coefficients of (7.35) are

$$A = S \sin^2 \theta + P \cos^2 \theta$$

$$B = RL \sin^2 \theta + SP(1 + \cos^2 \theta)$$

and

$$C = PRL$$

These coefficients are functions *only* of ω, θ, and the plasma parameters, so Eq. (7.35) is really biquadratic in (k/k_0). We must now determine how to obtain information from the dispersion relation (7.35). There are several ways to solve it.

First, if we solve (7.35) for the two roots of $(k/k_0)^2$, we get

$$\left(\frac{k}{k_0}\right)^2 = \frac{RL \sin^2 \theta + PS(1 + \cos^2 \theta) \pm [(RL - PS)^2 \sin^4 \theta + 4P^2 D^2 \cos^2 \theta]^{1/2}}{2S \sin^2 \theta + 2P \cos^2 \theta} \tag{7.36}$$

Note again that the identity $S^2 - D^2 = RL$ is used to obtain (7.36). We will examine this form of the solution shortly.

We can also solve (7.35) for θ. This will permit us to investigate additional properties of the dispersion relation. The solution for θ is

$$\tan^2 \theta = \frac{-P[(k/k_0)^2 - R][(k/k_0)^2 - L]}{[S(k/k_0)^2 - RL][(k/k_0)^2 - P]} \tag{7.37}$$

By specifying the angle θ in (7.37), we can tell what values of $(k/k_0)^2$ exist for given values of ω and the appropriate plasma parameters.

Equation (7.36) shows that there are two possible values of $(k/k_0)^2$. We call these two solutions, if both are propagating, the "slow" and "fast" waves. This notation refers to the phase velocity of the wave, which is defined as

$$v_{\text{phase}} = \frac{\omega}{k} = \frac{\omega}{(k/k_0)k_0} = \frac{c}{(k/k_0)} \tag{7.38}$$

The phase velocity is thus inversely proportional to k/k_0 and therefore the *larger* value of k/k_0 pertains to the *slow* wave, and likewise, the *smaller* value of k/k_0 is associated with the *fast* wave. Here c is the velocity of light.

Note that the values of k/k_0 for given values of ω are what is finally needed for understanding waves in cold plasmas. These may be obtained by taking the square root of each value of $(k/k_0)^2$ obtained from the dispersion relation. The process of taking the square root gives two values of (k/k_0) for *each* value of $(k/k_0)^2$. They are not different modes, but correspond only to opposite directions of propagation.

E. REPRESENTATIONS OF THE EIGENVALUES

A representation of the various forms of the solution to (7.35), (7.36), or (7.37) can take many forms. One possibility is to hold all quantities constant (θ, ω_{p_i}, ω_{p_e}, ω_{c_e}, ω_{c_i}) and vary only ω. We can then represent (k/k_0) as some function of ω and it may be plotted. We may also do the reverse; that is, use (k/k_0) as the independent variable and solve for and plot ω. Varying the density and/or magnetic field with ω fixed is yet another way of presenting the data. Obviously, then, there are a large number of schemes for presenting the dispersion relation, since there are so many parameters.

A very compact way to show many of the solutions to the dispersion relation is the *Clemmow–Mullaly–Allis* (*CMA*) *diagram.* This diagram is a plot of the forms of the magnitude of the phase velocity as a function of angle θ for normalized values of density and magnetic field.

We shall develop this diagram as follows. First, polar plots of *phase velocity* (not k/k_0) versus θ will be made, holding the plasma parameters fixed as shown in Fig. 7.3. In this figure the velocity of light is shown as a dashed circle. In most cases the fast wave has a phase velocity greater than c, while the slow wave's phase velocity is less than c. However, there are some exceptions, as will be seen shortly.

Because of the rotational property of the coordinate axes, we note that these "wave-normal surfaces" shown in Fig. 7.3 may be rotated about the z axis to produce a three-dimensional form. Both of the solutions for $(k/k_0)^2$ and hence both phase velocities (slow and fast) are superimposed on the same coordinate axes.

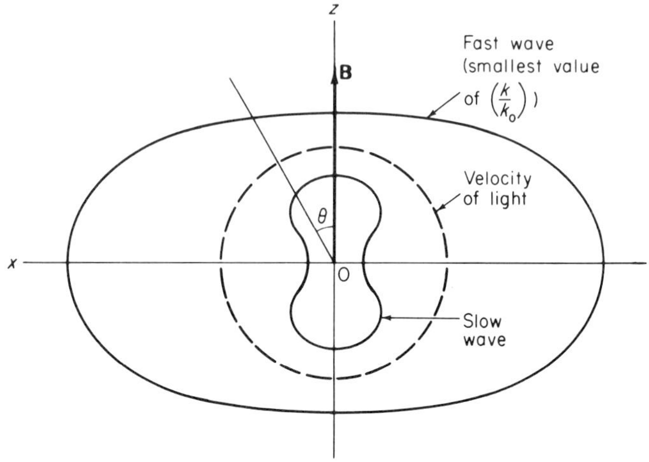

Fig. 7.3. Phase velocity, or wave-normal surface. These curves are the loci of points of constant phase, emitted from the origin.

Judging by the preliminary discussion, we would expect to find ranges of parameters where $(k/k_0)^2$ was either positive or negative. The former condition refers to propagating solutions whereas the latter refers to nonpropagating solutions. This can be seen directly when the square root of $(k/k_0)^2$ is taken, since a negative value of $(k/k_0)^2$ produces an imaginary (nonpropagating) value of (k/k_0). $(k/k_0)^2$ may change sign by going through zero, or by going through infinity. We shall define a zero crossing to be a cutoff (infinite phase velocity) and an infinity transition to be a resonance (zero phase velocity).

If we look at Eq. (7.36), we can determine where the cutoff and resonances occur at various angles as follows.

Cutoff exists, *regardless of angle*, when $P = 0$, $R = 0$, or $L = 0$. These occur whenever, in the plasma, the density, magnetic field, or frequency is varied to produce these conditions. Equation (7.37) also shows us that a resonance $[(k/k_0)^2 = \infty]$ occurs when

$$\tan^2 \theta = -\frac{P}{S} \tag{7.39}$$

It should be noted here that the forms of the solutions for k/k_0, and hence the phase velocity, should change as we pass *through* these cutoffs and resonances, and their form changes only when we cross these boundaries.

We call the resonances at $\theta = 0$ and $\theta = \pi/2$ *principal resonances.* $S = \infty$ is a principal resonance at $\theta = 0$, and $S = 0$ is a principal resonance at $\theta = \pi/2$. Also, another principal resonance can be found from Eq. (7.37) at $\theta = 0$. This condition occurs when $P = 0$.

The CMA diagram can now be displayed graphically on a coordinate system whose axes are proportional to plasma parameters, such as density and cyclotron frequency. This parameter space will be divided into regions formed by the cutoffs and principal resonances.

Table 7.1 lists these dividing boundaries.

Table 7.1 Dividing Boundaries for the CMA Diagram

Cutoffs	Principal resonances	
$P=0$	at $\theta=0$	and at $\theta=\pi/2$
$R=0$	$P=0$	$S=0$
$L=0$	$R=\infty$, $L=\infty$ $\rightarrow S=\infty$	

We now construct a two-dimensional parameter space required to produce the CMA diagram with the following axes:

Horizontal Axis

$$\alpha = \frac{\omega_{\mathrm{pi}}^2 + \omega_{\mathrm{pe}}^2}{\omega^2}$$

Vertical Axis

$$\frac{\Omega_{\mathrm{e}}^2}{\omega^2}$$

We will make the vertical axis logarithmic, and the horizontal axis linear. Figure 7.4 shows the labeling of the axes.

The cutoff and principal resonance lines are sketched in Fig. 7.5. We rewrite the appropriate quantities here so that Fig. 7.5 may be developed from them.

$$P = 1 - \left[\frac{\omega_{\mathrm{pe}}^2}{\omega^2} + \frac{\omega_{\mathrm{pi}}^2}{\omega^2}\right] = 1 - \alpha \tag{7.40}$$

$$R = 1 - \left[\frac{\omega_{\mathrm{pi}}^2}{\omega^2}\frac{\omega}{\omega+\Omega_{\mathrm{i}}} + \frac{\omega_{\mathrm{pe}}^2}{\omega^2}\frac{\omega}{\omega-\Omega_{\mathrm{e}}}\right] = 1 - \frac{\alpha\omega^2}{(\omega+\Omega_{\mathrm{i}})(\omega-\Omega_{\mathrm{e}})} \tag{7.41}$$

$$L = 1 - \left[\frac{\omega_{\mathrm{pi}}^2}{\omega^2}\frac{\omega}{\omega-\Omega_{\mathrm{i}}} + \frac{\omega_{\mathrm{pe}}^2}{\omega^2}\frac{\omega}{\omega+\Omega_{\mathrm{e}}}\right] = 1 - \frac{\alpha\omega^2}{(\omega-\Omega_{\mathrm{i}})(\omega+\Omega_{\mathrm{e}})} \tag{7.42}$$

$$S = \tfrac{1}{2}(R+L) \tag{7.43}$$

Equations (7.41) and (7.42) are true if we have charge neutrality; that is, if $n_{\mathrm{e}}q_{\mathrm{e}} = n_{\mathrm{i}}q_{\mathrm{i}}$.

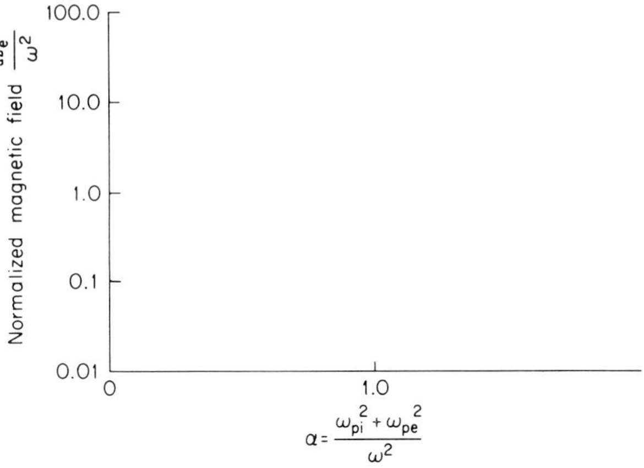

Fig. 7.4. The coordinates for the CMA diagram.

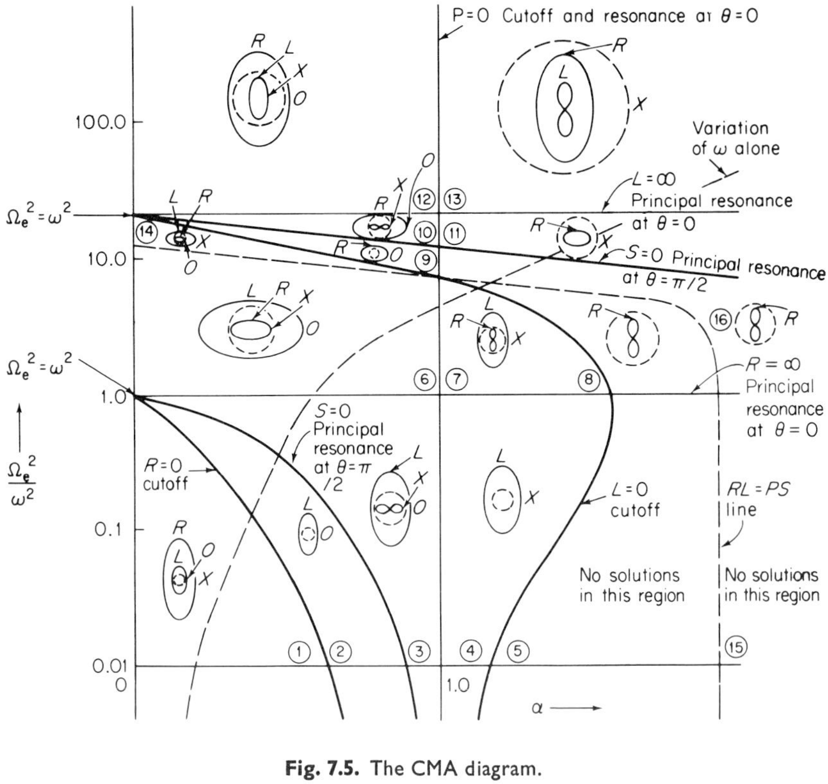

Fig. 7.5. The CMA diagram.

The line $RL = PS$ is also plotted in Fig. 7.5. The nature of this boundary will be explained subsequently. The parameter space in Fig. 7.5 will now be divided into numbered regions for ease of reference.

It should be noted here that if the true values of typical ion and electron masses were used, the vertical axis in Fig. 7.5 would be very much longer than it is. As a result, for simplification, m_i/m_e has been reduced to 4.

Each region in the CMA diagram has its possible phase velocity surfaces sketched in them. The dashed circle again corresponds to the velocity of light. The "slow" and "fast" wave notations now become quite apparent. Note also that in some regions of Fig. 7.5 *both* modes have phase velocities either less than or greater than the velocity of light. The labels R, L, O, and X need some explanation. R and L appear only on a phase velocity surface along the magnetic field axis. They refer to the polarization of the waves (right or left hand) in this direction. O and X appear only at 90° with respect to the magnetic field axis. These latter labels define *ordinary* and *extraordinary* waves, depending on whether or not the phase velocity depends on the magnetic field when the wave is propagating at 90° with respect to the field direction.

Figure 7.6 shows the four labeling designations depicted graphically. We note that the right and left circularly polarized waves rotate in the same direction as the electrons and positive ions, respectively. This means that resonant "heating" can take place for each species of particle if the appropriate circularly polarized wave is present at the particle's cyclotron frequency.

An examination of Fig. 7.5 shows that the boundaries in the CMA diagram form regions where certain modes are present and others are not. As the boundaries of these regions are crossed, the wave-normal surfaces for the modes appear, disappear, or change shape. For example, when moving from region 7 to 8 (increasing density) the fast wave disappears. If the parameters are changed to move along a path that goes from region 11 to 13, the slow wave appears as the boundary is crossed, and so on. The line $RL = PS$ also results in a change in the nature of the waves as this line is crossed, which can be seen by examining the CMA diagram.

The parameter space is set up in the CMA diagram so that free space and very low magnetic field appear in region 1 in the lower left corner of the diagram, and high density and high magnetic field appear in region 13 in the upper right corner, relative to the frequency of the waves. Since the coordinates of the graph are so normalized, then Alfvén waves, for example, which are of extremely low frequency, appear in region 13.

The dispersion relation may be solved directly for the two wave-normal surfaces in any of the regions by using the appropriate values of α and Ω_e/ω in equations (7.35), (7.36), or (7.37). Since the coordinates are normalized, it is easy to see that the same frequency may appear in several different modes, depending upon the values of density and magnetic field.

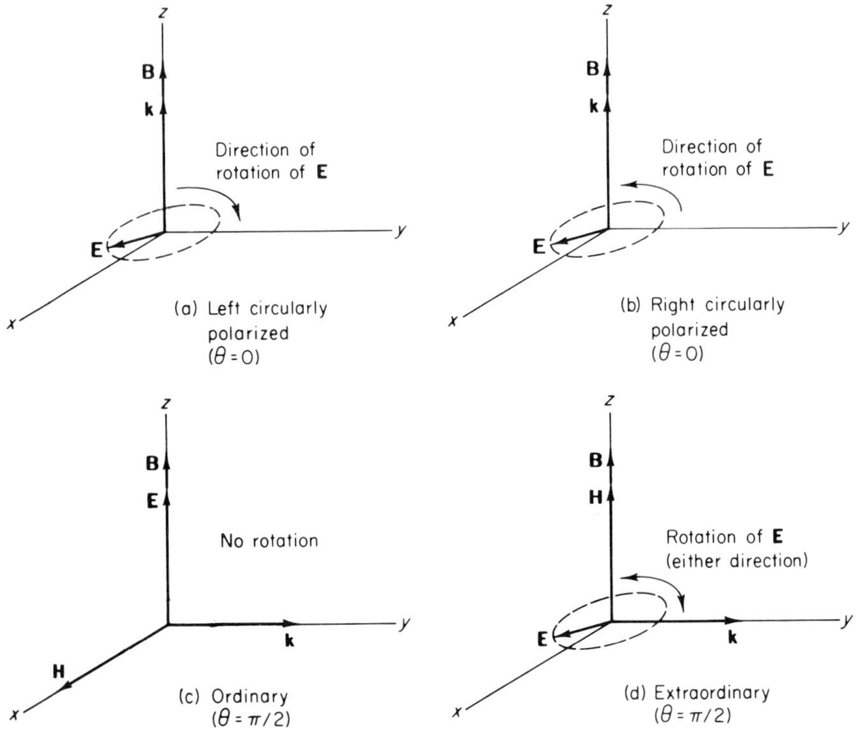

Fig. 7.6. Vector diagrams of the various polarizations possible at 0 and 90° with respect to the magnetic field. (a) and (b) are with **k** parallel to **B** and (c) and (d) are with **k** perpendicular to **B**.

The CMA diagram offers a very compact means of presenting much data regarding the solution to the dispersion relation expressed in Eq. (7.36). It supplies a very broad picture of the nature of the waves. Note again that although the characteristic shapes of the phase velocity surfaces remain the same inside each bounded region, their magnitudes may change. In all cases, the solution to the dispersion relation (7.35) gives the eigenmodes. Their names, number, and shape may change, but they all remain solutions of the same equation.

If it is desired to look at the picture in more detail, since the CMA diagram might appear to be a bit overwhelming, the following different representation can be used.

First, note that Eq. (7.36) is the solution for (k/k_0) as a function of ω, θ, and the plasma parameters. We may then hold θ and the plasma parameters constant, for example, and then plot k (not k/k_0) as a function of ω. We can move through the CMA diagram and plot the solutions in this form as we traverse boundaries. The solutions are for propagation in only one direction, so that the plots become line graphs.

Let us first examine the simple dispersion relation developed in Eq. (7.21), and plot the solution in this manner. We will, however, plot the graphs as ω versus k, rather than k versus ω. We will still allow ω to be the independent variable but we will plot it on the vertical axis. An ω versus k plot of the dispersion relation (7.21) is shown in Fig. 7.7.

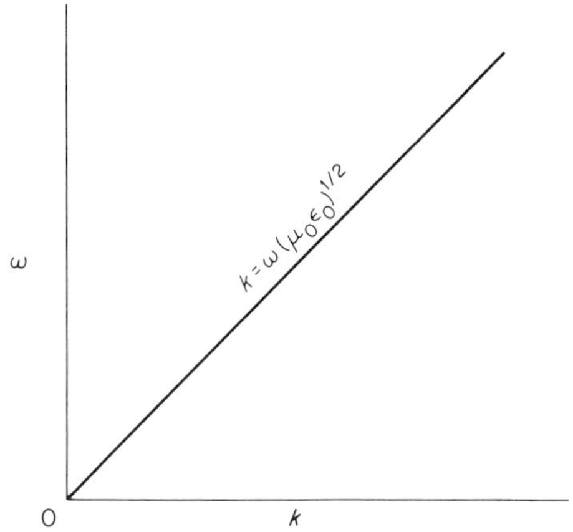

Fig. 7.7. ω versus k plot of the dispersion relation for plane monochromatic waves in free space.

Since there are no magnetic field or plasma effects, the dispersion plot is a straight line. The phase velocity is always the velocity of light and the group velocity, the slope of the line $[v_{\text{group}} = \partial\omega/\partial k]$ is also constant and equal to the velocity of light.

Now suppose ε becomes $\varepsilon_0[1 - (\omega_{\text{p}_e}^2/\omega^2)]$ (a cold plasma with no magnetic field and stationary ions). The dispersion relation (7.21) then becomes

$$k_z^{\,2} = \omega^2 \mu_0 \varepsilon_0 \left(1 - \frac{\omega_{\text{p}e}^2}{\omega^2}\right) \tag{7.44}$$

or

$$k_z = \pm\omega(\mu_0\,\varepsilon_0)^{1/2}\left(1 - \frac{\omega_{\text{p}e}^2}{\omega^2}\right)^{1/2} \tag{7.45}$$

Recall again that $\pm$ corresponds to propagation in two directions, not to two different modes. We shall consider the positive sign. A plot of ω versus k is given in Fig. 7.8. Note that the curve is only plotted for the real part of k_z as a function of ω. It can be seen from Eq. (7.45) that for $\omega < \omega_{\text{p}_e}$ the value

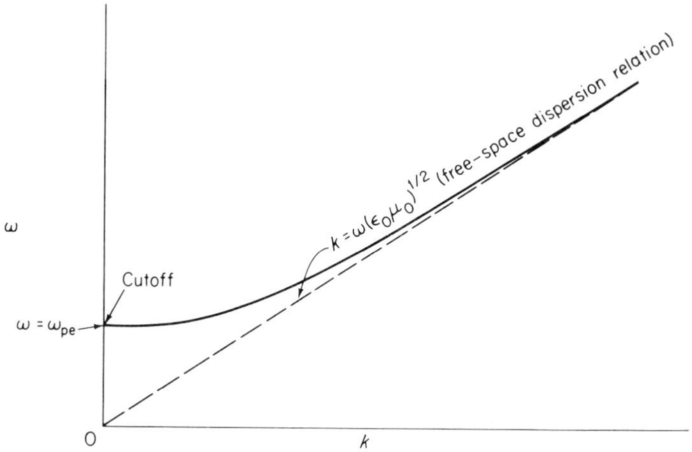

Fig. 7.8. ω versus k plot for plane monochromatic waves propagating in an isotropic cold plasma.

of k_z is purely imaginary. The curve also becomes asymptotic to the free space dispersion relation as $\omega \to \infty$, which is obvious from examination of Eq. (7.45). The phase velocity at $\omega = \omega_{pe}$ is infinite and the group velocity is zero, as can be seen by examining the figure. Cutoffs then, appear when the curves intersect the ω axis (infinite phase velocity). As we will see subsequently, resonances appear when the curves become parallel to the k axis, as $k \to \infty$ (zero phase velocity).

For this problem, since a plasma with no magnetic field is isotropic, Fig. 7.8 will have the same form for all k, regardless of the direction of propagation. This graph tells us what frequencies are excitable as eigenmodes and which ones are not.

We may also obtain the phase and group velocity from these plots. They are obtained from the graphs by using the following relations

$$v_{\text{phase}} = \frac{\omega}{k} = \frac{c}{[1 - (\omega_{pe}^2/\omega^2)]^{1/2}} \tag{7.46}$$

$$v_{\text{group}} = \frac{\partial \omega}{\partial k} = c\left(1 - \frac{\omega_{pe}^2}{\omega^2}\right)^{1/2} \tag{7.47}$$

The terms on the extreme right of (7.46) and (7.47) are the phase and group velocities for the plasma whose ω versus k plot is given in Fig. 7.8. Note that for this problem the product of $v_{\text{phase}}\, v_{\text{group}}$ is c^2, and that v_{phase} is always greater than c and v_{group} is always less than c. This is usually the situation for other dispersion plots of electromagnetic waves, although again there are some exceptions.

We may make similar plots for the dispersion relations of any problem, including that depicted in Eq. (7.36).

To make, therefore, an ω versus k plot from the dispersion relation, do the following.

1. Hold all variables (density, magnetic field, etc.) constant *except* ω and k.
2. Plot all solutions for k as a function of ω, or ω as a function of k, as desired.

Let us now investigate some further plots of dispersion relations for propagation at $\theta = 0$. Instead of using Eq. (7.36), however, let us use Eq. (7.37). We then note that, since $\tan^2 \theta = 0$ at $\theta = 0$, there can be three solutions giving that result

$$P = 0 \tag{7.48}$$

$$\left(\frac{k}{k_0}\right)^2 = R \tag{7.49}$$

and

$$\left(\frac{k}{k_0}\right)^2 = L \tag{7.50}$$

There are, therefore, three possible solutions (eigenmodes) to the dispersion relation at $\theta = 0$. Consider the first, Eq. (7.48). This is, in expanded form

$$P = 0 = 1 - \frac{\omega_{p_i}^2 + \omega_{p_e}^2}{\omega^2} \quad \text{or} \quad \omega^2 = \omega_{p_i}^2 + \omega_{p_e}^2 \tag{7.51}$$

Note that k does not appear. For fixed values of the plasma parameters ($\omega_{p_i}^2 + \omega_{p_e}^2 =$ constant) and ω versus k plot would be as shown in Fig. 7.9.

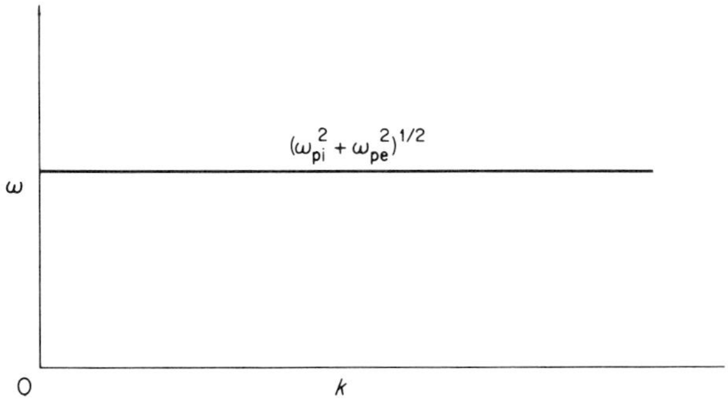

Fig. 7.9. ω versus k plot for the mode $P = 0$, which occurs at $\theta = 0°$.

Note that *all possible values* of k are permitted. This means that any wave number k and therefore any wavelength is possible for this mode. If $\omega_{p_i} \ll \omega_{p_e}$, then we find the characteristic oscillations at the electron plasma frequency. *They do not propagate.*

The second case $(k/k_0)^2 = R$ becomes

$$\left(\frac{k}{k_0}\right)^2 = 1 - \frac{\alpha\omega^2}{(\omega + \Omega_i)(\omega - \Omega_e)}$$

$$= \frac{\omega^2 - \omega\Omega_e + \omega\Omega_i - \Omega_e\Omega_i - \omega_{p_i}^2 - \omega_{p_e}^2}{(\omega + \Omega_i)(\omega - \Omega_e)} \tag{7.52}$$

or, after collecting terms,

$$\left(\frac{k}{k_0}\right)^2 = \frac{\omega^2 - \omega[\Omega_e - \Omega_i] - [\Omega_e\Omega_i + \omega_{p_i}^2 + \omega_{p_e}^2]}{(\omega + \Omega_i)(\omega - \Omega_e)} \tag{7.53}$$

The numerator of equation (7.53) may be factored to produce roots for ω. These roots are

$$\omega_{1,2} = \frac{(\Omega_e - \Omega_i) \pm \{(\Omega_e - \Omega_i)^2 + 4[\Omega_e\Omega_i + \omega_{p_i}^2 + \omega_{p_e}^2]\}^{1/2}}{2} \tag{7.54}$$

Note that ω_1 and ω_2 are both real if $\Omega_e > \Omega_i$, the usual case, but $\omega_1 < 0$ whereas $\omega_2 > 0$, so that Eq. (7.53) is

$$\left(\frac{k}{k_0}\right)^2 = \frac{(\omega + |\omega_1|)(\omega - \omega_2)}{(\omega + \Omega_i)(\omega - \Omega_e)} \tag{7.55}$$

If we solve this expression for k we obtain

$$k = \left[\frac{(\omega + |\omega_1|)(\omega - \omega_2)}{(\omega + \Omega_i)(\omega - \Omega_e)}\right]^{1/2} k_0$$

$$= \left[\frac{(\omega + |\omega_1|)(\omega - \omega_2)}{(\omega + \Omega_i)(\omega - \Omega_e)}\right]^{1/2} \omega(\varepsilon_0\mu_0)^{1/2} \tag{7.56}$$

We have again only considered the positive square root.

Figure 7.10 shows a plot of the real part of k as a function of ω, assuming $|\omega_1| < \Omega_e < \omega_2$, using the vertical axis for ω. Note that again, for $\omega \to 0$, the upper branch curve becomes asymptotic to the dispersion function for free space. However, there exists a resonance for the lower branch (the curve extends to ∞ parallel to the k axis) at $\omega = \Omega_e$ and a cutoff for the upper branch at $\omega = \omega_2$. There is also a gap in frequency between Ω_e and ω_2 where this mode does not propagate. This mode must, therefore, be the right circularly polarized wave that can resonate with the electron motion. Also, note that propagation is possible as $\omega \to 0$. This mode is sometimes called the "whistler" mode, due to its appearance as a variable low frequency signal when it was first observed in the ionosphere. Typically, whistlers in the ionosphere are observed in the band of frequencies between 0 and 10 kHz. It has been assumed that this mode is excited by lightning flashes in the atmosphere. The lightning discharge produces an electromagnetic pulse with a spectrum of frequencies. Since the phase velocity and group velocities for this mode are functions of frequency, as seen in Fig. 7.10, the electromagnetic spectral

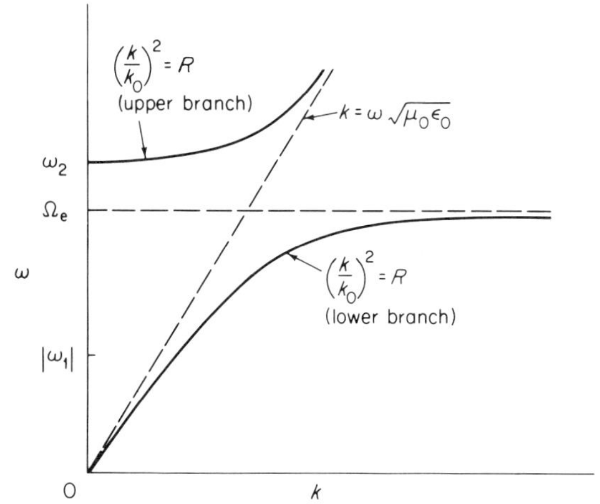

Fig. 7.10. Dispersion plot of the real part of k as a function of ω for the mode $(k/k_0)^2 = R$.

components of the pulse do not travel with the same speed. The result is that the pulse is distorted (dispersed) and may appear as a rising or falling signal. Note, too, that since the plasma parameters also change in the ionosphere, this may also cause changes in the phase and group velocities.

The third mode of propagation at $\theta = 0$, Eq. (7.50), becomes

$$\left(\frac{k}{k_0}\right)^2 = L = 1 - \frac{\alpha\omega^2}{(\omega - \Omega_i)(\omega + \Omega_e)} \tag{7.57}$$

or, in expanded form

$$\left(\frac{k}{k_0}\right)^2 = \frac{\omega^2 + \omega[\Omega_e - \Omega_i] - [\Omega_e\Omega_i + \omega_{pi}^2 + \omega_{pe}^2]}{(\omega - \Omega_i)(\omega + \Omega_e)} \tag{7.58}$$

We can solve the numerator for two values of ω:

$$\omega_{1,2} = \frac{-(\Omega_e - \Omega_i) \pm \{(\Omega_e - \Omega_i)^2 + 4[\Omega_e\Omega_i + \omega_{pi}^2 + \omega_{pe}^2]\}^{1/2}}{2} \tag{7.59}$$

These two frequencies have the *same magnitude but opposite signs* of ω_1 and ω_2, which were found in Eq. (7.54). So Eq. (7.58) becomes

$$\left(\frac{k}{k_0}\right)^2 = \frac{(\omega - \omega_1)(\omega + |\omega_2|)}{(\omega - \Omega_i)(\omega + \Omega_e)} \tag{7.60}$$

Solving for k, we get

$$k = \left[\frac{(\omega - \omega_1)(\omega + |\omega_2|)}{(\omega - \Omega_i)(\omega + \Omega_e)}\right]^{1/2} \omega(\mu_0 \varepsilon_0)^{1/2} \tag{7.61}$$

Plotting this relation in the (ω, k) plane results in the graph shown in Fig. 7.11. This looks slightly different from the dispersion plot shown in

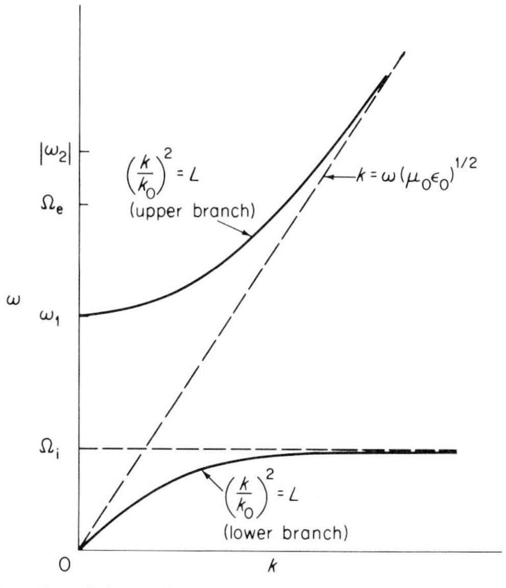

Fig. 7.11. Dispersion plot of the real part of k as a function of ω for the mode $(k/k_0)^2 = L$.

Fig. 7.10. There is a resonance at the *ion* cyclotron frequency, but a cutoff at $\omega = \omega_1$. Note again that propagation is possible as $\omega \to 0$. Also a gap between Ω_i and ω_1 appears. This mode might also be observed as a whistler in the ionosphere, but since it only exists below the *ion* cyclotron frequency, it usually appears at too low a frequency to be detected easily.

We now examine some cases where the angle θ is $\pi/2$. Again, Eq. (7.37) is the most convenient one to use. The two possible modes are

$$\left(\frac{k}{k_0}\right)^2 = P \tag{7.62}$$

and

$$\left(\frac{k}{k_0}\right)^2 = \frac{RL}{S} \tag{7.63}$$

These are the two possible solutions (modes) for all ranges of *plasma* parameters, with $\theta = \pi/2$. Equation (7.62) becomes

$$\left(\frac{k}{k_0}\right)^2 = 1 - \frac{\omega_{\mathrm{pi}}^2 + \omega_{\mathrm{pe}}^2}{\omega^2} \tag{7.64}$$

so that

$$k = \left[1 - \frac{\omega_{\mathrm{pi}}^2 + \omega_{\mathrm{pe}}^2}{\omega^2}\right]^{1/2} \omega(\mu_0 \varepsilon_0)^{1/2} \tag{7.65}$$

This is exactly the same relation as that depicted in Fig. 7.8, except that we have also included the effects of ions. It is *unaffected* by the magnetic field and is therefore labeled an ordinary wave. The intercept on the ω axis by the curve is at $\omega = (\omega_{\mathrm{pi}}^2 + \omega_{\mathrm{pe}}^2)^{1/2}$. This mode may conveniently be used for plasma diagnostics like the microwave interferometer.

Turning now to the other mode, we note that the solution of Eq. (7.63) can be written as

$$\left(\frac{k}{k_0}\right)^2 = \frac{RL}{S} = \frac{(\omega^2 - {\omega_1}^2)(\omega^2 - {\omega_2}^2)}{(\omega^2 - \omega_{LH}^2)(\omega^2 - \omega_{UH}^2)} \tag{7.66}$$

where ${\omega_1}^2$ and ${\omega_2}^2$ are the solutions to either Eq. (7.54) or (7.59), and

$$\omega_{UH}^2 = {\Omega_{\mathrm{e}}}^2 + \omega_{\mathrm{pe}}^2 \tag{7.67}$$

and

$$\frac{1}{\omega_{LH}^2} = \frac{1}{{\Omega_{\mathrm{i}}}^2 + \omega_{\mathrm{pi}}^2} + \frac{1}{\Omega_{\mathrm{i}}\Omega_{\mathrm{e}}} \tag{7.68}$$

ω_{UH} and ω_{LH} are called the upper and lower *hybrid resonance frequencies,* respectively. Note that several approximations are necessary to get (7.66) from (7.63). These are basically the neglect of terms involving the mass ratio of the electrons to ions $m_{\mathrm{e}}/m_{\mathrm{i}}$.

Equation (7.66), then, is the dispersion relation of an extraordinary wave, since here the phase velocity *does* depend upon the magnetic field. The solution for k becomes, after taking the square root of (7.66),

$$k = \omega(\mu_0 \varepsilon_0)^{1/2} \left[\frac{(\omega^2 - {\omega_1}^2)(\omega^2 - {\omega_2}^2)}{(\omega^2 - \omega_{LH}^2)(\omega^2 - \omega_{UH}^2)}\right]^{1/2} \tag{7.69}$$

A plot of this last equation in the (ω, k) plane is shown in Fig. 7.12. There are two cutoffs at $\omega = \omega_1$ and $\omega = \omega_2$ and two resonances at $\omega = \omega_{LH}$ and $\omega = \omega_{UH}$, respectively. This relation has three branches, as seen by the three separate curves in Fig. 7.12. Note it is still just *one* mode, but there are three branches.

If we hold all plasma parameters constant and increase the frequency, we will move along a path across Fig. 7.5, roughly along the dashed line marked

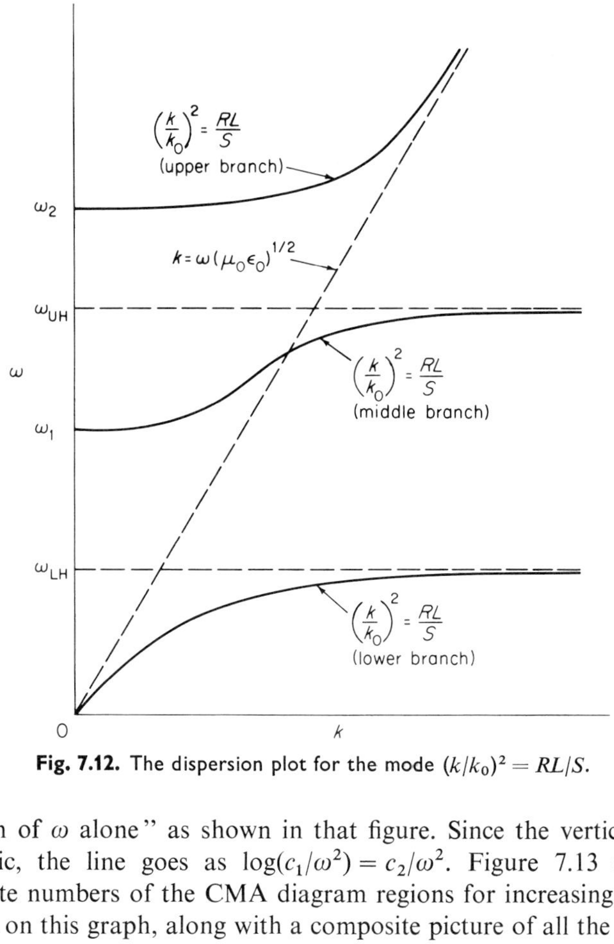

Fig. 7.12. The dispersion plot for the mode $(k/k_0)^2 = RL/S$.

"variation of ω alone" as shown in that figure. Since the vertical scale is logarithmic, the line goes as $\log(c_1/\omega^2) = c_2/\omega^2$. Figure 7.13 shows the appropriate numbers of the CMA diagram regions for increasing frequency as plotted on this graph, along with a composite picture of all the dispersion plots for $\theta = 0$ and $\theta = \pi/2$.

Figure 7.13 shows the many branches of the curves. However, by examining the appropriate regions in the CMA diagram, it can be seen that, since these graphs pertain to propagation only at $\theta = 0$ and $\theta = \pi/2$, the curves may be collected in at most two separate modes. For example, the lowest branches

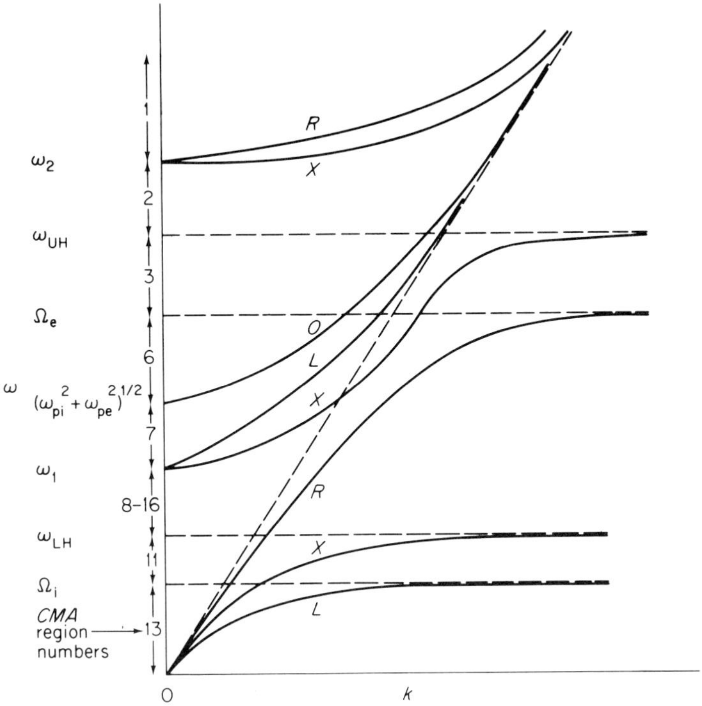

Fig. 7.13. Composite dispersion plot for the modes for $\theta = 0°$ and $90°$.

of the R and L modes of Fig. 7.13 are distinct, but the lowest X mode corresponds to the R mode at $\theta = \pi/2$.

Many other graphs of various forms of the cold plasma dispersion relation may be utilized. Sometimes the horizontal and vertical axes of the previous figures (ω and k) are interchanged. In this case, the resonances move up, toward ∞, the cutoffs go down to the horizontal axis, and the group velocity is the reciprocal of the slope of the curve.

F. NONZERO TEMPERATURE EFFECTS

We now briefly turn to investigating some waves that are formed in hot plasmas. Temperature effects may be considered in one of three ways:

1. We may include the Boltzmann equation and its associated velocity distribution as part of the system of equations to arrive at a dispersion relation.

2. We may take the moments of the Boltzmann equation, and use an appropriate "temperature" in the resultant momentum, energy, and so on, equations to be utilized in obtaining the dispersion relation.

3. We may include a momentum transfer collision frequency term in the expressions for conductivity and permittivity.

We shall consider some examples involving the latter two cases here and reserve the first case for a subsequent chapter. Let us look at the collision frequency terms first (third case). The development in Chapter 3, which led to Eq. (3.47), is applicable. The dispersion relations are of the same form except that m is now complex, that is, $m = m_0[1 + j(\nu_m/\omega)]$ when it appears in the dielectric or permittivity tensors.

The forms of permittivity tensor and the dispersion relation are exactly the same as before. Equations (7.36) and (7.37), however, are now functions of $\omega^2[1 + j(\nu_m/\omega)]$ rather than ω^2 alone. This has the effect of "blurring" the boundary regions of the CMA diagram and making cutoffs and resonances not quite so sharp.

The dispersion relation in Eq. (7.45) will now be developed as an example with the inclusion of ν_m. First, ε is now

$$\varepsilon = \varepsilon_0\left\{1 - \frac{\omega_{\text{pe}}^2}{\omega^2[1 + j(\nu_m/\omega)]}\right\} \tag{7.70}$$

Then, the dispersion relation is changed to

$$k^2 = \omega^2\mu_0\,\varepsilon_0\left\{1 - \frac{\omega_{\text{pe}}^2}{\omega^2[1 + j(\nu_m/\omega)]}\right\}$$

or, solving for k

$$k = \omega(\mu_0\,\varepsilon_0)^{1/2}\left\{1 - \frac{\omega_{\text{pe}}^2}{\omega^2[1 + j(\nu_m/\omega)]}\right\}^{1/2} \tag{7.71}$$

We again only desire solutions for k real.

Figure 7.14 shows the result for the real part of k for several values of ν_m. Note that now propagation is possible at frequencies below ω_{pe}. There is no sharp transition between propagating and nonpropagating regions as before. Since k is complex, a plot of the imaginary part of k will show the amount of spatial damping exhibited by this relation. Similar results may be obtained for other formulations of the dispersion relations.

At this point, however, we must be a bit more rigorous regarding the choice of independent variable for the dispersion plots. Basically, we have been choosing ω to be the independent variable, and allowing k to take on whatever values are determined for it by the dispersion relation appropriate to the problem.

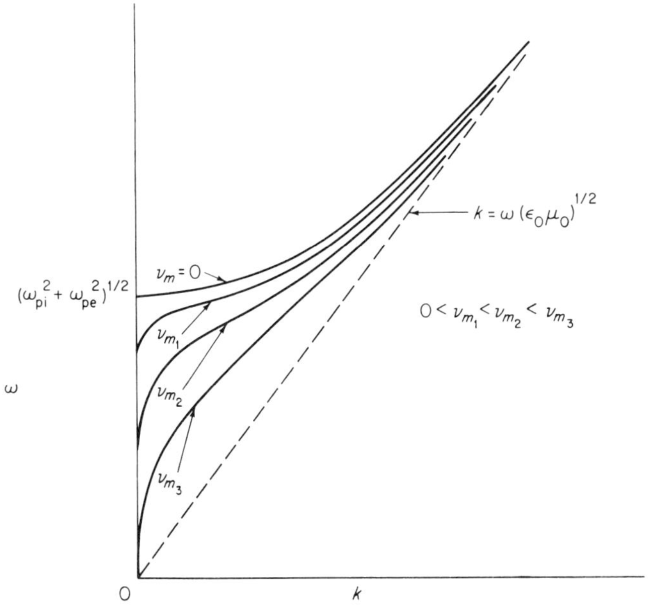

Fig. 7.14. Dispersion plot for a cold plasma with no magnetic field, considering the effects of collisions.

We may, in fact, produce essentially the same results, as long as k and ω are either purely real or purely imaginary, by allowing k to be the independent variable and ω to be the dependent variable.

However, if complex values appear, then, for example, the results should be specified for ω independent and always real, or k independent and always real. The former case may result in complex values for k, while the latter can produce complex ω.

The appearance of complex k values introduces spatial damping (or possibly spatial growth) and complex ω values produce temporal growth or decay of the wave. However, the mere existence of these complex terms does not guarantee the growth or damping predicted, and often additional conditions have to be met before a definite conclusion can be drawn. More will be said on this subject in the next chapter.

We now turn to the second method of introducing temperature effects. This method will appear in the taking of the moments of the Boltzmann equation. In so doing, we often get the result that a pressure gradient term appears in the momentum equation (6.25). The following is an example of this method.

We shall assume that the pressure can be represented by

$$p = \gamma_e n_e k T_e + \gamma_i n_i k T_i \tag{7.72}$$

where γ_e and γ_i are the ratio of the electron and ion specific heats, k is Boltzmann's constant, and T_e and T_i are the electron and ion "temperatures," respectively.

We now consider an "acoustic"-type perturbation disturbance, that is, one having only low frequency longitudinal oscillations and no magnetic field or gravitational effects. Equation (6.29) then becomes

$$\rho_m \frac{\partial \mathbf{v}}{\partial t} = \rho_E \mathbf{E} - \nabla_r p \tag{7.73}$$

The definitions of the quantities in Eq. (7.73) are

$$\rho_m = n_e m_e + n_i m_i \tag{7.74}$$

$$\mathbf{v} = \frac{n_e m_e \mathbf{v}_e + n_i m_i \mathbf{v}_i}{\rho_m} \tag{7.75}$$

and

$$\rho_E = n_e q_e + n_i q_i \tag{7.76}$$

We shall assume the plasma under consideration to have approximately equal numbers of singly charged ions and electrons. Therefore,

$$n_i \approx n_e \tag{7.77}$$

Also, $m_i \gg m_e$, so

$$\rho_m \approx n_i m_i \tag{7.78}$$

Now let us assume $T_e \gg T_i$, so that

$$p \approx \gamma_e n_e k T_e \tag{7.79}$$

If the electron temperature is constant throughout the plasma, then taking the gradient of Eq. (7.79), we have

$$\nabla_r p = \gamma_e k T_e \nabla_r n_e \tag{7.80}$$

for the pressure gradient.

With the use of Eq. (7.77) and (7.78), Eq. (7.75) is now

$$\mathbf{v} = \frac{n_e m_e}{n_i m_i} \mathbf{v}_e + \mathbf{v}_i \approx \mathbf{v}_i \tag{7.81}$$

Therefore, Eq. (7.73) may now be written as

$$n_i m_i \frac{\partial \mathbf{v}_i}{\partial t} = \rho_E \mathbf{E} - \gamma_e k T_e \nabla_r n_e \tag{7.82}$$

The equation of continuity for ions is

$$\frac{\partial(n_i m_i)}{\partial t} + \nabla_r \cdot (n_i m_i \mathbf{v}_i) = 0 \tag{7.83}$$

Taking the divergence of Eq. (7.82), we find

$$\nabla_r \cdot \left(n_i m_i \frac{\partial \mathbf{v}_i}{\partial t}\right) = \rho_E \nabla_r \cdot \mathbf{E} - \gamma_e k T_e \nabla_r^2 n_e \tag{7.84}$$

If we take the time derivative of (7.83) we get

$$\frac{\partial^2(n_i m_i)}{\partial t^2} + \frac{\partial}{\partial t}(\nabla_r \cdot n_i m_i \mathbf{v}_i) = 0 \tag{7.85}$$

If the time and spatial derivatives of (7.85) are exchanged and then the time derivative is expanded as a product, we get

$$\frac{\partial^2(n_i m_i)}{\partial t^2} + \nabla_r \cdot \left(\mathbf{v}_i \frac{\partial(n_i m_i)}{\partial t}\right) + \nabla_r \cdot \left(n_i m_i \frac{\partial \mathbf{v}_i}{\partial t}\right) = 0 \tag{7.86}$$

We now expand the middle term of Eq. (7.86). It is

$$\nabla_r \cdot \left(\mathbf{v}_i \frac{\partial(n_i m_i)}{\partial t}\right) = \left[\frac{\partial(n_i m_i)}{\partial t}\right] \nabla_r \cdot \mathbf{v}_i + \nabla_r\left(\frac{\partial(n_i m_i)}{\partial t}\right) \cdot \mathbf{v}_i \tag{7.87}$$

We will assume that $\nabla_r \cdot \mathbf{v}_i = 0$ and that the second term of the right-hand side of Eq. (7.87) may be neglected because both time and space operators are operating on $n_i m_i$, which should make this term of higher order than the others in Eq. (7.86). We may therefore neglect the middle term of Eq. (7.86) entirely.

We then substitute Eq. (7.86) without this middle term into Eq. (7.84), giving

$$-\frac{\partial^2(n_i m_i)}{\partial t^2} = \rho_E \nabla_r \cdot \mathbf{E} - \gamma_e k T_e \nabla_r^2 n_e \tag{7.88}$$

Use of Poisson's equation to eliminate $\nabla_r \cdot \mathbf{E}$ results in (7.88) becoming

$$-\frac{\partial^2(n_i m_i)}{\partial t^2} = \frac{\rho_E^2}{\varepsilon_0} - \gamma_e k T_e \nabla_r^2 n_e \tag{7.89}$$

Let us divide through by m_i. Then (7.89) becomes

$$\frac{\partial^2 n_i}{\partial t^2} = -\frac{\rho_E^2}{m_i \varepsilon_0} + \frac{\gamma_e k T_e}{m_i} \nabla_r^2 n_e \tag{7.90}$$

But $n_e \approx n_i$, so we now get, for Eq. (7.90),

$$\frac{\partial^2 n_i}{\partial t^2} = -\frac{\rho_E{}^2}{m_i \varepsilon_0} + \frac{\gamma_e k T_e}{m_i} \nabla_r{}^2 n_i \tag{7.91}$$

Now if we assume that the charge density ρ_E is approximately zero, since the ion motion is slow enough so that electrons can move quickly enough to neutralize the ions, then Eq. (7.91) may be written as

$$\frac{\partial^2 n_i}{\partial t^2} = \gamma_e \frac{k T_e}{m_i} \nabla_r{}^2 n_i \tag{7.92}$$

Equation (7.92) is a wave equation. It is homogeneous and has a dispersion relation. If plane waves for n_i propagating in the z direction ($\exp[j(kz - \omega t)]$) are assumed, then (7.92) is reduced to

$$\left[-\omega^2 + \frac{\gamma_e k T_e}{m_i} k_z{}^2\right] n_i = 0 \tag{7.93}$$

The dispersion relation, which is inside the bracket of Eq. (7.93), may be solved for $k_z{}^2$ and k_z. The solutions are

$$k_z{}^2 = \frac{\omega^2}{\gamma_e k T_e / m_i} \qquad \text{and thus} \qquad k_z = \frac{\omega}{(\gamma_e k T_e / m_i)^{1/2}} \tag{7.94}$$

A plot of ω versus k_z for (7.94) is shown in Fig. 7.15. This wave is known as an *ion-acoustic* wave. Note that it does not appear unless $T_e \neq 0$. The cold

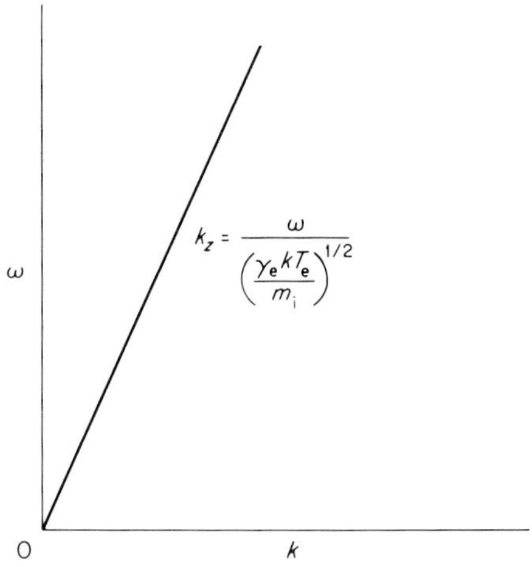

Fig. 7.15. ω versus k diagram for ion–acoustic waves.

plasma dispersion relation does not predict the existence of such a mode, and it is only because of the presence of a nonzero temperature that it does appear. Typically, these waves appear at frequencies of a few kilohertz to a few megahertz. Physically, these frequencies must be low enough so that the electrons can rearrange themselves continuously around the ions to permit the approximation of zero net electrical charge to be made.

It has been the goal of this chapter to acquaint the reader with some of the basic approaches to waves in plasmas. It should become apparent that the dispersion relation (in whatever form) can give much insight into observed plasma phenomena. Once a dispersion relation has been written, it is usually possible to produce a CMA diagram or ω versus k plots of the eigenfrequencies. From this it can be determined what ranges of k and ω are possible for various parameters (e.g., magnetic field, density, temperature). Sometimes, the structure of the electric and magnetic fields may also be obtained conveniently.

It should again be emphasized that the existence of a suitable eigenmode does not guarantee its presence in a real plasma. Careful consideration must be given to the means by which the plasma is excited.

For example, suppose that the region of parameter space where the desired mode is to be excited lies in the middle of a plasma with a variable density in a spatially variable dc magnetic field. Other modes may have to be excited and somehow couple to the desired modes across the appropriate boundaries in the CMA diagram. In addition, when crossing these boundaries, the wave may be totally absorbed or reflected at the boundary.

The eigenmodes are therefore very important as a means of identifying and producing wave phenomena in plasmas. Subsequently we shall consider the important effects that appear when the Boltzmann equation is used in considering wave behavior.

Suggested Reading

Allis, W. P., Buchsbaum, S. J., and Bers, A., "Waves in Anisotropic Plasmas," M.I.T. Press, Cambridge, Massachusetts, 1963. This book includes a good discussion of bounded waves, as well as much material covered in this chapter.

Gartenhaus, S., "Elements of Plasma Physics," Holt, New York, 1964.

Schmidt, G., "Physics of High Temperature Plasmas—An Introduction," Academic Press, New York, 1966.

Stix, T. H., "The Theory of Plasma Waves," McGraw-Hill, New York, 1963. This is an excellent treatment of waves in both hot and cold plasmas. Caution: it uses cgs units throughout.

Vandenplas, P. E., "Electron Waves and Resonances in Bounded Plasmas," Wiley (Interscience), New York, 1968. This is a broad collection of the subject matter.

Problems

1 Construct the wave-normal surfaces for region 7 in the CMA diagram. Determine the actual equation used to construct the surfaces. Make the surfaces of correct relative magnitude, if possible. Locate the point at which the surface is to be plotted at

$$\frac{\Omega_e^{\,2}}{\omega^2} = 2 \qquad \text{and} \qquad \alpha = 1.1$$

(a) Use $m_i/m_e = 4$.
(b) Use $m_i/m_e = 1836$ (approximately the real value).

2 Consider the boundaries between regions 3 and 4 of the CMA diagram. Plot the behavior of the two wave-normal surfaces in the direction 90° with respect to the magnetic field and then in the direction parallel to the magnetic field as the line $\alpha = 1$ is crossed in moving from region 3 to region 4. Label any resonances and cutoffs that may appear.

3 Locate the region in the CMA diagram where Alfvén waves appear. Write the dispersion relation for these waves, assuming that the magnetic field **B** is extremely large and the plasma density is also very large. Show that two modes appear. Identify them as slow, fast, ordinary, extraordinary, right and left circularly polarized, and so on. Show how the identification is made.

4 Develop a k versus ω plot assuming that k is always real for the modes described in Eqs. (7.44) and (7.49). Is there any difference in the diagrams for this method of solution? (Numerical evaluation of (7.49) may be helpful for this problem.)

5 Upon examination of the ω versus k plots, construct a plot of the *group* velocity versus ω. Do so for the modes in Figs. 7.10 and 7.11. Are there any examples where the group velocity is *negative*? What is the physical interpretation of such a result?

6 Find an eigenvalue (dispersion relation) for the different electromagnetic wave modes propagating between two infinite perfectly conducting parallel plates whose spacing is a. Plot some of these modes on an ω versus k diagram.

7 What happens to the modes in Problem 6 if a plasma fills the space between the plates whose dielectric constant is $\varepsilon/\varepsilon_0 = 1 - (\omega_{pe}^2/\omega^2)$ as ω_{pe}^2 goes from 0 to ∞?

8 Do Problem 6 for a rectangular cavity whose dimensions are a, b, c. Plot some of the modes as an ω versus k_z plot. Note that in this problem k is set by the dimensions of the cavity to be a set of values, rather than constants, so the ω versus k plot is a series of points.

9 Suppose that the cavity of Problem 8 is filled with a cold uniform plasma. What happens to the eigenfrequencies if it is assumed that the density increases from zero and that k is unaffected by the presence of the plasma? From these problems it can be seen that another possible way to measure plasma density is by use of cavities of this type. Explain how this could be done.

10 Find the dispersion relation for plane monochromatic electromagnetic waves in a plasma if the effects of ions are included, but the plasma is cold.

11 Find the wave equation in a medium in which:

(a) ε is a function of position;
(b) ε is a function of time;
(c) ε is a function of both position and time.

12 Show that, for a cold plasma

$$v_{\text{phase}}\, v_{\text{group}} = c^2$$

and that

$$v_{\text{phase}} > c \qquad v_{\text{group}} < c$$

Does this violate relativity?

13 Examine the mode $(k/k_0)^2 = P$. Determine the direction of the electric field vector, relative to the dc magnetic axis. Determine whether this mode is O, X, R, or L. How can a wave guide be oriented so as to launch this mode for a propagation experiment. Assume the guide is operated in its principal mode. What is the mode labeling if $\theta = \pi/2$? Find the region or regions in the CMA diagram in which this mode exists.

8 Waves (Stable and Unstable) in Hot Plasmas

In this chapter we examine in more detail the problem only briefly mentioned at the end of the last chapter, that is, the effects of finite temperatures on waves when the velocity distribution function is considered. The plasma will now influence the electric field in the wave equation through the collisionless Boltzmann equation and Poisson's equation. We will examine the behavior of electrostatic oscillations in a plasma having a velocity distribution and obtain a dispersion relation. An extremely interesting result will be found: in the *absence of collisions* energy may be transferred from a wave to a plasma.

The general method used to obtain a dispersion relation involving velocity distributions is developed, as well as a general set of criteria for distinguishing the various types of waves and instabilities in plasmas.

The fundamental example to be studied is that of electrostatic electron oscillations in a plasma. This means that the waves in this example will be longitudinal. We will assume that the *collisionless* Boltzmann equation for electrons applies. This equation is restated here.

$$\frac{\partial f}{\partial t} + \mathbf{v} \cdot \nabla_r f + \frac{\mathbf{F}_e}{m_e} \cdot \nabla_v f = 0 \tag{8.1}$$

$\mathbf{F}_e$ will be assumed to be only $q_e \mathbf{E}$. We will assume that the ions remain fixed in position.

We now reintroduce a perturbation condition on f, which is

$$f = f_0 + f_1 \tag{8.2}$$

If we let f_0 be an equilibrium distribution that is a function only of velocity, then f_1 can be a small perturbation that can vary with space and time as well as velocity. In equilibrium, f_1 will be zero. The electric field **E** will also be zero in equilibrium, since it is assumed to be produced by the perturbations of f only, and vice versa. Therefore, both f_1 and **E** are considered to be perturbed quantities of the same order, and **v** will point in the direction of the fluctuations.

A. TRANSFORM THEORY

Now, in general, **E** and f_1 are functions of position and time. We assume that the position and time dependence of both of these functions may be represented as

$$\exp(j(kz - \omega t))$$

We realize that assuming this variation really means that, unless we have only a plane monochromatic wave, the desired function has some kind of spectral representation; for example, Fourier and/or Laplace transforms are possible.

We recapitulate the definitions of these transforms. The usual methods of introduction show the following.

Fourier Transform

$$F(k) = \int_{-\infty}^{+\infty} f(z)e^{-jkz}\, dz \tag{8.3}$$

Inverse Fourier Transform

$$f(z) = \frac{1}{2\pi}\int_{-\infty}^{+\infty} F(k)e^{jkz}\, dz \tag{8.4}$$

Laplace Transform

$$F(s) = \int_{0}^{\infty} f(t)e^{-st}\, dt \tag{8.5}$$

where $s = \sigma + j\beta$.

Inverse Laplace Transform

$$f(t) = \frac{1}{2\pi j}\int_{\sigma_1 - j\infty}^{\sigma_1 + j\infty} F(s)e^{st}\, ds \tag{8.6}$$

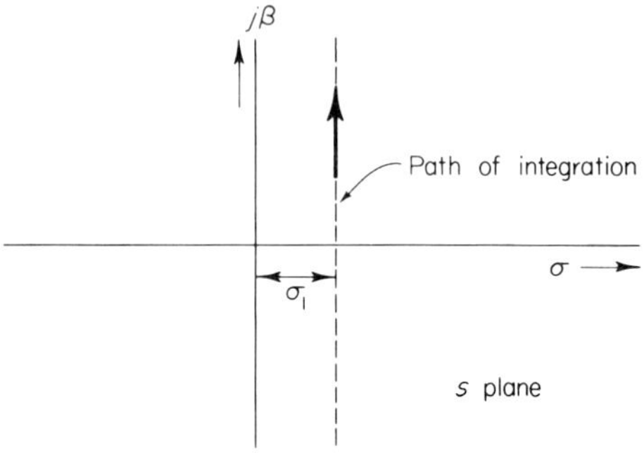

Fig. 8.1. Integration path in the s plane for the inverse Laplace transform.

Figure 8.1 shows the path of integration required for the integral (8.6) in the complex s plane. The path of integration must be to the right of σ_1; that is, it must be far enough to the right so that all the singularities (poles) of $F(s)$ are to the left of the contour.

There are restrictions on the nature of the functions utilized in the Fourier and Laplace transforms. For the Fourier transform, the restriction is

$$\int_{-\infty}^{+\infty} |f(t)|\, dt < \infty$$

[$f(t)$ must be absolutely integrable.] However, if $f(t)$ is not absolutely integrable, the Fourier transform *may* exist as a limit if the following integral exists.

$$\int_{-\infty}^{+\infty} |f(t)|\, dt = \lim_{T\to\infty} \int_{-T}^{+T} |f(t)|\, dt < \infty \tag{8.7}$$

For the Laplace transform, the restriction that ensures existence of the transform is

$$\int_{0}^{+\infty} |f(t)|e^{-\sigma_1 t}\, dt < \infty \tag{8.8}$$

The natures of the two transforms are very similar. Basically, the restriction for the Laplace transform defined in Eq. (8.8) permits us to obtain the Laplace transform of functions that may not have a Fourier transform. Important cases of this are those functions $f(t)$ which do not have a well-defined value at $t = \infty$, such as $\sin t$. Remember that σ is the real part of the complex frequency s. The transform is defined, therefore, for all values of $\sigma > \sigma_1$. σ_1 *may* or may not be zero, depending on the nature of the function. In addition, the Laplace transforms used here begin at $t = 0$, so that initial conditions in time may be used.

We shall now make a change of variable in the Laplace transform. Specifically, let

$$s = \sigma + j\beta = -j\omega = -j(\omega_r + j\omega_{\mathrm{i}})$$
$$s = \omega_{\mathrm{i}} - j\omega_r \tag{8.9}$$

so

$$\sigma = \omega_{\mathrm{i}} \qquad \beta = -\omega_r \qquad \text{and} \qquad ds = -j\,d\omega$$

The defining integrals for the modified Laplace transform and its inverse now become the following.

Modified Laplace Transform

$$F(\omega) = \int_0^\infty f(t)e^{j\omega t}\,dt \tag{8.10}$$

Modified Inverse Laplace Transform

$$f(t) = \frac{1}{2\pi}\int_{-\infty + j\omega_{\mathrm{i}_1}}^{+\infty + j\omega_{\mathrm{i}_1}} F(\omega)e^{-j\omega t}\,d\omega \tag{8.11}$$

where $\omega_{\mathrm{i}_1} = \sigma_1$. The restriction on the function $f(t)$ for the existence of the transform is now

$$\int_0^\infty |f(t)|e^{-\omega_{\mathrm{i}_1}t}\,dt < \infty \tag{8.12}$$

The contour of integration for the inverse modified transform is shown in Fig. 8.2. The path of integration must be above ω_{i_1}.

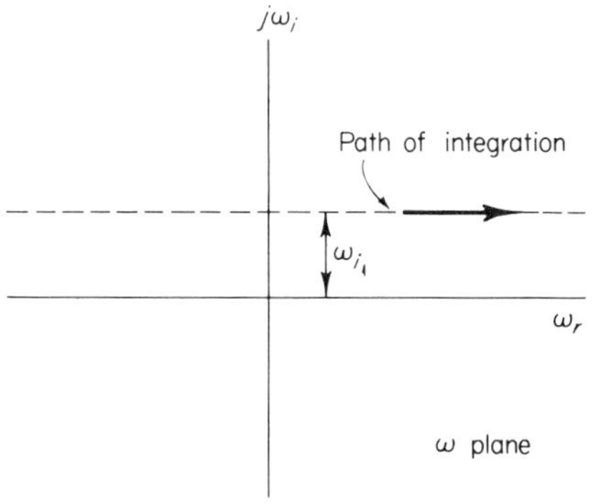

Fig. 8.2. Integration path in the ω plane for the inverse modified Laplace transform.

When Fourier or Laplace transforms are used to solve differential equations, the entire equation is first transformed and then solved algebraically for the dependent variable. The inverse transform is taken to obtain the solution in terms of the real variables.

It should be noted here that, in general, the transformed solution for the dependent variable becomes a quotient of polynomials. The polynomials may be factored, and the resultant infinities (poles or singularities) and zeros of the solution will appear in the complex frequency plane.

The inverse Laplace transformation path must be above all the poles of the quotient of polynomials in the solution for the dependent variable. A similar development results for the poles and zeros in the solution if the Fourier transform is used. In this case, the inverse integration path, however, lies directly *on* the real k axis. This implies that some poles *may* be on either side of the integration path. More will be said about this later. In addition, the Fourier transform is taken from $-\infty < z < +\infty$ so initial conditions at $z = 0$ do not appear.

We now will solve an ordinary *RLC* circuit by means of the modified Laplace transform, so that the methods just presented can be utilized in a familiar problem.

The fundamental circuit equation becomes, after differentiating,

$$L\frac{d^2 i}{dt^2} + R\frac{di}{dt} + \frac{i}{C} = Y(t) \tag{8.13}$$

where $Y(t)$ is the time derivative of a source function.

We are looking for the solution to Eq. (8.13). If we now take the modified Laplace transform, so that

$$I(\omega) = \int_0^\infty i(t)e^{j\omega t}\, dt \tag{8.14}$$

then we may also find the modified Laplace transform for the derivative di/dt of the current to be

$$I'(\omega) = \int_0^\infty \frac{di}{dt} e^{j\omega t}\, dt$$

Integrating by parts, we get

$$I'(\omega) = ie^{j\omega t}\Big]_0^\infty - j\omega \int_0^\infty ie^{j\omega t}\, dt$$

Therefore, the modified Laplace transform of the first derivative is

$$I'(\omega) = -i(0) - j\omega I(\omega) \tag{8.15}$$

We have assumed that $ie^{j\omega t} \to 0$ as $t \to \infty$ and $i(0)$ is the value of the current at $t = 0$. The transform of the second derivative $I''(\omega)$, can also be defined as shown below:

$$I''(\omega) = \int_0^\infty \frac{d^2 i}{dt^2} e^{j\omega t}\, dt = \frac{di}{dt} e^{j\omega t}\Big]_0^\infty - j\omega \int_0^\infty \frac{di}{dt} e^{j\omega t}\, dt$$

$$= -i'(0) - j\omega I'(\omega) = -i'(0) - j\omega[-i(0) - j\omega I(\omega)]$$

So,

$$I''(\omega) = -i'(0) + j\omega i(0) - \omega^2 I(\omega) \tag{8.16}$$

$i'(0)$ is the value of di/dt at $t = 0$. Putting these definitions into Eq. (8.13), we get the complete transformed equation

$$\frac{I(\omega)}{C} + R[-i(0) - j\omega I(\omega)] + L[-i'(0) + j\omega i(0) - \omega^2 I(\omega)] = Y(\omega) \tag{8.17}$$

$Y(\omega)$ is the modified Laplace transform of the source function. Collecting terms in Eq. (8.17) results in

$$\left[\omega^2 L + j\omega R - \frac{1}{C}\right] I(\omega) = -Li'(0) - Ri(0) + j\omega Li(0) - Y(\omega) \tag{8.18}$$

We may solve (8.18) for $I(\omega)$, and get

$$I(\omega) = \frac{-Li'(0) - Ri(0) + j\omega Li(0) - Y(\omega)}{[\omega^2 L + j\omega R - (1/C)]} \tag{8.19}$$

Factoring the denominator yields

$$I(\omega) = \frac{-Li'(0) + i(0)[j\omega L - R] - Y(\omega)}{(\omega - \omega_1))(\omega - \omega_2)} \tag{8.20}$$

where ω_1 and ω_2 are defined as

$$\omega_{1,2} = -j\frac{R}{2L} \pm \left[-\left(\frac{R}{2L}\right)^2 + \frac{1}{LC}\right]^{1/2} \tag{8.21}$$

We now plot these last two frequencies in the complex ω plane. If $1/LC > (R/2L)^2$, the roots are complex and are shown in Fig. 8.3. Otherwise they lie on the imaginary ω axis.

In order to obtain the solution for $i(t)$ we must take the inverse Laplace transform of Eq. (8.20). As long as the inverse integration path is above ω_1 and ω_2, and any poles of the source function $Y(\omega)$, the integral will give the correct result. We note that the initial conditions on i and its derivative with respect to time must be given, that is, $i(0)$ and $i'(0)$ must

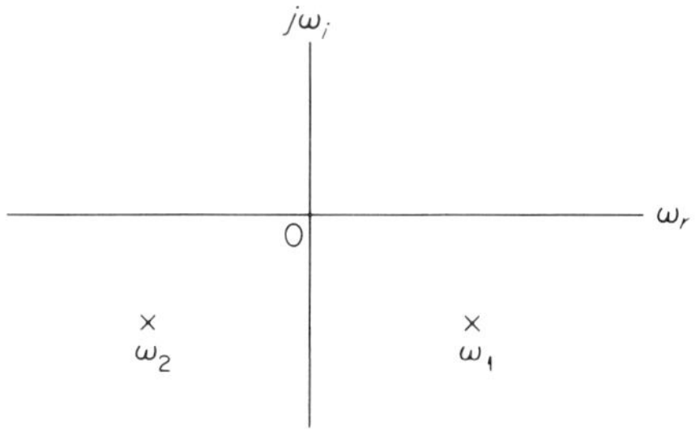

Fig. 8.3. Eigenvalues of the series *RLC* circuit.

be specified for the integration to be completed. These latter quantities are not transforms, but the *actual* values of i and i' at $t = 0$.

If we desire a "dispersion relation," then $i(0)$, $i'(0)$, and $Y(t)$ are set to zero. We have used quotation marks around the words dispersion relation because in this case, there is really no dispersion, since we only have one variable, ω. When we consider partial differential equations, then actual dispersion can appear. Equation (8.18), for the "dispersion relation," is then simply

$$\left[\omega^2 L + j\omega R - \frac{1}{C}\right] I(\omega) = 0 \tag{8.22}$$

The desired "dispersion relation" is the bracketed quantity in Eq. (8.22) set equal to zero. Note that this quantity also appears in the denominator of the solution for $I(\omega)$ with sources and initial conditions.

We may now extend this method to a partial differential equation.

B. APPLICATIONS OF TRANSFORM THEORY TO THE WAVE EQUATION AND BOLTZMANN EQUATION

The one-dimensional wave equation for plane monochromatic electromagnetic waves in free space will be the first partial differential equation considered. This equation is

$$\frac{\partial^2 E(z, t)}{\partial z^2} = \varepsilon\mu \frac{\partial^2 E(z, t)}{\partial t^2} \tag{8.23}$$

We have dispensed with the vector notation, since this is a one-dimensional problem. We now take the *Fourier* transform in *space* and the *Laplace* transform in *time* of the dependent variable, $E(z, t)$.

The spatial Fourier transform of $E(z, t)$ becomes

$$E(k, t) = \int_{-\infty}^{+\infty} E(z, t)e^{-jkz}\, dz \tag{8.24}$$

We also define the Fourier transform of the derivative of E with respect to z to be

$$\begin{aligned} E_z(k, t) &= \int_{-\infty}^{+\infty} \frac{\partial E(z, t)}{\partial z} e^{-jkz}\, dz \\ &= E(z, t)e^{-jkz}\Big]_{-\infty}^{+\infty} + jk \int_{-\infty}^{+\infty} E(z, t)e^{-jkz}\, dz \\ &= jkE(k, t) \end{aligned} \tag{8.25}$$

if $E(z, t) = 0$ at $z = \pm\infty$. Note that initial conditions (at $t = 0$) do not appear in (8.25). The Fourier transform of the second derivative of E with respect to z becomes

$$\begin{aligned} E_{zz}(k, t) &= \int_{-\infty}^{+\infty} \frac{\partial^2 E}{\partial z^2} e^{-jkz}\, dz = \left[\frac{\partial E}{\partial z} e^{-jkz}\right]_{-\infty}^{+\infty} + jk \int_{-\infty}^{+\infty} \frac{\partial E}{\partial z} e^{-jkz}\, dz \\ &= \left[\frac{\partial E}{\partial z} e^{-jkz}\right]_{-\infty}^{+\infty} + jkEe^{-jkz}\Big]_{-\infty}^{+\infty} - k^2 \int_{-\infty}^{+\infty} Ee^{-jkz}\, dz = -k^2 E(k, t) \end{aligned} \tag{8.26}$$

if $E(z, t) = \partial E(z, t)/\partial z = 0$ at $z = \pm\infty$. Note again that the time initial conditions (at $t = 0$) do *not* appear in the Fourier transform of any of the derivative terms. Only the Laplace transform will carry these initial conditions.

The modified Laplace transformation is now carried out on $E(z, t)$. That is, for example, the modified Laplace transform of the second derivative of E with respect to t will be

$$E_{tt}(z, \omega) = -E_t(z, 0) + j\omega E(z, 0) - \omega^2 E(z, \omega) \tag{8.27}$$

Note that the initial conditions (at $t = 0$) of E and $\partial E/\partial t$ must be known *for all z*. We may now *Fourier* transform Eq. (8.27), the result of which will be the Laplace transform in time and the Fourier transform in space of the second derivative of E with respect to time. This is

$$E_{tt}(k, \omega) = -E_t(k, 0) + j\omega E(k, 0) - \omega^2 E(k, \omega) \tag{8.28}$$

We have been required to take the Fourier transform of the initial conditions in Eq. (8.28). We may now *Laplace* transform Eq. (8.26) to obtain the Laplace transform in time and the Fourier transform in space of the second derivative of E with respect to z, namely,

$$E_{zz}(k, \omega) = -k^2 E(k, \omega) \tag{8.29}$$

The complete transformed wave equation can be obtained by using Eqs. (8.28) and (8.29) together. The result is

$$-k^2E(k, \omega) = -\varepsilon\mu\omega^2E(k, \omega) - \varepsilon\mu E_t(k, 0) + j\omega\varepsilon\mu E(k, 0) \tag{8.30}$$

We can solve Eq. (8.30) for $E(k, \omega)$ since Eq. (8.30) is algebraic. We obtain

$$E(k, \omega) = \frac{-\varepsilon\mu E_t(k, 0) + j\omega\varepsilon\mu E(k, 0)}{(\omega^2\varepsilon\mu - k^2)} \tag{8.31}$$

We can, if desired, take both the *inverse* Fourier and Laplace transforms of (8.31) to get $E(z, t)$ again. If the inverse Laplace transform is taken first, then the denominator of (8.31) may be factored and the path for the inverse Laplace transform can then be placed above all the poles of the function $E(k, \omega)$ in the ω plane.

If the dispersion relation alone is sought, the denominator of Eq. (8.31), when set equal to zero, gives the desired relation.

This development has been *identical* to that leading to Eq. (8.22), except that we have used two transform variables, ω and k.

We can now extend this process to the collisionless Boltzmann equation and find a dispersion relation from it. We will Fourier transform the perturbed collisionless Boltzmann equation in space and Laplace transform it in time. As before, both perturbed quantities $f_1(v, z, t)$ and $E(z, t)$, which are used in the Boltzmann equation, must be transformed. Recall that f_1 is the perturbation in the distribution function, which results in the electric field $E(z, t)$ and fluctuations $f_1(v_z, z, t)$ (longitudinal waves). The transforms for E and f_1 are

$$E(\omega, k) = \int_0^{\infty}\int_{-\infty}^{+\infty} E(z, t)\exp[-j(kz - \omega t)]\,dz\,dt \tag{8.32}$$

$$F_1(v_z, \omega, k) = \int_0^{\infty}\int_{-\infty}^{+\infty} f_1(v_z, z, t)\exp[-j(kz - \omega t)]\,dz\,dt \tag{8.33}$$

Remember that v points only in the z direction in f_1, since the fluctuations only move in this direction. In most cases, the order of taking the transforms does not matter. The inverse transformations become

$$E(z, t) = \frac{1}{2\pi}\int_{-\infty+j\sigma}^{+\infty+j\sigma}\int_{-\infty}^{+\infty} E(\omega, k)\exp[j(kz - \omega t)]\,dk\,d\omega \tag{8.34}$$

and

$$f_1(v_z, z, t) = \frac{1}{2\pi}\int_{-\infty+j\sigma}^{+\infty+j\sigma}\int_{-\infty}^{+\infty} F_1(v_z, \omega, k)\exp[j(kz - \omega t)]\,dk\,d\omega \tag{8.35}$$

Note again that when taking the transform of the derivative terms, the only initial conditions needed are those in time for the Laplace transform, that is, $f_1(v_z, z, 0)$ and $E(z, 0)$.

C. THE ELECTROSTATIC AND LONG-WAVELENGTH APPROXIMATIONS

At this juncture, we will discuss two approximations to the wave equation that are often made. They are (1) the electrostatic (slow-wave) and (2) the long-wavelength approximations.

The *electrostatic* or slow-wave approximation implies that we consider only waves that have an electric field that is produced by charge separation. That is, the electric field must be in the same direction as the movement of the local fluctuations in plasma density. In addition, since the electric field is produced only by charges, the field values may be determined by use of the methods of electrostatics. This implies that **E** can be written as the negative gradient of a scalar potential, namely

$$\mathbf{E} = -\nabla_r \phi$$

This situation also affects Maxwell's equations. The curl of **E** will be identically zero, since we may say

$$\nabla_r \times \mathbf{E} = \nabla_r \times (-\nabla_r \phi) \equiv 0$$

This means that the wave magnetic field is small enough so that it may be neglected. Since we have assumed that waves can propagate as $\exp[j(\mathbf{k} \cdot \mathbf{r} - \omega t)]$ then

$$\nabla_r \times \mathbf{E} = j(\mathbf{k} \times \mathbf{E}) = 0$$

so that **E** must be parallel to **k**, the propagation vector. Hence, electrostatic waves must be longitudinal.

The conditions under which the electrostatic approximation is valid must be noted. First, electrostatic waves *cannot* exist in free space, since no charges exist there. Then, the phase velocity of these electrostatic waves must be low enough so that *electromagnetic* effects do not become significant. We may express this as an inequality, namely,

$$v_{\text{phase}} = \frac{\omega}{k} \ll c$$

Under these two conditions, the electrostatic model may be applicable.

It is possible to define the scalar and vector potentials making up the wave electric field in general as

$$\mathbf{E} = -\nabla_r \phi - \frac{\partial \mathbf{A}}{\partial t}$$

such that

$$\nabla_r \cdot \mathbf{A} \equiv 0 \qquad \text{(transverse gauge)}$$

Then, if **E** is separated into two parts, such that

$$\mathbf{E}_1 = -\nabla_r \phi \qquad \text{and} \qquad \mathbf{E}_2 = -\frac{\partial \mathbf{A}}{\partial t}$$

it is seen that $\nabla_r \cdot \mathbf{E}_2 = 0$ and $\nabla_r \times \mathbf{E}_1 = 0$. The result is that $\mathbf{E}_1$ is a purely electrostatic field and $\mathbf{E}_2$ is purely electromagnetic in origin. The relative magnitudes of $\mathbf{E}_1$ and $\mathbf{E}_2$ determined from the particular problem under consideration can show the conditions under which the electrostatic approximation is valid.

The *long-wavelength* approximation may be used whenever the wavelength in the assumed direction of propagation is very long compared to the distances over which a solution is considered. Under these conditions we may let the *wave number* associated with the propagation vector approach zero.

If this is true, then the equation

$$\nabla_r \times \mathbf{E} = -\frac{\partial \mathbf{B}}{\partial t}$$

becomes

$$\nabla_r \times \mathbf{E} = j(\mathbf{k} \times \mathbf{E}) \equiv 0$$

Since we have let **k** approach zero, then **E** may have a component in the direction of **k** (TM wave). It may also have components perpendicular to **k**. In addition, **E** is still derivable as the sum of a negative gradient of a scalar potential plus an arbitrary constant.

In both approximations the methods of electrostatics may be used to solve for the electric field. In both approximations the effects of the wave magnetic field can be neglected.

The essential difference between these approximations is that in the electrostatic approximation **k** is defined to be parallel to **E**, but frequency and wave number may be comparable to the scales of the particular problem, provided $\omega/k \ll c$. In the long-wavelength approximation, the propagation vector must be extremely small compared to the dimensions of interest and, as a result, **E** can have components both perpendicular and parallel to the direction of propagation.

It should be noted, as a word of caution, that many authors have referred to either of these approximations as *quasi-static* approximations, since they do appear to be somewhat similar, but as can be seen from the foregoing discussion, there are some differences between the approximations. The example under consideration in this chapter uses the electrostatic approximation.

D. THE DISPERSION RELATION FOR LONGITUDINAL ELECTROSTATIC WAVES: I

We now must obtain the transforms for each of the terms in the collisionless Boltzmann equation. The perturbed collisionless Boltzmann equation, for longitudinal electrostatic waves propagating in the z direction under the assumptions given above, becomes, as developed previously,

$$\frac{\partial f_1(v_z, z, t)}{\partial t} + v_z \frac{\partial f_1(v_z, z, t)}{\partial z} + \frac{q_e E(z, t)}{m_e} \frac{\partial f_0(v)}{\partial v_z} = 0 \tag{8.36}$$

We again do not need vector notation because of the one-dimensional nature of the problem. The first term in the collisionless Boltzmann equation (8.36) may be *Laplace* transformed to become

$$F_{1t}(\omega, z, v_z) = \int_0^\infty \frac{\partial f_1}{\partial t} e^{j\omega t}\, dt = -j\omega F_1(\omega, z, v_z) - f_1(0, z, v_z) \tag{8.37}$$

We can now *Fourier* transform Eq. (8.37) to obtain

$$F_{1t}(\omega, k, v_z) = -j\omega F_1(\omega, k, v_z) - g_1(k, v_z) \tag{8.38}$$

where g_1 is the Fourier transform of $f_1(0, z, v_z)$—the initial conditions on the perturbation of the velocity distribution. It is also possible to include initial conditions on **E** or the effects of a source function as well if required. The second term in Eq. (8.36) becomes, after taking both of the transforms

$$jkv_z F_1(\omega, k, v_z) = \int_0^\infty \int_{-\infty}^{+\infty} v_z \frac{\partial f_1}{\partial z} \exp[-j(kz - \omega t)]\, dz\, dt \tag{8.39}$$

The third term in Eq. (8.36) is, in transformed form,

$$\frac{q_e}{m_e}\frac{\partial f_0(v)}{\partial v_z} E(k, \omega) = \frac{q_e}{m_e}\frac{\partial f_0(v)}{\partial v_z} \int_0^\infty \int_{-\infty}^{+\infty} E(z, t) \exp[-j(kz - \omega t)]\, dz\, dt \tag{8.40}$$

We may now combine all of the four terms to obtain, from (8.36), the transformed collisionless Boltzmann equation as

$$-j\omega F_1(\omega, k, v_z) - g_1(k, v_z) + jkv_z F_1(\omega, k, v_z) + \frac{q_e}{m_e}\frac{\partial f_0(v)}{\partial v_z} E(k, \omega) = 0 \tag{8.41}$$

We may collect terms and solve for $F_1(\omega, k, v_z)$ algebraically to give

$$F_1(\omega, k, v_z) = \frac{(q_e/m_e)(\partial f_0(v)/\partial v_z)E(k, \omega) - g_1(k, v_z)}{j(\omega - kv_z)} \tag{8.42}$$

Although (8.42) is algebraic, it is still an equation in two unknowns. We must eliminate $F_1(\omega, k, v_z)$ to obtain the dispersion relation. We can do so if we utilize Poisson's equation,

$$\nabla_r \cdot \mathbf{E} = \frac{\rho_E}{\varepsilon_0} \tag{8.43}$$

We may define ρ_E, the net electric charge density, to be

$$\rho_E(z, t) = n q_e \int_{-\infty}^{+\infty} \int_{-\infty}^{+\infty} \int_{-\infty}^{+\infty} f_1(v_z, z, t)\, dv_x\, dv_y\, dv_z \tag{8.44}$$

assuming again that only the electrons that are perturbed produce a net localized charge. n is the density normalization constant. If the assumption of allowing only longitudinal (z-direction) variations in f_1 and ρ_E is applied to Poisson's equation (8.43), it becomes, after substituting in (8.44),

$$\frac{\partial E(z, t)}{\partial z} = n \frac{q_e}{\varepsilon_0} \int_{-\infty}^{+\infty} \int_{-\infty}^{+\infty} \int_{-\infty}^{+\infty} f_1(v_z, z, t)\, dv_x\, dv_y\, dv_z \tag{8.45}$$

If we now Fourier and Laplace transform this equation, we get the following.

$$jkE(k, \omega) = n \frac{q_e}{\varepsilon_0} \int_{-\infty}^{+\infty} \int_{-\infty}^{+\infty} \int_{-\infty}^{+\infty} F_1(v_z, k, \omega)\, dv_x\, dv_y\, dv_z \tag{8.46}$$

We may now integrate the solution for $F_1(v_z, k, \omega)$, which is Eq. (8.42), over velocity space. This becomes

$$\begin{aligned} n \int_{-\infty}^{+\infty} \int_{-\infty}^{+\infty} \int_{-\infty}^{+\infty} & F_1(\omega, k, v_z)\, dv_x\, dv_y\, dv_z \\ &= n \frac{q_e}{m_e} \frac{E(k, \omega)}{j} \int_{-\infty}^{+\infty} \int_{-\infty}^{+\infty} \int_{-\infty}^{+\infty} \frac{\partial f_0(v)}{\partial v_z} \frac{1}{(\omega - kv_z)}\, dv_x\, dv_y\, dv_z \\ &\quad + jn \int_{-\infty}^{+\infty} \int_{-\infty}^{+\infty} \int_{-\infty}^{+\infty} \frac{g_1(k, v_z)}{\omega - kv_z}\, dv_x\, dv_y\, dv_z \end{aligned} \tag{8.47}$$

If both sides of Eq. (8.47) are multiplied by q_e/ε_0, then the left-hand side of (8.47) is the right-hand side of (8.46) and $F_1(\omega, k, v_z)$ may be eliminated. The resultant equation now is in one unknown and is

$$\begin{aligned} jkE(k, \omega) &= \frac{n q_e^2}{m_e \varepsilon_0} \frac{E(k, \omega)}{j} \int_{-\infty}^{+\infty} \int_{-\infty}^{+\infty} \int_{-\infty}^{+\infty} \frac{\partial f_0(v)/\partial v_z}{\omega - kv_z}\, dv_x\, dv_y\, dv_z \\ &\quad + j \frac{n q_e}{\varepsilon_0} \int_{-\infty}^{+\infty} \int_{-\infty}^{+\infty} \int_{-\infty}^{+\infty} \frac{g_1(k, v_z)}{\omega - kv_z}\, dv_x\, dv_y\, dv_z \end{aligned} \tag{8.48}$$

We can collect terms and obtain

$$\left[1 + \frac{nq_e^2}{km_e \varepsilon_0} \int_{-\infty}^{+\infty} \int_{-\infty}^{+\infty} \int_{-\infty}^{+\infty} \frac{\partial f_0(v)/\partial v_z}{\omega - kv_z}\, dv_x\, dv_y\, dv_z\right] E(k, \omega)$$

$$= \frac{nq_e}{k\varepsilon_0} \int_{-\infty}^{+\infty} \int_{-\infty}^{+\infty} \int_{-\infty}^{+\infty} \frac{g_1(k, v_z)}{\omega - kv_z}\, dv_x\, dv_y\, dv_z \tag{8.49}$$

Now Eq. (8.49) is similar in form to Eqs. (8.30) and (8.31), and so we may solve it for $E(k, \omega)$. In doing so, we obtain

$$E(k, \omega) = \frac{\dfrac{nq_e}{k\varepsilon_0} \displaystyle\int_{-\infty}^{+\infty} \int_{-\infty}^{+\infty} \int_{-\infty}^{+\infty} \frac{g_1(k, v_z)}{\omega - kv_z}\, dv_x\, dv_y\, dv_z}{1 + \dfrac{\omega_{pe}^2}{k} \displaystyle\int_{-\infty}^{+\infty} \int_{-\infty}^{+\infty} \int_{-\infty}^{+\infty} \frac{\partial f_0(v)/\partial v_z}{\omega - kv_z}\, dv_x\, dv_y\, dv_z} \tag{8.50}$$

since

$$\omega_{pe}^2 = \frac{nq_e^2}{m_e \varepsilon_0}$$

If the dispersion relation alone is desired, the denominator of (8.50) is set equal to zero, just as in the case for Eq. (8.31). If $E(z, t)$ is desired, the inverse Fourier and Laplace transforms of (8.50) may be taken. The restriction on the path for the inverse Laplace transform is the same as in previous cases; it must be above all the singularities of the expression (8.50), which defines $E(k, \omega)$.

This process may be applied to many different types of linear partial differential and coupled equations. If the transforms of the equations can exist, then, in most cases, a dispersion relation can be obtained.

It should be noted that, although the waves fluctuate only in the z direction, it is necessary to integrate with respect to all three coordinates in velocity space. Normally, this results only in a multiplicative constant for each term that is integrated in this way.

We may also solve the dispersion relation for longitudinal electrostatic waves (the denominator of (8.50) set equal to zero) for k^2. In doing so, we get

$$k^2 = \omega_{pe}^2 \int_{-\infty}^{+\infty} \int_{-\infty}^{+\infty} \int_{-\infty}^{+\infty} \frac{\partial f_0(v)/\partial v_z}{[(\omega/k) - v_z]}\, dv_x\, dv_y\, dv_z = \omega_{pe}^2 \sqrt{\pi}\, Z\left(\frac{\omega}{k}\right)$$

If $f_0(v)$ is an isotropic Maxwellian, then the function $Z(\omega/k)$ is the Fried and Conte plasma dispersion function, the values of which have been tabulated for complex values of ω/k. $Z(\omega/k)$ is then

$$Z\left(\frac{\omega}{k}\right) = \frac{1}{\sqrt{\pi}} \int_{-\infty}^{+\infty} \int_{-\infty}^{+\infty} \int_{-\infty}^{+\infty} \frac{\partial f_0(v)/\partial v_z}{[(\omega/k) - v_z]}\, dv_x\, dv_y\, dv_z$$

or, using the normalized form of the Maxwellian,

$$Z(\eta) = \frac{1}{\sqrt{\pi}} \int_{-\infty}^{+\infty} \frac{e^{-x^2}}{x - \eta}\, dx$$

ω versus k plots may be made utilizing the methods of Chapter 7.

The inverse integration necessary to obtain $E(z, t)$ must now be examined in greater detail in order to learn how to find the solutions and what conclusions may be drawn from them. Equation (8.50) is of a form that requires complex variable contour integrations, in both the k and ω planes. Before this takes place, however, we must make sure that $E(k, \omega)$ is defined correctly to perform the integration. Then we shall investigate the inverse integration in the ω plane. After this an important digression into the properties of the dispersion relation will be made. Finally a concluding discussion of both inverse integrations and the properties of the solution will be presented.

We may write the inverse Laplace integration of E in terms of ω as

$$E(k, t) = \frac{1}{2\pi} \int_{-\infty + j\omega_i}^{+\infty + j\omega_i} E(k, \omega) e^{-j\omega t}\, d\omega \tag{8.51}$$

In general, $E(k, \omega)$ has poles and zeros in both the k and ω planes. In order for the integral to be valid, ω_i must be *above* all of the singularities (poles) of $E(k, \omega)$ in the ω plane. This is shown in Fig. 8.4. We temporarily neglect branch lines in this analysis.

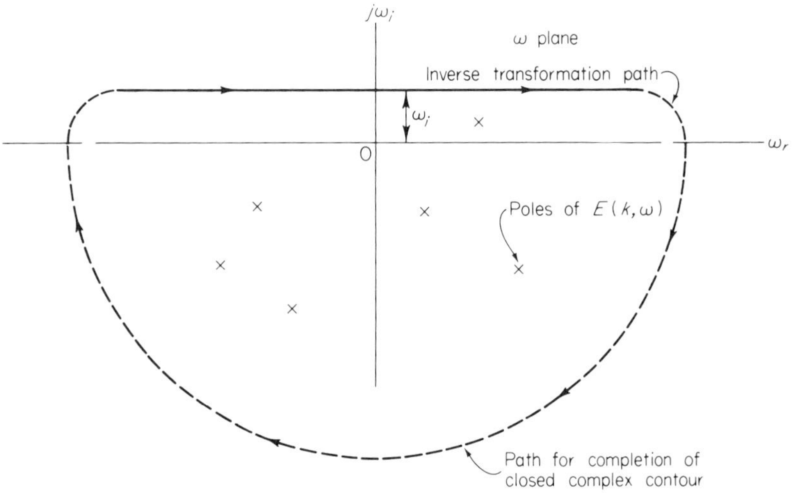

Fig. 8.4. Integration path for the inverse Laplace transformation of $E(k, \omega)$.

E. COMPLEX CONTOUR INTEGRATION AND ANALYTIC CONTINUATION

The integral in Eq. (8.51) can be a portion of a closed *contour* integral. This means that we may use a fundamental theorem of complex variables to help evaluate it. This theorem is the *Cauchy integral theorem*. It is expressible as an integral and has the following three different values, depending on conditions.

Case I

$$2\pi j f(a) = \oint_c \frac{f(z)}{z - a}\, dz \qquad \text{if } a \text{ is inside the contour} \tag{8.52}$$

Case II

$$0 = \frac{1}{2\pi j}\oint_c \frac{f(z)}{z - a}\, dz \qquad \text{if } a \text{ is outside the contour} \tag{8.53}$$

Case III

$$\pi j f(a) = \mathrm{P}\oint_c \frac{f(z)}{z - a}\, dz \qquad \text{if } a \text{ is } on \text{ the contour} \tag{8.54}$$

The symbol P will be defined subsequently. $\oint_c$ means a closed contour integral in the complex plane.

This theorem provides a way of evaluating contour integrals. A brief explanation of the theorem is developed as follows. Let z be a complex variable and a be a fixed complex number. The positive direction of several arbitrary closed contour integrals is shown in Fig. 8.5. The path is always counterclockwise in the complex plane as shown.

Any *closed* contour may be used, regardless of its shape. Now $f(z)$ is a function of the complex variable z. It must have no poles (singularities) inside or on the contour. If this is so, we say that $f(z)$ is *analytic* inside and on the contour. If $f(z)$ happens to be analytic everywhere in the complex plane, it is called an *entire* function of z. If a, the fixed complex number, is inside the contour, then Case I applies and the integral has the value $2\pi j f(a)$. $f(a)$ is the value of $f(z)$ at point a. If a is outside the contour, Case II applies and the integral has the value 0. Figure 8.6 shows these situations.

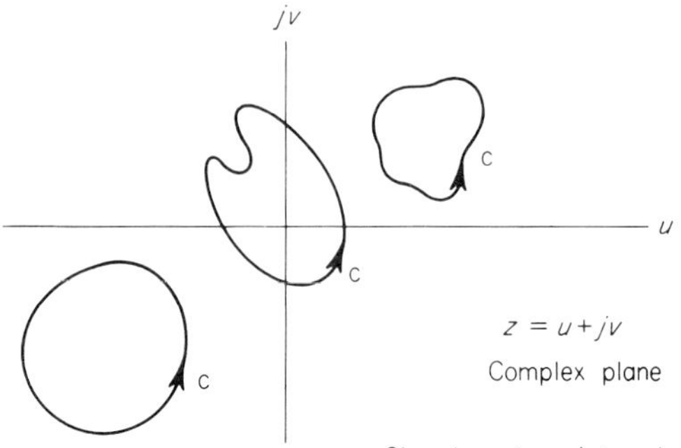

Fig. 8.5. Closed complex contour integrals. The "positive" sense is in the counterclockwise direction as shown.

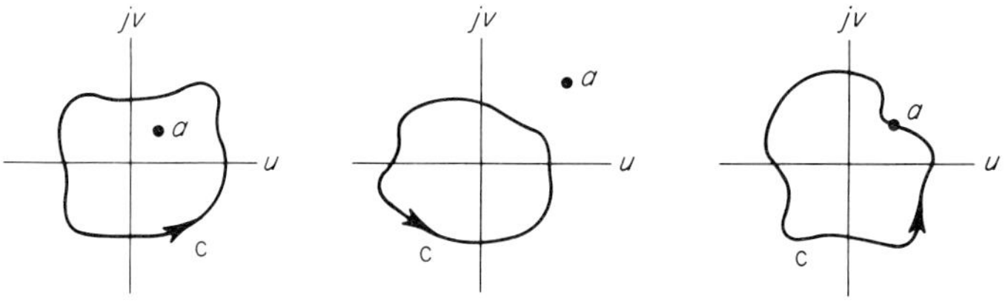

Fig. 8.6. Three possible locations of the point a with respect to a closed contour integral.

If a lies *on* the contour (Case III), a limiting process must be used, since the integral will "blow up" as it crosses the singularity. We will define the *Cauchy principal value* of the integral to be

$$\mathrm{P}\oint_{c} \frac{f(z)}{z-a}\,dz = \lim_{\delta\to 0}\left[\not\oint_{c} \frac{f(z)}{z-a}\,dz\right] \tag{8.55}$$

Note that the symbol P must appear in front of the integral to note that it is a principal value integral. The contour required for this integral is shown in Fig. 8.7 in an expanded view. δ is the radius of the path around the point a. $\not\oint_c$ is the original closed contour integral except for the small semicircular part of radius δ at the singularity. The theorem predicts that the principal value of the integral is $\pi j f(a)$. In summary, then, the three cases are as follows.

Case I. a inside c

$$2\pi j f(a) = \oint_{c} \frac{f(z)\,dz}{z-a} = \oint_{c} g(z)\,dz$$

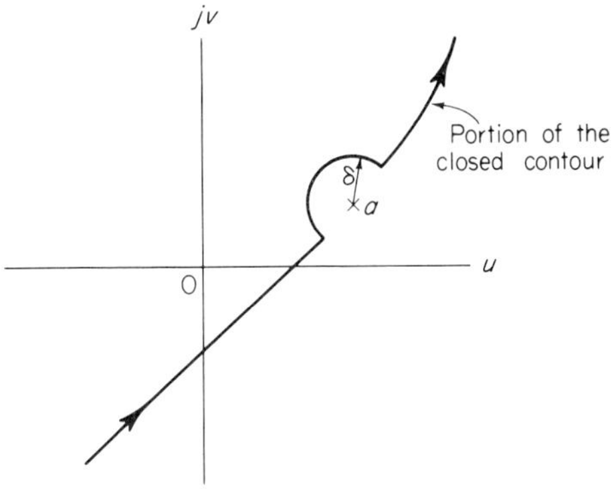

Fig. 8.7. A portion of the *closed* contour for a principal value integral.

Case II. a outside c

$$0 = \oint_c \frac{f(z)\,dz}{z-a} = \oint_c g(z)\,dz$$

Case III. a on c

$$\pi j f(a) = \mathrm{P}\oint_c \frac{f(z)\,dz}{z-a} = \mathrm{P}\oint_c g(z)\,dz$$

In the expressions above, we have defined a new function $g(z)$ such that

$$g(z) = \frac{f(z)}{z-a}$$

$g(z)$ has one singularity (pole) at $z = a$. We will define the *residue* of $g(z)$ at the pole $z = a$ to be

$$\operatorname{res} g(z)\Big|_{z=a} = \lim_{z\to a}[(z-a)g(z)] = f(a) \tag{8.56}$$

If $g(z)$ has more singularities inside the contour, the integral has the value, by successive applications of Cauchy's theorem, of $2\pi j$ times the sum of the residues at each singularity. That is,

$$\begin{aligned}\oint_c g(z)\,dz = 2\pi j \sum_{\text{res}} g(z) \qquad & \textit{inside} \text{ the contour} \\ + \pi j \sum_{\text{res}} g(z) \qquad & \textit{on} \text{ the contour}\end{aligned} \tag{8.57}$$

if the principal values are included. In many cases $g(z)$ has *simple* poles such that

$$g(z) = \frac{f(z)}{z - a} \tag{8.58}$$

but often we may find functions in which the poles are of higher order, for example,

$$g(z) = \frac{f(z)}{(z - a)^2} \quad \text{which is a pole of order 2, etc.} \tag{8.59}$$

In these cases, the residues for poles of order m can still be evaluated, by use of the expression

$$\text{res } g(z)\bigg|_{z=a} = \frac{1}{(m-1)!} \lim_{z \to a} \left[\frac{d^{m-1}}{dz^{m-1}} \frac{(z-a)^m f(z)}{(z-a)^m} \right] \tag{8.60}$$

In general form $g(z)$ may be expressed as:

$$g(z) = \frac{(z - z_1)^a (z - z_2)^b \cdots}{(z - z_n)^k (z - z_0)^h (z - z_p)^i \cdots} \tag{8.61}$$

if there are many different poles of various orders. Note that the factors of the numerator do not produce poles.

Each residue can be calculated as follows.

$$\text{res } g(z)\bigg|_{z=z_n} = \frac{1}{(k-1)!} \lim_{z \to z_n} \left[\frac{d^{k-1}}{dz^{k-1}} (z - z_n)^k g(z) \right] \tag{8.62}$$

which is the residue for the pole at $z = z_n$. We will apply the foregoing results to the contour integration of Eq. (8.50) using Cauchy's theorem to evaluate it. Remember that, as long as the same poles are enclosed, the shape of the contour *makes no difference*. The integral has the same value, provided the integrand is still analytic and defined on the new contour.

The inverse Laplace transform path of integration is normally *above* all of the poles of the function $E(k, \omega)$. However, it is not a *closed* contour. To use contour integration, we must construct a closed contour. We will do this by taking a large arc joining the ends of the Laplace contour and enclosing all the poles of the function $E(k, \omega)$. If the arc is pushed out toward infinity, it usually does not contribute to the integral† and the value of the closed contour integration is the same as the integral along the inverse Laplace path alone. If we wish to deform the contour so that we still enclose all of the poles, but allow *parts* of the contour to be below some or all of the poles of $E(k, \omega)$, we may use the deformed contour shown in Fig. 8.8.

† See *Jordan's lemma* in most books on complex variable theory.

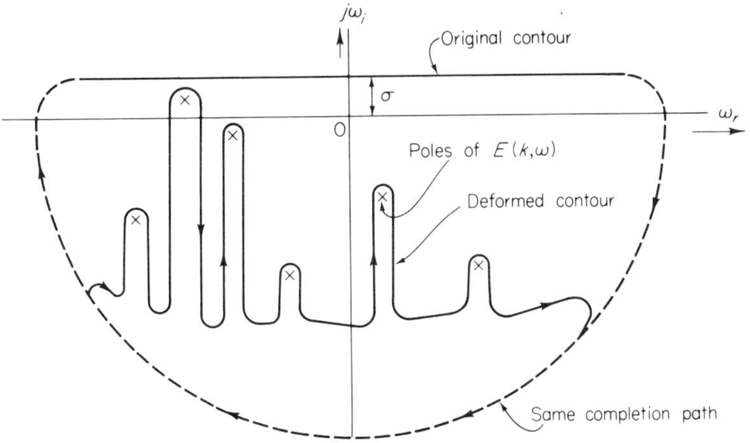

Fig. 8.8. The original and deformed inverse Laplace contours.

The two contour integrals around the original and the deformed contours *will have the same value*, since we still have a closed contour going in the same direction enclosing the same number of poles. However, we must know $E(k, \omega)$ for $\omega_i < \sigma$. This function has not as yet been defined in this region because the Laplace transform only exists when $\omega_i > \sigma$. To define the $E(k, \omega)$ in this region, we will make use of a method called *analytic continuation.*

This method is developed as follows. Let one complex function be analytic in a region of the complex plane, and another function be analytic in another region, which could possibly even be the same function. If the two functions can be shown to be equal in some overlapping portion of the two regions, or along some line common to both regions, then the functions are said to be the analytic continuations of each other.

For example, let us show that two functions

$$f_1(z) = \cosh^2(z) \qquad \text{and} \qquad f_2(z) = 1 + \sinh^2(z)$$

are analytic continuations of each other and, in fact, are equal everywhere in the complex plane, where $z = x + jy$. First, we note that if z is purely real, then $f_1(x) = \cosh^2(x)$ and $f_2(x) = 1 + \sinh^2(x)$. These functions are equal along the real axis. Since neither $f_1(z)$ or $f_2(z)$ have any singularities in the z plane, then, by analytic continuation

$$f_1(z) = f_2(z)$$

for all z, and $f_1(z)$ is the analytic continuation of $f_2(z)$ and vice versa.

Let us take another example. The function $f(z) = 1/(1 - z)$ has two expansions, depending on whether $|z| \geq 1$ or $|z| \leq 1$. They are, for $|z| \leq 1$,

$$f(z) = \frac{1}{1 - z} = 1 + z + z^2 + z^3 + \cdots \tag{8.63}$$

and for $|z| \geq 1$,

$$f(z) = \frac{1}{1 - z} = -\frac{1}{z}\frac{1}{\left(1 - \frac{1}{z}\right)} = -\frac{1}{z} - \frac{1}{z^2} - \frac{1}{z^3} - \frac{1}{z^4} - \cdots \tag{8.64}$$

Both expansions are analytic everywhere in their respective regions except at the point where the value of z is purely real and equal to 1, where they each have a pole. Also, expansion (8.63) is not valid in the region where $|z| > 1$, and expansion (8.64) is not valid in the region where $|z| < 1$.

However, both functions have the same values along the circle where $|z| = 1$, except at the point $z = 1$ where the singularities appear. Therefore (8.63) is the analytic continuation of (8.64) and vice versa, since the functions are equal along a common line (the circle).

F. THE DISPERSION RELATION FOR LONGITUDINAL ELECTROSTATIC WAVES: II

We now return to Eq. (8.50) and examine its behavior while deforming the contour as shown in Fig. 8.8. The integral is the sum of the residues of the integrand of Eq. (8.51), determined at the singularities of $E(k, \omega)$. However, in order to utilize the deformed contour, we must analytically continue $E(k, \omega)$ in the portion of the complex plane where it has not been defined, since the contour now passes in this region. Note that the contour for completion is assumed to be so far away from all the poles that it has a negligible contribution to the integral.

The reason for deforming the contour should now be stated. What we are interested in is the response of the system as time gets large, that is, the steady state response, if any. The integrand of Eq. (8.51) is $E(k, \omega)e^{-j\omega t}$ and must always be evaluated on the contour, deformed or not. However, for values of ω far below the real axis, the imaginary part of ω is very negative and damps out the response as time progresses. The result is that as time increases, the last remaining response (asymptotic response) must be that due to those locations on the deformed contour with the smallest negative value of the imaginary part of ω (or for those growth responses where the imaginary part of ω is greater than zero). Therefore, the asymptotic time response, for stable systems, must be due to the singularity closest to the real axis (the highest singularity). We shall disregard poles above the axis for the present. The deformed contour integral may be evaluated for the asymptotic response by finding the residue for the highest singularity or singularities only. If several poles are close to the axis, the asymptotic response will be the sum of their individual responses.

We must, however, still be able to evaluate $E(k, \omega)$ at its singularities, which are below $\omega_i = \sigma$, in order to calculate the residues to complete the integration. Analytic continuation provides the answer. Since $E(k, \omega)$ is expressed as a quotient, we can analytically continue the numerator and denominator separately. The numerator is

$$\frac{n\mathscr{q}_e}{k\varepsilon_0}\int_{-\infty}^{+\infty}\int_{-\infty}^{+\infty}\int_{-\infty}^{+\infty}\frac{g_1(k, v_z)}{\omega - kv_z}\,dv_x\,dv_y\,dv_z \tag{8.65}$$

and is a function of ω and k. $g_1(k, v_z)$ is the Fourier transform of the initial disturbance in v_z and z. Also, remember that the numerator of Eq. (8.50) [which is given here as Eq. (8.65)] may include sources, as well. We will assume that there will be no values of v_z for which g_1 goes to ∞. Furthermore, for the mathematics involved, we will assume g_1 may be a function of a complex v_z. This is allowed by analytic continuation, since if there are no singularities of g_1 anywhere in the complex v_z plane, and since $g_1(v_{\text{real}}) = g_1(v_{\text{complex}})$ on the real v_z axis, then $g_1(v_{\text{complex}})$ is the analytic continuation of $g_1(v_{\text{real}})$. We now define an integral along the real v_z axis that is the same as (8.65). It is, except for a constant,

$$I\left(\frac{\omega}{k}\right) = \int_{-\infty}^{+\infty}\frac{g_1(k, v_z)\,dv_z}{v_z - (\omega/k)} \tag{8.66}$$

We will drop the triple integral notation for the next steps by assuming that the integrations with respect to v_x and v_y have been made, and result only in normalization constants. If $g_1(k, v_z)$ is everywhere analytic in the complex v_z plane, then the integral in (8.66) may be evaluated by means of contour integration. The only singularity of the integrand is found where $v_z = \omega/k$ and the residue of this integrand at this singularity is

$$g_1\left(k, \frac{\omega}{k}\right)$$

Figure 8.9 shows the required closed contour in the complex v_z plane when the imaginary part of $\omega > 0$, which is the normal case for the contour in the inverse Laplace transform, that is, when the inverse *Laplace* contour lies on its original undeformed path.

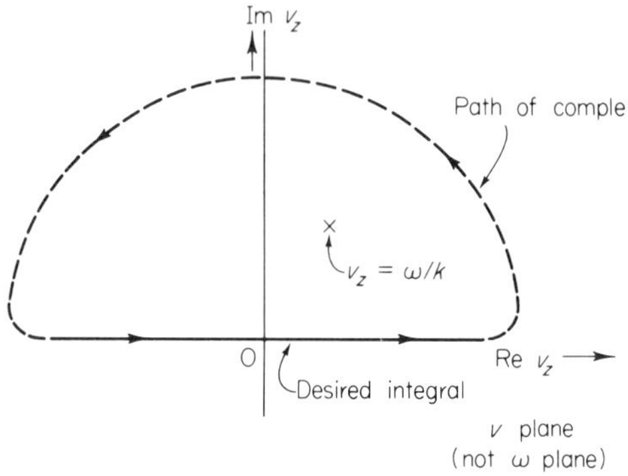

Fig. 8.9. Complex v_z plane contour. Note that this is not a Laplace transform contour, but is a complex integration used to help evaluate the integral in (8.66).

This contour integral has the value

$$\oint_c \frac{g_1(k, v_z)}{v_z - (\omega/k)}\, dv_z = 2\pi j g_1\left(k, \frac{\omega}{k}\right) \tag{8.67}$$

The desired linear integral (Eq. (8.66)) is then, in terms of the contour integral value,

$$I\left(\frac{\omega}{k}\right) = 2\pi j g_1\left(k, \frac{\omega}{k}\right) - \int_{\text{around the outside circle}} \frac{g_1(k, v_z)\, dv_z}{(v_z - \omega/k)} \tag{8.68}$$

This is the integral required to evaluate the numerator of $E(k, \omega)$ when the imaginary part of $\omega > 0$. This integral exists. The next question to be asked is whether this function is analytic *for all* ω. The answer is yes, because $g_1[k, (\omega/k)]$ was assumed to be analytic in ω, and the "integral around the outside circle" is the integral of a function that has no poles on this path, that is, the integrand is

$$\frac{g_1(k, v_z)}{v_z - (\omega/k)}$$

and is therefore analytic. The integrand is thus analytic on the contour in Fig. 8.9, since we have not intercepted the singularity at the point $v_z = \omega/k$.

The case where the imaginary part of $\omega < 0$ must now be investigated. Here we use analytic continuation. Figure 8.10 shows the situation in the complex v_z plane when the imaginary part of $\omega < 0$. In this case, *if we deform the contour* for the integral (8.66) around the pole which is now located in the lower part of the v_z plane, we may again write an equation of the same form as Eq. (8.68) in exactly the same way as before, where the contour for the integral I, instead of going straight along the real axis, now dips down to

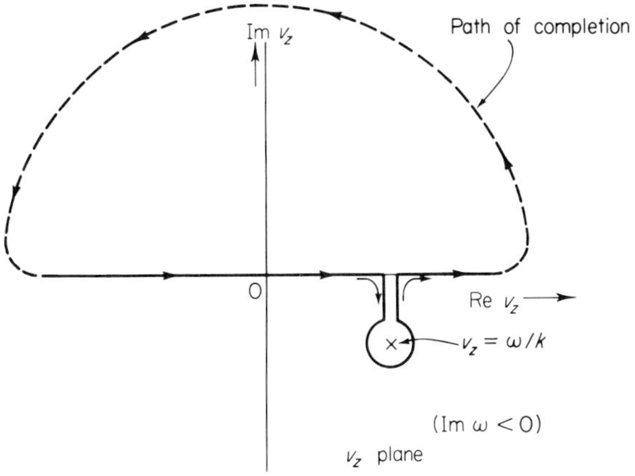

Fig. 8.10. Complex v_z plane showing the deformation of the integration path along the real v_z axis when Im $\omega < 0$.

encircle the pole. The result is that the left-hand side of Eq. (8.68) is *still analytic* in ω for the imaginary part of $\omega < 0$. The same arguments can be used to show this as before. It should be noted again, however, that the integral is no longer taken only along the real v_z axis, *but is on the deformed contour* shown in the figure. If we wish to *find the integral along the real axis*, we may do so, but in order to get it, we must now *subtract* the contribution from the residue at $v_z = \omega/k$. That is, in summary,

$$I\left(\frac{\omega}{k}\right)_{\text{along axis}} = 2\pi j g_1\left(\frac{\omega}{k}, k\right) - \int_{\text{outside circle}} \frac{g_1(k, v_z)\, dv_z}{v_z - (\omega/k)} \tag{8.69}$$

for Im $\omega > 0$. The integral along the deformed axis, when Im $\omega < 0$, is the same, namely,

$$I\left(\frac{\omega}{k}\right)_{\text{along deformed axis}} = 2\pi j g_1\left(\frac{\omega}{k}, k\right) - \int_{\text{outside circle}} \frac{g_1(k, v_z)\, dv_z}{v_z - (\omega/k)}$$

We can write an equation for the integral along the *undeformed* axis when Im $\omega < 0$ to be

$$\begin{aligned} I\left(\frac{\omega}{k}\right)_{\text{along axis}} &= -\int_{\text{outside circle}} \frac{g(k, v_z)\, dv_z}{v_z - (\omega/k)} \\ &= I\left(\frac{\omega}{k}\right)_{\text{along deformed axis}} - 2\pi j g_1\left(\frac{\omega}{k}, k\right) \end{aligned} \tag{8.70}$$

since the closed contour for this integral now has the value of zero when

$$\text{Im}\, \omega < 0,$$

if the integral $I(\omega/k)$ on the left-hand side of (8.69) and (8.70) is *always defined along the real* v_z *axis*. If we allow the contour to be deformed, then Eq. (8.69) applies, regardless of the sign of the imaginary part of ω. The analytic continuation of the function occurs because there is a region of overlap of the two functions, namely, the upper half ω plane. If Im $\omega > 0$, the contour in the v_z plane is not deformed, the singularity is above the real v_z axis, and both functions are identical.

Hence, we have now determined that the numerator of the integral for $E(k, \omega)$ is an analytic function for all values of the imaginary part of ω and, in fact, for all values of ω. This means that there are no singularities in the numerator which will cause a response to be present in the inverse Laplace transform integral.

The denominator of Eq. (8.50) must now be investigated. Its form is very similar to the numerator. We may use analytic continuation and a series of similar arguments to show that the denominator of the integrand is analytic in the entire ω plane. The only singularities in the inverse transformation integral, then, are due to the *zeros* of the denominator, since they make poles for the integrand.

The denominator may be written in two forms, as was the numerator, for Im $\omega > 0$ and Im $\omega < 0$. These are

$$1 + \frac{\omega_{\text{pe}}^2}{k} \int_{-\infty}^{+\infty} \frac{\partial f_0(v_z)}{\partial v_z} \frac{1}{\omega - kv_z}\, dv_z \qquad \text{for} \quad \text{Im}\, \omega > 0 \tag{8.71}$$

and

$$1 + \frac{\omega_{\text{pe}}^2}{k} \int_{-\infty}^{+\infty} \frac{\partial f_0(v_z)}{\partial v_z} \frac{1}{\omega - kv_z}\, dv_z - \frac{2\pi j \omega_{\text{pe}}^2}{k|k|} \frac{\partial f_0(\omega/k)}{\partial v_z} = 0 \qquad \text{for} \quad \text{Im}\, \omega < 0 \tag{8.72}$$

$f_0(\omega/k)$ is $f_0(v_z)|_{v_z=\omega/k}$. In both cases the integration with respect to v_z is made along the real v_z axis.

The denominator, when set equal to zero, is the dispersion relation and it is valid in either of its two forms for the appropriate values of ω. *Both integrals are evaluated along the real* v_z *axis*. If the path along the v_z axis is deformed so that it passes *through* the singularity at $v_z = \omega/k$, then a single dispersion relation may be written, which is valid for all ω. It can be written in terms of the *principal value* of the integral and its residue as

$$1 + \frac{\omega_{\text{pe}}^2}{k} \,\mathrm{P} \int_{-\infty}^{+\infty} \frac{\partial f_0(v_z)}{\partial v_z} \frac{1}{\omega - kv_z}\, dv_z - \frac{\pi j \omega_{\text{pe}}^2}{k|k|} \frac{\partial f_0(\omega/k)}{\partial v_z} = 0 \tag{8.73}$$

Note that the integral in (8.73), although it extends from $-\infty$ to $+\infty$, *must* intercept the singularity.

It should again be noted that, in deriving Eqs. (8.71), (8.72), and (8.73), poles of the function for $E(k, \omega)$ have *not* been assumed to lie above the real ω axis. Otherwise, instabilities and violent growth of the disturbances would occur and the linearized equations we have used might not be valid.

G. THE NATURE OF THE EIGENMODES—LANDAU DAMPING

We now further investigate the dispersion relations in (8.71) through (8.73) to get some idea of the nature of the modes that may exist, by expanding the integral with respect to v_z in powers of (kv_z/ω); that is, we change the integrands of these equations to the form

$$\frac{1}{\omega}\frac{\partial f_0(v_z)}{\partial v_z}\frac{1}{1-(kv_z/\omega)} = \frac{\partial f_0(v_z)}{\partial v_z}\frac{1}{\omega}\left(1+\frac{kv_z}{\omega}+\frac{k^2 v_z{}^2}{\omega^2}+\frac{k^3 v_z{}^3}{\omega^3}\cdots\right) \tag{8.74}$$

This truncated expansion is only valid if, for most of the particles, $kv_z/\omega \ll 1$. In other words, the phase velocity of the wave has to be much greater than the average thermal speed of the particles. When the series (8.74) is used for the integrands, we may integrate each term in the series by parts. We continue to remove the triple integral notation for simplicity. The first integrated term in the series of (8.74) is quite simple. It is

$$\frac{1}{\omega}\int_{-\infty}^{+\infty}\frac{\partial f_0(v_z)}{\partial v_z}\,dv_z = \left.\frac{f_0(v_z)}{\omega}\right]_{-\infty}^{+\infty} = 0 \tag{8.75}$$

The integral of the second term in the series gives the result

$$\frac{k}{\omega^2}\int_{-\infty}^{+\infty}\left(\frac{\partial f_0(v_z)}{\partial v_z}\right) v_z\,dv_z = \left.\frac{k}{\omega^2}f_0(v_z)v_z\right]_{-\infty}^{+\infty} - \frac{k}{\omega^2}\int_{-\infty}^{+\infty} f_0(v_z)\,dv_z = -\frac{k}{\omega^2} \tag{8.76}$$

The foregoing statements are true since $\int_{-\infty}^{+\infty} f_0(v_z)\,dv_z \equiv 1$ and we assume that $f_0(v_z) \to 0$ at $v_z = \pm\infty$ very rapidly (exponentially). The integrated third term is

$$\begin{aligned}\frac{k^2}{\omega^3}\int_{-\infty}^{+\infty}\frac{\partial f_0(v_z)}{\partial v_z}v_z{}^2\,dv_z &= \left.\frac{k^2}{\omega^3}f_0(v_z)v_z{}^2\right]_{-\infty}^{+\infty} - \frac{k^2}{\omega^3}2\int_{-\infty}^{+\infty} f_0(v_z)v_z\,dv_z \\ &= -\frac{2k^2}{\omega^3}\langle v_z\rangle\end{aligned} \tag{8.77}$$

which has the value zero if $f_0(v_z)$ is an even function of v_z. This is usually the case.

The integral of the fourth term becomes

$$\frac{k^3}{\omega^4}\int_{-\infty}^{+\infty}\frac{\partial f_0(v_z)}{\partial v_z}v_z^{\,3}\,dv_z = \frac{k^3}{\omega^4}f_0(v_z)v_z^{\,3}\Big]_{-\infty}^{+\infty} - \frac{3k^3}{\omega^4}\int_{-\infty}^{+\infty}f_0(v_z)v_z^{\,2}\,dv_z$$

$$= \frac{3k^3}{\omega^4}\langle v_z^{\,2}\rangle \tag{8.78}$$

where $\langle v_z^{\,2}\rangle$ is the mean of the square of the thermal speed. Utilizing all these terms in the dispersion relation, we now obtain, for the series of (8.74) after the integration,

$$1 + \frac{\omega_{\text{pe}}^2}{k}\left[0 - \frac{k}{\omega^2} + 0 - \frac{3k^3}{\omega^4}\langle v_z^{\,2}\rangle + \cdots\right] = 0 \quad \text{for} \quad \text{Im}\,\omega > 0 \tag{8.79}$$

and the corresponding series for $\text{Im}\,\omega < 0$:

$$1 + \frac{\omega_{\text{pe}}^2}{k}\left[0 - \frac{k}{\omega^2} + 0 - \frac{3k^3}{\omega^4}\langle v_z^{\,2}\rangle + \cdots\right] - \frac{2\pi j\omega_{\text{pe}}^2}{k\,|k|}\frac{\partial f_0(\omega/k)}{\partial v_z} = 0 \tag{8.80}$$

We may write the expanded terms for the dispersion relation in terms of the principal value integral using Eqs. (8.73) and (8.74). The result is

$$1 + \frac{\omega_{\text{pe}}^2}{k}\left[-\frac{k}{\omega^2} - \frac{3k^3}{\omega^4}\langle v_z^{\,2}\rangle + \cdots\right] - \frac{\pi j\omega_{\text{pe}}^2}{k\,|k|}\frac{\partial f_0(\omega/k)}{\partial v_z} = 0 \tag{8.81}$$

We may now solve these dispersion relations for ω as a function of k, and so on, as with any of the other previous algebraic dispersion relations. Note that the derivative of the velocity distribution at the phase velocity $v_z = \omega/k$ is still present in the dispersion relation for $\text{Im}\,\omega < 0$. It is now possible to examine this in more detail. For $k \to 0$ (infinitely long wavelength) first of all, the dispersion relations all reduce to

$$1 - \frac{\omega_{\text{pe}}^2}{\omega^2} = 0 \tag{8.82}$$

which is the dispersion relation for longitudinal electrostatic oscillations at the plasma frequency. This is true *regardless* of the form of the distribution function, because the term involving $\partial f_0(v_z)/\partial v_z|_{v_z=\omega/k}$ goes to zero, as $k \to 0$, faster than $k \to 0$ itself, so that the limit of this term is zero. Equation (8.82) must be, therefore, the "zero-order" dispersion relation for electrostatic waves. As a result, these plasma oscillations appear in the limit of infinitely long wavelength for the dispersion relations just developed. We will now compute a first-order correction to (8.82). To do this, we will neglect $\langle v_z^{\,2}\rangle$

and higher-order terms in the dispersion relations. Equation (8.81) may then be solved for ω as follows. First, we can obtain ω^2, namely,

$$\omega^2 = \frac{\omega_{p_e}^2}{1 - \dfrac{\pi j \omega_{p_e}^2}{k\,|k|}\dfrac{\partial f_0(\omega/k)}{\partial v_z}} \tag{8.83}$$

We may expand the denominator of (8.83) for small values of the term

$$\frac{\pi j \omega_{p_e}^2}{k\,|k|}\frac{\partial f_0(\omega/k)}{\partial v_z}$$

giving

$$\omega^2 \simeq \omega_{p_e}^2\left(1 + \frac{\pi j \omega_{p_e}^2}{k\,|k|}\frac{\partial f_0(\omega/k)}{\partial v_z}\right)$$

Using the first two terms of the binomial theorem, we may approximate the square root of the foregoing equation yielding

$$\omega \approx \omega_{p_e}\left(1 + \frac{\pi j \omega_{p_e}^2}{2k\,|k|}\frac{\partial f_0(\omega/k)}{\partial v_z}\right) \tag{8.84}$$

Note that ω *is complex*. This shows that a "damping" of the wave can occur. If terms of the order of $\langle v_z{}^2\rangle$ are also considered, a similar result is obtained. The damping process is named after Landau, who first derived it. Note that it is a *collisionless* process, that is, the velocity distribution itself may be used to show addition or subtraction of energy from the wave, depending upon the sign of the slope of the velocity distribution function at $v_z = \omega/k$. In this approximation, ω/k is very large, so that almost invariably the derivative of the distribution function at $v_z = \omega/k$ has a negative sign. This means that, in Eq. (8.84), the imaginary part of ω, although small, is negative. This imaginary term will result in damping of waves that have oscillatory parts as used in this development. This shows that energy has gone *from the wave to the plasma.*

The physical mechanism for such damping may be looked at in the following way. The wave is a disturbance that propagates through the plasma. As a result, traveling potential waves are set up. These potential waves tend to "trap" particles in them. Those particles whose velocities are exactly equal to the wave phase velocity are, of course, trapped automatically. Those particles whose velocities are slightly greater than the wave phase velocity must attempt to surmount a wave potential hill. If the particle does not have enough energy to go over the top of the hill, it will "roll back" down the hill, up the other side, oscillate between the two crests, and eventually be trapped. Since the particle had been moving faster than the wave, it has now been decelerated, on the average, and has therefore lost some of its energy to the wave. The wave energy is thus increased and grows in amplitude because of the gain in energy obtained from this particle. However, at the same time, those particles moving slightly *slower* than the wave phase velocity are speeded up, that is, they get some more energy from the wave, which results in a damping effect on the wave. In the region of velocity space where the distribution function has a slope that is negative there are slightly *more* particles moving at the speed that produces damping than at the speed that produces growth of the wave. The net result is that, when the slope of the distribution function is negative at $v_z = \omega/k$, damping occurs. When the slope is positive at this point, then growth of the wave amplitude occurs. Figure 8.11 shows particles being trapped by the wave to produce Landau damping.

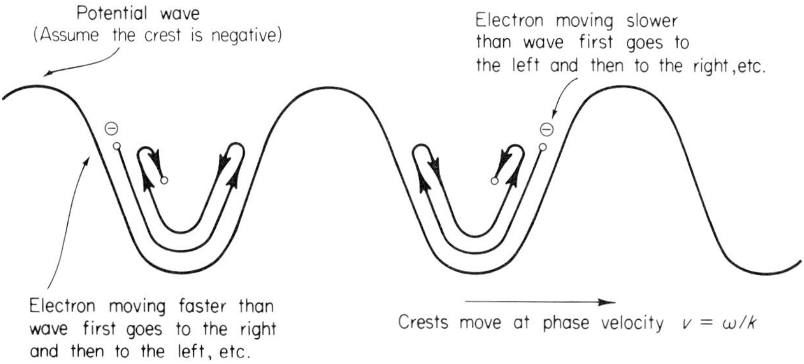

Fig. 8.11. Potential wave showing trapping of charged particles. The particle motion described is relative to the wave phase velocity ω/k.

H. WAVES AND INSTABILITIES—CRITERIA

This work on the dispersion relation has found the *normal modes* for the system under consideration. We have now completed a general discussion of the inverse Laplace integration of Eq. (8.50). To find the complete response, the inverse Fourier transforms of Eq. (8.50) must be taken. There are, then, two contour integrations for the complete solution. The first, the inverse Laplace transform, is performed in ω space, while the inverse Fourier integration is in k space. These two contours are shown in Fig. 8.12.

Equation (8.50), as initially written, has no sources; only an initial condition is present. It is possible, however, to include the effects of sources, in a similar manner to the *RLC* circuit methods developed earlier in this chapter. If a source function is considered, it will, in general, have a spatial and time

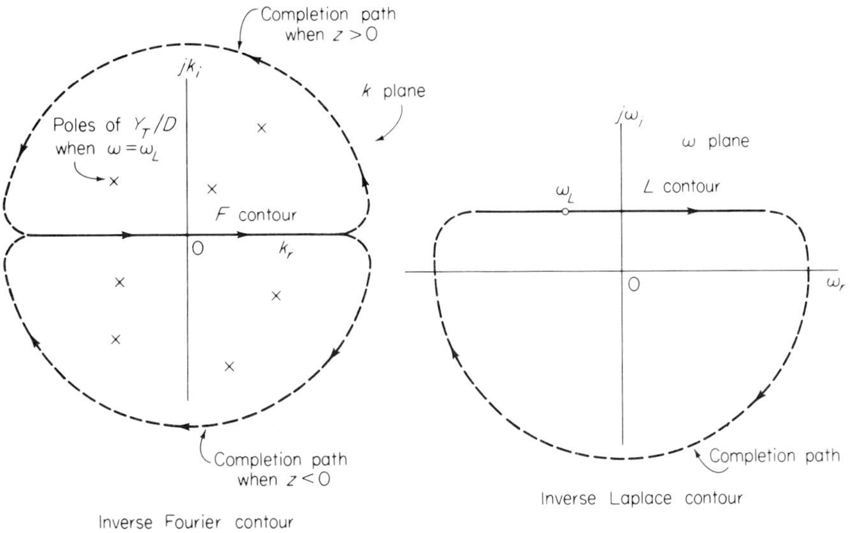

Fig. 8.12. Inverse Fourier and Laplace contours necessary for the solution of Eq. (8.50). Note the different path of completion in k space when $z < 0$.

dependence $Y(z, t)$, and will be permitted to have both a Fourier and Laplace transform, $Y(k, \omega)$.

To utilize source functions in Eq. (8.50), the transformed source function must be properly added in the numerator. We may then combine the source function and the initial condition function to produce a numerator function $Y_T(k, \omega)$. The denominator, the dispersion relation, we shall call $D(k, \omega)$. Equation (8.50) now becomes, simply,

$$E(k, \omega) = \frac{Y_T(k, \omega)}{D(k, \omega)} \tag{8.85}$$

The two required inverse integrations of Eq. (8.85), written together are,

$$\frac{1}{2\pi} \int_{\mathrm{L}\,-\infty + j\omega_{i1}}^{+\infty + j\omega_{i1}} \left[\frac{1}{2\pi} \int_{\mathrm{F}\,-\infty}^{+\infty} \frac{Y_T(k, \omega)}{D(k, \omega)} \exp(jkz)\, dk \right] \exp(-j\omega t)\, d\omega \tag{8.86}$$

The integrations must now be investigated. The symbol L stands for the inverse Laplace path and F stands for the inverse Fourier path. We may perform either integral first. We choose to do the inverse Fourier integration first. Then the Laplace contour must lie above any singularities of the *inverse Fourier integral* of the function $Y_T(k, \omega)/D(k, \omega)$ that are in the ω plane. Initially, note that k is real when ω is on the undeformed Laplace contour. The inverse Fourier integral can then be evaluated along the real k axis. That is, if we do *not* deform the inverse Laplace contour, then k is always real. For a given frequency ω_L that lies *on* the *Laplace contour*, the function Y_T/D may exhibit singularities in the k plane. The direction in which the inverse Fourier transform integral contour may be closed must now be decided. The *Laplace contour* is completed as shown in Fig. 8.12 as before.

Both inverse integrals, Fourier and Laplace, are assumed to be of the form that will vanish at very great distances from their respective origins of coordinates. We must close the inverse Fourier contour in the upper half k plane for values of $z > 0$, since otherwise the function will blow up for large values of z. Similarly, if we are at values of $z < 0$, we must close the contour in the lower half plane to avoid blowing up of the integral. This can be seen by remembering that the z variations go as $\exp(jkz)$. If $z > 0$, then the imaginary part of k must always be kept positive as $|k| \to \infty$ to keep the function from blowing up. The reverse is true if $z < 0$.

The inverse Fourier integral is then in one of two forms, depending upon whether $z > 0$ or $z < 0$. It may be evaluated by using Cauchy's theorem and summing over the residues of the function Y_T/D. The inverse Fourier integral has the value

$$F(\omega_L, z)_{z>0} = 2\pi j \sum_{\text{res}} \left[\frac{Y_T(\omega_L, k)}{D(\omega_L, k)} e^{jkz} \right]_{\text{upper half plane}, z>0} \tag{8.87}$$

or

$$F(\omega_L, z)_{z<0} = -2\pi j \sum_{\text{res}} \left[\frac{Y_T(\omega_L, k)}{D(\omega_L, k)} e^{jkz} \right]_{\text{lower half plane}, z<0} \tag{8.88}$$

The minus sign is necessary in the second equation, because in this case we are moving in a direction opposite to the positive sense of the contour for Cauchy's theorem when the inverse Fourier transform integration is made, due to the fact that the contour is closed in the lower half plane.

Since we have closed the contour by assuming that the solutions do not blow up in space, all of the functions $F(\omega_L, z)$ drop to zero for large magnitudes of z. Note again that this is always true for all values of ω_L.

The question to be answered here is one of the nature of the time-dependent solutions. Do they grow, decay, or propagate uniformly? What kinds of growth and decay are possible? To examine these questions we first present a few definitions and concepts regarding growth (instability) and decay (evanescence) of disturbances.

There are two kinds of instabilities to be considered: *absolute* and *convective*. Figure 8.13 shows a diagram of *three* disturbances. The *stationary disturbance* grows in time and its center never moves in space. It never decays at a fixed point in space. An *absolute* instability is a disturbance that grows in time and does not decay at any fixed point, but is allowed to move. This means that an observer at any fixed point will see the disturbance begin from

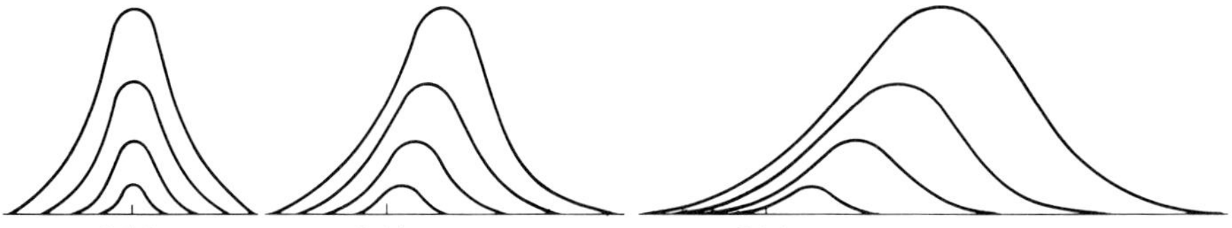

Fig. 8.13. From left to right: stationary, absolute, and convective disturbances, respectively.

zero and then increase without limit. The *convective* instability also grows and is permitted to move, but an observer sitting at one point will see a *growth and then an eventual decay* as the disturbance passes the observation point. It can be seen from Fig. 8.13 that the absolute instability can move as well, but not as fast as the convective instability, relative to the velocity of the peak of the disturbance. It should also be noted that the drawings in Fig. 8.13 may be transformed into one another. A moving disturbance may appear stationary if the observer of the disturbance moves with the "group velocity," that is, the velocity of the peak of the disturbance. The mechanism for changing from an absolute (the extreme case is the stationary disturbance) to the convective disturbance is that the relative velocity between the observer and the peak of the disturbance changes so that the disturbance either grows and never dies down (absolute) or grows and eventually decays to zero (convective) at a given point.

Bers and Briggs have developed a criterion for distinguishing between these types of instabilities (or noninstabilities as well), using as a basis the plot in the k plane of the singularities of the function Y_T/D as a function of ω.

If ω is allowed to move directly downward from ω_L in the ω plane of Fig. 8.12 (corresponding to the deformation of the Laplace contour) the poles of the function Y_T/D move in the k plane. The Briggs and Bers criterion states that an *absolute* instability occurs if two poles, one from the upper half k plane and one from the lower half k plane, *merge* through the F contour as ω goes from any given value of ω_L down (parallel to) the imaginary ω axis in the ω plane. Note that this merging process results in a *branch line* beginning at the merged poles. This may not permit us to close the Fourier contour, so that the residue theorem cannot be used. However, it may be possible to evaluate the inverse Fourier integral by other means if desired. A discussion of branch lines will appear shortly.

If no poles merge as the frequency is changed, the system may then have only convective or evanescent modes. If poles *cross* the real k axis as the frequency is moved to the real ω axis, they are convective instabilities. If they do not cross this axis, they are evanescent, and if the poles are on the axis, the waves are propagating.

The upper part of Fig. 8.14 shows the motion of several poles in the k plane as ω is changed indicating absolute, convective, evanescent, and propagating disturbances. Note that the F contour must be deformed if poles cross the real k axis so that it still encircles the same number of poles. It must also "pass through" merged poles, as shown in the figure. The lower part of Fig. 8.14 shows the ω plane with the frequency going from ω_L to the real ω axis.

In general, it is only necessary to move the frequency down to the real ω axis for identification of the modes, because when ω goes below the real axis, all solutions will eventually decay away in time.

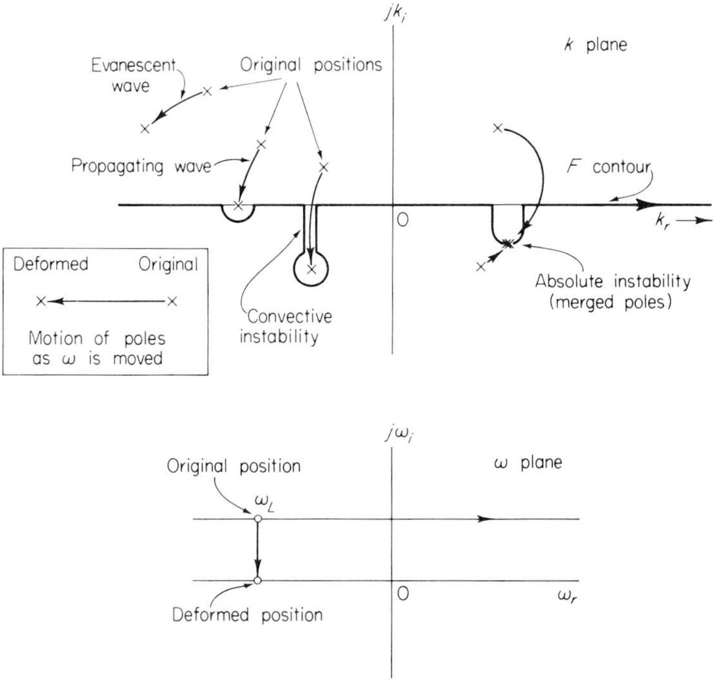

Fig. 8.14. Examples of movement of poles in the k plane.

I. BRANCH LINES

The previous development has omitted a direct discussion of one form of singularity in complex variables, namely, the branch line. A branch line appears in the complex plane whenever a function is multivalued, that is, for one value of the independent variable, the function has two or more values.

The representation of the function then "branches" in two or more paths, hence the name branch line.

Integration through branch lines is not permitted. Only by deforming a contour around the line can the integration proceed. Figure 8.15 shows an example of a branch line and a contour of integration.

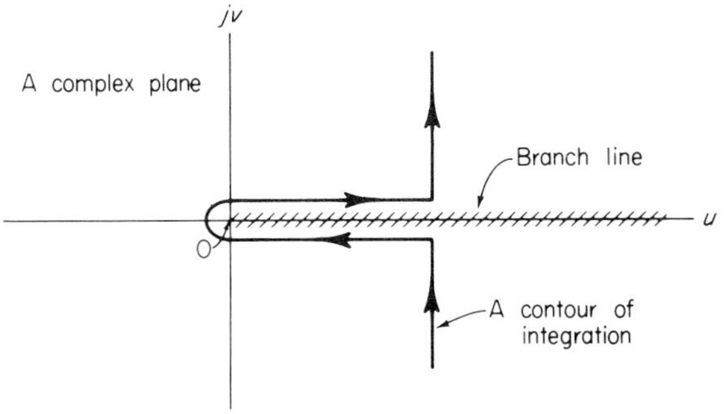

Fig. 8.15. Contour and branch line.

A typical multivalued function which shows a branch line is

$$f(z) = z^{1/2} \tag{8.89}$$

This means, that for every value of z, except $z = 0$, there are two values for $f(z)$. We may plot both values on separate graphs and then place the graphs over each other. They may be joined along a branch line which may be placed, as shown in Fig. 8.15, radially outward from the origin. The branch line may be along the real axis, as shown, or in any other radial direction desired. In order to perform integration, complex contours must always be on the same branch of the function. This is especially important if a closed contour is desired. It is often the case that multiple (high-order) poles of a function are the beginnings and/or the ends of branch lines. If a contour of integration can be deformed outside of the branch line, then the residue at the multiple singularity does not contribute to the integral.

The multivalued nature of a typical dispersion function can be seen by looking at Fig. 8.12. Note that a completely different set of singularities will appear in the integral for the inverse Fourier transform, depending upon whether $z > 0$ or $z < 0$. In addition, branch lines may appear in the ω or k planes if other forms of the dispersion relation are used, which happen to be multivalued.

Also, the analytic continuation process used to arrive at the appropriate dispersion functions assumed a positive value for k, so that the singularity at

$v_z = \omega/k$ was located at the appropriate place in the complex v_z plane depending on the sign of ω. When k is negative, the location of the singularity is reversed, so the dispersion relations are also reversed. Hence, a branch line of the dispersion function can appear along the real k axis.

A brief summary of this chapter is now in order. Starting from the linearized collisionless Boltzmann equation, we have obtained dispersion relations for electrostatic waves that showed collisionless absorption when the velocity distribution was considered. This implied that the *nonuniformity* in the velocity distribution has resulted in Landau damping. The distribution function is obviously not a constant with respect to velocity, hence the nonuniformity.

J. POSITION-SPACE DAMPING

The problem to be considered now is the behavior of this type of absorption if the effects of the velocity distribution are removed by integration over velocity space (the moments of the Boltzmann equation) and a nonuniformity in *position* space is considered instead of the previous nonuniformity in velocity space.

The 0th moment of the collisionless Boltzmann equation for the electrons is

$$\frac{\partial n_e(\mathbf{r}, t)}{\partial t} + \nabla_r \cdot (n_e(\mathbf{r}, t) \cdot \langle \mathbf{v}_e \rangle) = 0 \tag{8.90}$$

where $\langle \mathbf{v}_e \rangle$ is the average velocity.

Poisson's equation will be written as

$$\nabla_r \cdot \mathbf{E} = \frac{q_e}{\varepsilon_0} [n_e(\mathbf{r}, t) - n_i(\mathbf{r}, t)] \tag{8.91}$$

assuming that n_e is the electron density and n_i the ion density. Here we have taken $q_e = -q_i$. Note that the electric field *does not appear* in Eq. (8.90) so we cannot solve these two equations together. We will need a third equation, which in this case will be the first moment of the collisionless Boltzmann equation. It is written in Eq. (8.92).

$$\frac{\partial}{\partial t} n_e(\mathbf{r}, t) m_e \langle \mathbf{v}_e \rangle + \nabla_r \cdot [n_e(\mathbf{r}, t) m \langle \mathbf{v}_e \mathbf{v}_e \rangle] - n_e(\mathbf{r}, t) \langle \mathbf{F}_e \rangle = 0 \tag{8.92}$$

We will assume that

$$\langle \mathbf{F}_e \rangle = \frac{q_e \mathbf{E}}{m_e} \qquad \text{(electrostatic waves)}$$

We will now consider a perturbation solution. We do this by assuming that n_e is the sum of two terms, an equilibrium, but spatially nonuniform term, and a perturbation. In other words, we may write

$$n_e(\mathbf{r}, t) = n_0(\mathbf{r}) + n_1(\mathbf{r}, t) \tag{8.93}$$

for the perturbation. This is the same kind of expansion that was previously made for f. We will also consider $\langle \mathbf{v}_e \rangle$ to be a perturbed quantity, that is to say, $\langle \mathbf{v}_e \rangle$ is zero in equilibrium. Also, the plasma is to be cold, so $\langle v_e{}^2 \rangle = 0$ as well. The ions will be assumed to be stationary and to act only as a neutralizing background. Under these conditions and assumptions, the three equations needed for the solution become, after neglecting higher order terms,

$$\frac{\partial n_1(\mathbf{r}, t)}{\partial t} + \nabla_r \cdot (n_0(\mathbf{r})\langle \mathbf{v}_e \rangle) = 0 \tag{8.94}$$

$$\nabla_r \cdot \mathbf{E} = \frac{q_e}{\varepsilon_0} n_1(\mathbf{r}, t) \tag{8.95}$$

and

$$\frac{\partial \langle \mathbf{v}_e \rangle}{\partial t} - \frac{q_e}{m_e} \mathbf{E} = 0 \tag{8.96}$$

These three equations are solved simultaneously in an attempt to obtain the dispersion relation, if it exists.

We can find a single equation in $\mathbf{E}$ alone to be

$$\nabla_r \cdot \left[\frac{\partial^2 \mathbf{E}}{\partial t^2} + \frac{n_0(\mathbf{r}) q_e{}^2 \mathbf{E}}{\varepsilon_0 m_e} \right] = 0 \quad \text{or} \quad \nabla_r \cdot \left[\frac{\partial^2 \mathbf{E}}{\partial t^2} + \omega_{pe}^2(\mathbf{r}) \mathbf{E} \right] = 0 \tag{8.97}$$

We could now proceed to take Fourier and Laplace transforms as before. However, note that, because of the presence of the variable plasma frequency, the second term in Eq. (8.97) is not a constant times $\mathbf{E}$. The result is that we will *not* be able to find a dispersion relation. This can be seen very easily by noting that, for each value of $\mathbf{r}$, a different dispersion relation can be obtained.

If ω_{pe}^2 is *not* a function of $\mathbf{r}$, then the dispersion relation *can* be obtained from Eq. (8.97). We can develop this relation using the methods described in this chapter. The Fourier and Laplace transforms of (8.97) can be taken, if $\omega_{pe}^2(\mathbf{r}) = \text{constant}$, giving

$$j\mathbf{k} \cdot [-\omega^2 + \omega_{pe}^2]\mathbf{E} = 0 \tag{8.98}$$

From (8.98) we can see that the dispersion relation itself becomes

$$\omega_{pe}^2 = \omega^2 \tag{8.99}$$

which is the dispersion relation for electrostatic plasma oscillations. Note that the frequency is purely real and no absorption appears.

We may now, however, instead of a dispersion relation, find the response of the system (when $\omega_{p_e}^2$ is a function of $\mathbf{r}$) to an initial condition with or without a source function by taking the appropriate transforms and doing the inverse integrations. This will permit us to solve for $\mathbf{E}(\mathbf{r}, t)$. By looking at this quantity, it may be possible to ascertain whether absorption is taking place when the density is nonuniform.

We will consider a one-dimensional problem. That is, only variations in z will be permitted for the perturbed quantities, as well as for the total density. We will start the solution by having an initial uniform displacement of all the particles in the z direction at time equal zero. The initial velocity $\langle \mathbf{v}_e \rangle$ will be assumed to be zero at time equal zero. $\mathbf{E}$ will therefore only be a function of z and t and only have a component in the z direction.

Equation (8.97) now becomes

$$\frac{\partial}{\partial z}\left[\frac{\partial^2 E(z, t)}{\partial t^2} + \omega_{p_e}^2(z)E(z, t)\right] = 0 \tag{8.100}$$

After taking the Laplace transform with respect to time, Eq. (8.100) becomes

$$\frac{\partial}{\partial z}\left[-\omega^2 + \omega_{p_e}^2(z)\right] E(z, \omega) = \frac{\partial}{\partial z}\left[\frac{\partial E(z, 0)}{\partial t} - j\omega E(z, 0)\right] \tag{8.101}$$

We can take the Laplace transform because the problem is linear in time. The right-hand side of Eq. (8.101) is the initial conditions on E at time equal zero. These can be evaluated from the initial displacement, if desired.

However, it is easier to evaluate the problem by using the initial conditions directly in terms of the displacement of the density and the initial velocity, if any, before the three equations are solved simultaneously. It will then not be necessary to know E or its derivative with respect to time at $t = 0$. This can be seen by examining Eqs. (8.94)–(8.96). In no place is found a derivative of E with respect to time. This means that if these three equations are transformed *first* and then solved for E, we may avoid the necessity of knowing E and its derivatives at $t = 0$.

The Laplace transform of the three equations is

$$-j\omega n_1(z, t) - n_1(z, 0) + \frac{\partial}{\partial z}(n_0(z)\langle v_e(z, \omega)\rangle) = 0 \tag{8.102}$$

$$\frac{\partial E(z, \omega)}{\partial z} = \frac{q_e}{\varepsilon_0} n_1(z, \omega) \tag{8.103}$$

$$-j\omega\langle v_e(z, \omega)\rangle - \langle v_e(z, 0)\rangle - \frac{q_e}{m_e} E(z, \omega) = 0 \tag{8.104}$$

which includes the initial conditions on n_1 and v_e. We can now solve these three equations for $E(z, \omega)$. It should be noted that sources for the electric field may also be added into Eqs. (8.103) and (8.104), since the electric fields can be added by superposition. The single equation for $E(z, \omega)$, with the initial conditions on n_1 and $\langle v_e \rangle$ is

$$\frac{\partial}{\partial z}[-\omega^2 + \omega_{pe}^2(z)]E(z, \omega) = \frac{\partial}{\partial z}\left[\frac{n_0(z)\langle v_e(z, 0)\rangle q_e}{\varepsilon_0}\right] - j\frac{\omega q_e}{\varepsilon_0} n_1(z, 0) \quad (8.105)$$

Note that there are still two initial conditions in Eq. (8.105), as there were in (8.101). We may now integrate Eq. (8.105) with respect to z and solve for $E(z, \omega)$. The solution is

$$E(z, \omega) = \frac{[q_e n_0(z)/\varepsilon_0]\langle v_e(z, 0)\rangle - \int_{-\infty}^{z}(j\omega q_e/\varepsilon_0)n_1(z, 0)\, dz}{\omega_{pe}^2(z) - \omega^2} \quad (8.106)$$

Note that, if we were to apply the analog of the previous work, the denominator of Eq. (8.106) should be the dispersion relation. Again, however, since ω_{pe}^2 is a function of z, we cannot do this. We *can* take the inverse Laplace transform of Eq. (8.106) to find the electric field as a function of time.

We may also integrate the electric field over all z to obtain the voltage across the plasma. Let us do this integration on Eq. (8.106) before we take the inverse Laplace transform. To do this, we must specify $n_0(z)$, $n_1(z, 0)$ and $\langle v_e(z, 0)\rangle$. Let

$$n_0(z) = \frac{n_c}{1 + z^2}$$

where n_c is the density at the center. This relation obviously specifies the way ω_{pe}^2 varies with z as well.

Also, we will let

$$n_1(z, 0) = -\delta\, \frac{dn_0(z)}{dz}$$

and

$$\langle v_e(z, 0)\rangle = 0$$

These last two conditions correspond to an initial small displacement δ of all the particles. This can be seen if it is noted that $-\delta\, dn_0(z)/dz$ is the first term in a Taylor expansion for $n_0(z)$ about the point z. This term is set equal to $n_1(z, 0)$.

Under these conditions, we can easily integrate Eq. (8.106) with respect to z to obtain the total voltage.

We will first, however, perform the integration in the *numerator* of Eq. (8.106), which, although it is also integration with respect to z, must be done before the additional integration. The first term in the numerator of (8.106) is zero because $\langle v_e(z, 0)\rangle = 0$. The integral of the numerator is then

$$-\int_{-\infty}^{z} \frac{j\omega q_e}{\varepsilon_0} n_1(z, 0)\, dz = j\frac{\delta\omega q_e}{\varepsilon_0} n_0(z) \tag{8.107}$$

The integration of (8.106) with respect to z to obtain the potential across the entire plasma may now be performed. This integral is

$$-\int_{-\infty}^{+\infty} E(z, \omega)\, dz = \int_{-\infty}^{+\infty} \frac{(j\delta\omega q_e/\varepsilon_0)n_0(z)}{\omega_{p_e}^2(z) - \omega^2}\, dz \tag{8.108}$$

The negative sign is used in (8.108) because of the definition of the potential. Now

$$\omega_{p_e}^2(z) = \frac{q_e{}^2 n_0(z)}{m_e \varepsilon_0} = \frac{q_e{}^2}{m_e \varepsilon_0}\frac{n_c}{1 + z^2} = \frac{\omega_{p_c}^2}{1 + z^2}$$

where ω_{p_c} is the plasma frequency at $z = 0$. We may change Eq. (8.108), where we will let $\mathscr{V}(\omega)$ be the voltage across the plasma, to the form

$$\begin{aligned}\mathscr{V}(\omega) &= -\frac{j\omega q_e \delta n_c}{\varepsilon_0}\int_{-\infty}^{+\infty} \frac{dz}{(\omega_{p_c}^2 - \omega^2) - \omega^2 z^2} \\ &= \frac{j\omega q_e \delta n_c}{\varepsilon_0}\int_{-\infty}^{-\infty} \frac{dz}{\left[z + \left(\dfrac{\omega_{p_c}^2}{\omega^2} - 1\right)^{1/2}\right]\left[z - \left(\dfrac{\omega_{p_c}^2}{\omega^2} - 1\right)^{1/2}\right]}\end{aligned} \tag{8.109}$$

This expression may be integrated by use of residues, assuming that we can use a complex variable z. Analytic continuation shows that the complex z function will match the real z function along the real z axis. The only singularities are located at two values of z, which are at

$$z = \pm\left(\frac{\omega_{p_c}^2}{\omega^2} - 1\right)^{1/2} \tag{8.110}$$

The singularities are off the real z axis if $\omega_{p_c}^2/\omega^2 < 1$, and on the real axis if $\omega_{p_c}^2/\omega^2 > 1$. Figure (8.16) shows the two types of singularities, together with the appropriate contours of integrations.

We first consider the case when $\omega_{p_c}^2/\omega^2 \le 1$. The integral is then $2\pi j$ times the residue at $z = j[1 - (\omega_{p_c}^2/\omega^2)]^{1/2}$. The integral (8.109) has the value

$$\mathscr{V}(\omega) = 2\pi j\left(\frac{j\omega q_e \delta n_c}{\varepsilon_0}\right)\frac{1}{2j(1 - \omega_{p_c}^2/\omega^2)^{1/2}}$$

or

$$\mathscr{V}(\omega) = \frac{q_e n_c \delta}{\varepsilon_0}\frac{\pi}{(\omega_{p_c}^2 - \omega^2)^{1/2}} \tag{8.111}$$

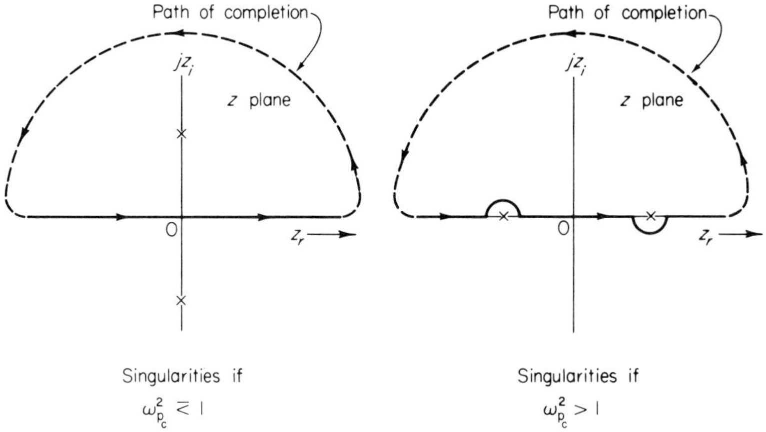

Fig. 8.16. The singularities of Eq. (8.109).

We will now take the *inverse Laplace* transform of Eq. (8.111), to get the time response. The inverse Laplace transform of (8.111) shows the time response of the voltage across the plasma to be

$$\mathscr{V}(t) = -\frac{q_e n_c \delta\pi}{\varepsilon_0} J_0(\omega_{p_c} t) \tag{8.112}$$

J_0 is a Bessel function of the first kind of order zero. As can be seen from a plot of this function versus time, the voltage oscillates, *and* damps out to zero. This result is rather surprising, since there does not seem to be a mechanism for absorption of the energy. It perhaps may be likened to the position space analogue of Landau damping, which is also a collisionless process.

When the singularities lie on the z contour, the integral may be evaluated by noting that we have a principal value problem, since the contour now goes through both singularities. We will keep the same "sense" of the integral, in that the upper singularity, which goes to the right of the imaginary axis, is still kept inside the contour as the center density is increased to make $\omega_{p_c} > \omega$. The left-hand root still remains below (outside) the contour.

In this case, we have two principal values of opposite sign to be evaluated. The integral then has the value of

$$\mathscr{V}(\omega) = \pi j\left(\frac{j\omega q_e \delta n_c}{\varepsilon_0}\right)\left[\frac{1}{2[(\omega_{p_c}/\omega)^2 - 1]^{1/2}} + \frac{1}{2[(\omega_{p_c}/\omega)^2 - 1]^{1/2}}\right]$$

or

$$\mathscr{V}(\omega) = -\frac{\pi q_e \delta n_c}{\varepsilon_0}\frac{1}{(\omega_{p_c}^2 - \omega^2)^{1/2}} \tag{8.113}$$

This is, interestingly, exactly the same function as that found when $\omega_{p_e} \leq \omega$. The inverse Laplace transform is also the same as before, that is, a Bessel function of argument $(\omega_{p_e} t)$ that again gives damping.

The physical mechanism for such damping is not very clear. It only appears when the plasma is nonuniform in density. It does seem to be similar in mathematical structure to Landau damping. Perhaps it is an absorption of electrostatic energy (that energy set up when the changes were initially displaced) into motional (kinetic) energy of the particles. It may not be reversible (recoverable as potential energy again) because of phase mixing of oscillations that take place over the range of plasma frequencies specified by $\omega_{p_e}(z)$.

K. CONFINEMENT AND EQUILIBRIUM OF A PLASMA—HYDRODYNAMIC AND MICROINSTABILITIES

A general summary of the various confinement and instability conditions will be made. It is made here because the developments on instabilities, which can now be related to confinement problems, have been introduced.

To confine a plasma we must first note that, if electric and magnetic fields are used for the confinement, the plasma will really never be in equilibrium. Therefore, once equilibrium is achieved, the plasma is no longer confined.

We note that the Maxwellian velocity distribution should be of the form

$$f \sim \exp\left(-\frac{\mathscr{E}}{kT}\right)$$

where $\mathscr{E} = H = \frac{1}{2}mv^2 + q\phi$ for charged particles. Here ϕ is a potential function which could be used for confinement. However, it will only confine one species of charge, that is, either electrons or ions, but *not both*. Therefore, for successful results, the confinement of one species of particle cannot produce an equilibrating effect on the other species. If this effect is produced, then the result is an approach to equilibrium and a consequent *loss* of confinement. Collisions are one mechanism that brings the plasma to equilibrium, but for high energies the rate is usually slow enough so that many experiments can be performed with an essentially confined plasma. However, sometimes collisions themselves produce instabilities, often termed *resistive instabilities*, which move the plasma toward equilibrium much faster than the collisional processes themselves, and confinement is lost very quickly.

If the plasma can be considered collisionless, then we can investigate confinement from two aspects. The first will be the fluid-mechanical approach.

If the plasma can be assumed to be a perfect conductor, then the magnetic field lines are frozen into the plasma, and Ohm's Law

$$\mathbf{J} = \sigma(\mathbf{E} + \mathbf{v} \times \mathbf{B})$$

becomes

$$\mathbf{E} + \mathbf{v} \times \mathbf{B} = 0$$

Motions in this regime, such as Alfvén waves, are usually of very low frequency and long wavelength. This means that any instabilities that could appear in this range are large scale (macroscopic) and very destructive of plasma confinement. Often, these instabilities are predicted by an energy calculation. This method computes the potential energy of a hydrodynamic plasma in a given magnetic field configuration. If a configuration of a lower energy is available, the prediction is that the plasma can move to this state. If this motion follows a path away from confinement, the plasma will become unstable hydrodynamically.

A mirror confined plasma is unstable in this way, for the radial confinement force $B^2/2\mu_0$ produced by the magnetic field decreases with increasing radius, so that once a portion of the plasma moves away from the magnetic field axis, it tends to keep moving away from the confinement region and hence tends to become unstable.

However, other arrangements of magnetic fields can result in macroscopically stable configurations. These field arrangements only provide for stability when we consider the magnetic field lines frozen into the plasma, which is a direct implication of the assumption of infinite conductivity.

This condition is easily violated, especially at high frequencies and short wavelengths. More complex forms of plasma motion, involving the velocity distribution, must be used. Instabilities resulting from these considerations are termed *microinstabilities*. These are much more difficult to guard against because high plasma conductivity and the adiabatic invariance of the magnetic moment, which tend to produce confinement, will not hold in these regimes.

The discussion before, which resulted in a dispersion relation of the form shown in (8.71) and (8.72), shows one treatment of a possible microinstability. Usually, since the microinstabilities are of short wavelength, compared to the characteristic dimensions of the plasma, the growth of the amplitude of the instability rapidly makes the problem nonlinear. Such unstable disturbances might therefore be modified so that enhanced (anomalous) diffusion might occur instead of a cataclysmic event. This has been proposed as a mechanism for Bohm diffusion.

In most cases involving either hydromagnetic instabilities or microinstabilities, we must use some sort of perturbation expansion from an initial state. This usually implies a Fourier representation of the fluctuating quantities, which can permit the development of a dispersion relation. The eigenvalues of this dispersion relation may be examined to determine the nature of the waves or instabilities that may appear. The methods of this chapter are thus directly applicable to a wide range of problems.

The discussions in this chapter have served as an introduction into the nature of instabilities, waves and wave-type phenomena propagating in plasmas, when the velocity or the density distribution of particles making up the plasma must be used. The nature and functional form of the distribution functions must be considered. Knowledge of these distribution functions is important to discover the behavior of the plasma–wave system. If a dispersion relation can be found, it may determine the asymptotic time response of such a system.

Subsequently we shall consider, in greater depth than with the two-particle elastic cross section theory previously described, how collections of charged particles interact as well as radiation and correlation effects on these collective phenomena.

Suggested Reading

Allis, W. P., Buchsbaum, S. J., and Bers, A., "Waves in Anisotropic Plasmas," M.I.T. Press, Cambridge, Massachusetts, 1963. This work is similar to Stix's "The Theory of Plasma Waves," suggested in this list but uses mks units; it includes a good discussion on bounded waves as well.

Baldwin, D. E., and Ignat, D. W., Resonant Absorption in Zero Temperature Non-Uniform Plasma, *Phys. Fluids* **12**, 697 (1969). Similar to the Bers and Schneider report in this list but discusses the forced oscillation of a nonuniform plasma column that leads to collisionless damping.

Bers, A., and Schneider, H. M., M.I.T. Res. Lab. of Electronics, Quart. Progr. Rept. No. 89, 1968. This is a discussion of collisionless damping in spatially nonuniform plasmas.

Briggs, R. J., "Electron Stream Interactions with Plasma," M.I.T. Press, Cambridge, Massachusetts, 1964. This is an excellent discussion of instabilities in plasmas.

Churchill, R., "Introduction to Complex Variables and Applications," Prentice-Hall, Englewood Cliffs, New Jersey, 1954. This book treats contour integration and analytic continuation very well.

Denisse, J. F., and Delcroix, J. L., "Plasma Waves," Wiley (Interscience), New York, 1963.

Fried, B. D., and Conte, S. D., "The Plasma Dispersion Function," Academic Press, New York, 1961. This is a "data book" of the plasma dispersion function for large ranges of the complex variable.

Landau, L., *J. Phys.* (*USSR*) **10**, 25 (1946). This is the original work on collisionless damping.

Montgomery, D., Validity of the Electrostatic Approximation, *Phys. Fluids* **13**, 1401 (1970).

Schmidt, G., "Physics of High Temperature Plasmas—An Introduction," Academic Press, New York, 1966. This is a wide-ranging coverage with much material in depth.

Stix, T. H., "The Theory of Plasma Waves," McGraw-Hill, New York, 1963.

Vandenplas, P. E., "Electron Waves and Resonances in Bounded Plasmas," Wiley (Interscience), New York, 1968. This presents a broad collection of the subject matter.

Problems

1 If the dispersion relation in (8.50) has, as its velocity distribution, a form as $f_0(v) = \delta(v - u) + \delta(v + u)$ then we may consider that the plasma

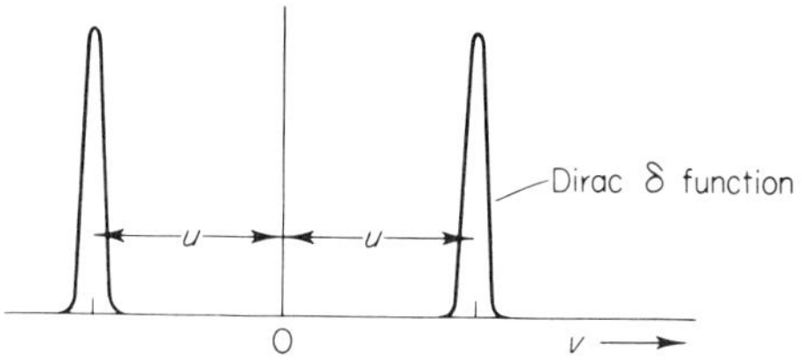

is composed of two interpenetrating beams moving with relative velocity $2u$ with respect to each other. If this is the case, the dispersion relation in the denominator of Eq. (8.50) may be integrated by parts. Show that the result is

$$1 - \omega_{\text{pe}}^2 \left[\frac{1}{(\omega - ku)^2} + \frac{1}{(\omega + ku)^2}\right] = 0$$

This dispersion relation is that of *the two-stream instability.*

2 Plot the roots of the two-stream dispersion function developed in Problem 1 in the complex ω plane and show what happens to the modes as k increases from zero for a fixed value of u.

3 Apply the Bers and Briggs method to determine the nature of the modes in

(a) the dispersion relation for waves in free space;
(b) the dispersion relation for waves in a cold isotropic plasma;
(c) the dispersion relation for waves in the two-stream instability.

Be sure that all possible values of k are considered that might cause a change in the modes. Numerical evaluation may be helpful for case (c).

4 The development of this chapter determined a dispersion relation for longitudinal electrostatic waves, considering that the ions were fixed. Determine a dispersion relation for *transverse* waves. That is, the fluctuations in **E** are perpendicular to the direction of propagation of the wave. In this case, Poisson's equation will always be zero so the waves are *not* electrostatic. Show that the dispersion relation is, for long wavelengths,

$$\omega^2 - k_z{}^2c^2 + \omega\omega_{\text{pe}}^2 \int_{-\infty}^{+\infty}\int_{-\infty}^{+\infty}\int_{-\infty}^{+\infty} \frac{f_0(v)}{k_z v_z - \omega}\, dv_x\, dv_y\, dv_z = 0$$

$$\text{for} \quad \text{Im}\,\omega > 0$$

Assume again that the ions are fixed and the wave propagates in the z direction. Find the dispersion relation for Im $\omega < 0$, as well.

5 Use the results of Problem 4 to compute the long-wavelength approximation to the dispersion relation.

6 Find the analytic continuations of the function $1/(1-z)^2$ for $z > 1$ and $z < 1$.

7 Develop the dispersion relations in (8.71), (8.72), and (8.73) but assume that $k < 0$. Be consistent with the definitions of the Fourier and Laplace transforms.

8 Find the dispersion relation using (8.71) and (8.72) if

$$f_0(v) = \frac{\alpha}{\pi} \frac{1}{v^2 + \alpha^2}$$

where α is a constant.

9 Determine the *impedance* seen across a plasma of the type described in the section on nonuniform density damping seen by a voltage source placed completely across the z dimension of the plasma $(-\infty < z < +\infty)$. Assume the current in the z direction is the same for all values of z. Is the impedance real or complex? What can you conclude from this?

10 For a plane monochromatic wave in free space find ϕ and $\mathbf{A}$, assuming the Lorentz gauge

$$\nabla_r \cdot \mathbf{A} = \frac{1}{c^2} \frac{\partial \phi}{\partial t}$$

Now, change the gauge to $\nabla_r \cdot \mathbf{A} = 0$ and determine ϕ and $\mathbf{A}$. Show that $\nabla_r \cdot \nabla_r \phi \equiv 0$ in free space.

11 We wish to use the long-wavelength approximation to solve for the electromagnetic field around a uniform plasma column parallel to the z axis when a plane electromagnetic wave of frequency ω is incident upon it. Under what conditions may we use the approximation? Find the

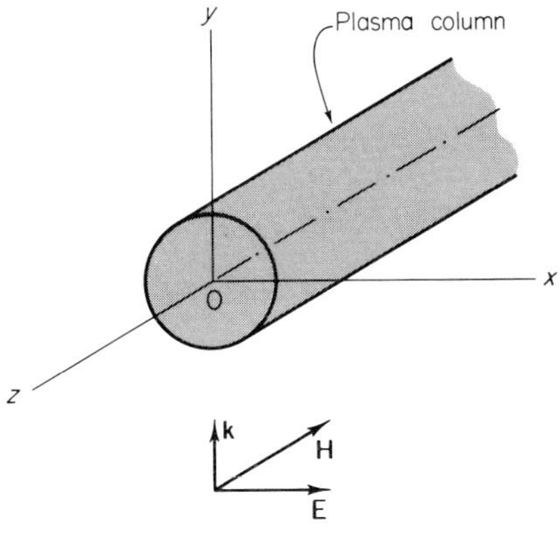

potential and electric fields everywhere in space, assuming the plasma column is infinitely long in the z direction. Solve for all unknown coefficients in terms of the amplitude and frequency of the incident electric field and the plasma frequency.

12 Assume that an electric field source $Y(t)$ is included in the free space wave equation. Let the source be located only at $z = 0$, but exist for all time. Find the wave equation, including the source, and then determine the solution, assuming that the source term produces an electric field at $z = 0$ of $E_0 \cos(\omega t)$. Does the wave equation change its form?

13 Suppose electrostatic waves are to be investigated when the velocity distribution is such that $\langle v_z \rangle \neq 0$. Does damping similar to that shown in (8.84) appear? Derive an expression for the frequency ω which gives evidence for the existence or nonexistence of damping. Does the sign of $\langle v_z \rangle$ affect the results? What is the result if $\langle v_z \rangle = 0$ and $\langle v_z^2 \rangle \neq 0$?

14 Show that the denominator of Eq. (8.50) is analytic in the entire ω plane for both $k > 0$ and $k < 0$.

15 Assume a source function $E_s(z, t)$. Place it properly in Eq. (8.50).

16 Plot a k versus ω diagram for Eq. (8.84) and the results of problem 13.

9 Plasma Kinetic Theory—Nonequilibrium Statistical Mechanics

Up to this point we have successfully used the collisionless Boltzmann equation in looking at plasma problems. We have attempted to make headway into the problems of collisions and have ended up with some assumptions and approximations, such as the truncation of collision interactions for lengths greater than the Debye length.

We have also obtained the Fokker–Planck equation, which, in principle, provides one possible formulation for the collision term of the Boltzmann equation. In fact, the Boltzmann equation itself is based on a large number of assumptions, which were given in Chapter 4.

It is the purpose of this chapter to begin with the Fokker–Planck equation and evaluate it for a plasma. Next, a consideration of the domains of validity of the single-particle, Boltzmann–Fokker–Planck, and fluid models will be made.

Finally, we will return to the Liouville theorem and produce a method for treating the behavior of the N-particle distribution function. This function f_N is much more representative of plasma behavior than f, because particle interactions *other* than simple collisions are implicitly "located" inside the function.

This kinetic approach is quite general, and should be reasonably independent of the distribution function itself. For example, it will not always be necessary to assume an equilibrium condition on the plasma. The equilibrium assumption (total energy of the plasma remains constant) produced the Maxwellian distribution, which is one possible solution to the Boltzmann–Vlasov equation. However, some nonequilibrium distribution functions are *also* solutions to the Boltzmann–Vlasov equation. (The collisionless Boltzmann equation is the Boltzmann–Vlasov equation.) Previously, such functions were "assumed" and then tested to see if they satisfied the collisionless

Boltzmann equation. Now, however, nonequilibrium statistical mechanics can provide a more rigorous method for determining the nature of nonequilibrium distribution functions. The goal of kinetic theory is to obtain an equation that determines the rate of change of the distribution function f_N as it progresses toward a final "equilibrium" state.

A. COLLISIONAL VELOCITY AVERAGES

We begin this chapter by returning to the Fokker–Planck equation, which serves as a starting point for an examination of the kinetic theory of plasmas in general and nonequilibrium statistical mechanics in particular. We begin by restating the Fokker–Planck equation (5.73).

$$\Delta t \frac{\partial f}{\partial t} = -\frac{\partial}{\partial \mathbf{v}}(f\langle \Delta \mathbf{v} \rangle) + \frac{1}{2}\frac{\partial^2}{\partial \mathbf{v}\,\partial \mathbf{v}}(f\langle \Delta \mathbf{v}\,\Delta \mathbf{v} \rangle) \tag{5.73}$$

This is a representation of the net change in f, the single-particle distribution function, over a collision time Δt. Recall that $\langle \Delta \mathbf{v}\,\Delta \mathbf{v} \rangle$ is a dyadic. If Eq. (5.73) is divided through by Δt, we can then obtain the formulation necessary to place this quantity on the right-hand side of the Boltzmann equation.

The first term on the right-hand side of Eq. (5.73) is the expression for dynamical friction. This process tends to slow down or speed up particles to drag them toward the average velocity ($\Delta \mathbf{v} = 0$). The second term, velocity space diffusion, results in a spreading of the velocities of the particles, in opposition to the dynamical friction. Normally, both quantities are present, and, in equilibrium, they should be of equal value, so that the left-hand side of Eq. (5.73) is zero.

We will reexamine the elastic collision of two particles in a velocity space coordinate system. Figure 9.1 shows a diagram of the velocity coordinates of such a collision. $\mathbf{G}$ is the velocity of the center of mass, $\mathbf{v}_1$ and $\mathbf{v}_2$ are the velocities of the two particles before the collision, and $\mathbf{v}_1'$ and $\mathbf{v}_2'$ are the velocities after the collision. We shall also define a relative velocity vector $\mathbf{g}_{21}$ as shown in the figure. We have already shown in Chapter 2 that $|\mathbf{g}_{21}| = |\mathbf{g}_{21}'|$. That is, the relative velocity changes direction, but not magnitude. Suppose that we now define a new velocity space coordinate system in which the z axis is directed along $\mathbf{g}_{21}$. Such a system is shown in Fig. 9.2. It can be seen from this figure that if $\Delta \mathbf{g}_{21}$, the change in the relative velocity vector, is small, then it is perpendicular to $\mathbf{g}_{21}$.

We can also write

$$\Delta g_{21} \cong 2 g_{21} \sin\left(\frac{\chi}{2}\right) \tag{9.1}$$

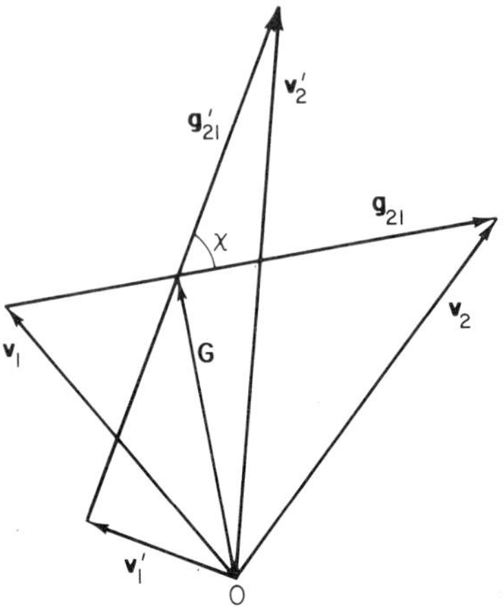

Fig. 9.1. Velocity coordinates for two elastically colliding particles.

as an approximation for the magnitude of the change in velocity. Recall that χ is the deflection angle. The components of $\Delta \mathbf{g}_{21}$ are

$$(\Delta g_{21})_z = -2g_{21} \sin^2\left(\frac{\chi}{2}\right) \tag{9.2}$$

$$(\Delta g_{21})_x = 2g_{21} \sin\left(\frac{\chi}{2}\right) \cos\left(\frac{\chi}{2}\right) \cos \varphi \tag{9.3}$$

$$(\Delta g_{21})_y = 2g_{21} \sin\left(\frac{\chi}{2}\right) \cos\left(\frac{\chi}{2}\right) \sin \varphi \tag{9.4}$$

in terms of the spherical angles defined in Fig. 9.2. We wish to compute the average values of the quantities defined in Eq. (9.2), (9.3), and (9.4). In order

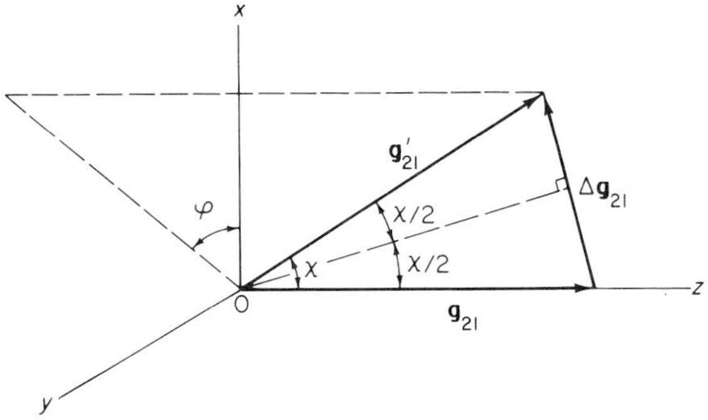

Fig. 9.2. Spherical relative velocity coordinate system.

to do this we need to note the definitions of $\langle \Delta\mathbf{v} \rangle$ and $\langle \Delta\mathbf{v}\, \Delta\mathbf{v} \rangle$ defined previously. They are, from Chapter 5,

$$\langle \Delta\mathbf{v} \rangle = \int \text{Prob}(\mathbf{v}, \Delta\mathbf{v})\, \Delta\mathbf{v}\, d(\Delta\mathbf{v}) \tag{5.70}$$

and

$$\langle \Delta\mathbf{v}\, \Delta\mathbf{v} \rangle = \int \text{Prob}(\mathbf{v}, \Delta\mathbf{v})\, \Delta\mathbf{v}\, \Delta\mathbf{v}\, d(\Delta\mathbf{v}) \tag{5.71}$$

Prob($\mathbf{v}$, $\Delta\mathbf{v}$) is defined as the probability that a particle, given a velocity $\mathbf{v}$, will acquire a velocity $\mathbf{v} + \Delta\mathbf{v}$ in a collision time Δt. We have already determined $\Delta\mathbf{v}$ in Chapter 2 for a particle that has a relative velocity g_{21} and a deflection angle χ. Now, however, we must obtain the averages of $\Delta\mathbf{v}$ and $\Delta\mathbf{v}\, \Delta\mathbf{v}$, first over all possible deflection angles, and then over the complete range of allowed relative velocities. We can perform the averaging by integration. This should result in an expression that has the form shown in Eqs. (5.70) and (5.71).

We first examine the integration over allowed deflection angles, given a fixed value of g_{21}. The probability of a deflection through angle χ, given a relative velocity g_{21} is expressed as the differential collision cross section, which was developed previously. The expression $d\sigma/d\Omega$ is, for Coulomb collisions,

$$\frac{d\sigma}{d\Omega} = \frac{n_1 k^2}{4 g_{21}^4} \frac{1}{\sin^4(\chi/2)} \tag{9.5}$$

where

$$k = \frac{q_1 q_2}{4\pi\varepsilon_0 m_r}, \qquad m_r = \frac{m_1 m_2}{m_1 + m_2}$$

and n_1 is the density of scattering particles per unit volume of position space. We shall now call the colliding particle the test particle, and the particles with which it collides the target particles.

The incremental deflections of the test particle are first multiplied by $g_{21}(d\sigma/d\Omega) f_{\mathrm{T}}(\mathbf{v})$ where $f_{\mathrm{T}}(\mathbf{v})$ is the velocity distribution of the target particles, to obtain the total number of deflections that the test particle undergoes in collision time Δt. This expression is then integrated over all possible deflection angles. We will thus be able to obtain the average value for the deflections of a test particle that is colliding with target particles of velocity $\mathbf{v}$ in collision time Δt. This expression can be used in the dynamical friction term for the angular average of the component of $\Delta\mathbf{g}_{21}$. In the z direction the deflection is

$$\overline{\Delta g_z} = \int_0^{2\pi} \int_{\chi_{\min}}^{\chi_{\max}} (\Delta g)_z \frac{d\sigma}{d\Omega} g^2 \sin\chi\, d\chi\, d\varphi\, f_{\mathrm{T}}(\mathbf{v}) \tag{9.6}$$

The subscript 21 of g is understood; $f_T(\mathbf{v})$ is, again, the velocity distribution of the target particles. The limits on χ will be discussed shortly. Figure 9.2 shows the coordinate system for this integration. Equation (9.6) becomes, after substitution of Eq. (9.5) and (9.2),

$$\overline{\Delta g_z} = -2\left(\frac{q_1 q_2}{8\pi\varepsilon_0 m_r g^2}\right)^2 f_T(\mathbf{v}) 4\pi g^2 n_1 \int_{\chi_{\min}}^{\chi_{\max}} \frac{\sin^2(\chi/2)\sin\chi}{\sin^4(\chi/2)}\,d\chi \tag{9.7}$$

The integration with respect to φ was performed to arrive at (9.7). We will now change the variable of integration from χ to $\chi/2$. Equation (9.7) then becomes

$$\overline{\Delta g_z} = -\left(\frac{q_1 q_2}{2\varepsilon_0 m_r g^2}\right)^2 f_T(\mathbf{v})\frac{g^2}{\pi} n_1 \int_{\chi_{\min}}^{\chi_{\max}} \frac{\cos(\chi/2)\,d(\chi/2)}{\sin(\chi/2)} \tag{9.8}$$

We now examine the angular averages for Δg_x and Δg_y. It will become obvious that the averages of Δg_x and Δg_y should be zero because it is equally likely that a collision could deflect a particle in the positive or negative x and y directions. This result can be seen from an examination of the integrals for Δg_x and Δg_y as shown below.

The integral for $\overline{\Delta g_x}$ is

$$\overline{\Delta g_x} = \left(\frac{q_1 q_2}{8\pi\varepsilon_0 m_r g^2}\right)^2 f_T(\mathbf{v})\,2g^2 n_1 \int_0^{2\pi}\int_{\chi_{\min}}^{\chi_{\max}} \sin(\chi/2)\cos(\chi/2)\sin\chi\cos\varphi\,d\varphi\,d\chi$$

$$\equiv 0 \qquad \text{since} \qquad \int_0^{2\pi}\cos\varphi\,d\varphi = 0 \tag{9.9}$$

and similarly for $\overline{\Delta g_y}$. This means that the change in relative velocity averaged over angles occurs *only* in the direction of g *itself.*

The limits on χ now have to be set. We shall assume that the minimum deflection angle $\chi_{\min}$ occurs when a particle has an impact parameter of value $b = \lambda_D$. Equation (2.34) gives the relationship between b and χ. This is

$$\tan\left(\frac{\chi}{2}\right) = \frac{q_1 q_2}{4\pi\varepsilon_0 g^2 b m_r}$$

so

$$\tan\left(\frac{\chi_{\min}}{2}\right) = \frac{q_1 q_2}{4\pi\varepsilon_0 g^2 m_r \lambda_D} \tag{9.10}$$

Completing the integration of Eq. (9.8) results in the net angular average of Δg_z to be

$$\overline{\Delta g_z} = -\left(\frac{q_1 q_2}{2\varepsilon_0 m_r g^2}\right)^2 n_1 f_T(\mathbf{v})\frac{g^2}{\pi}\ln(\sin\chi/2)\Big]_{\chi_{\min}}^{\chi_{\max}} \tag{9.11}$$

If $\chi_{\max}$ is assumed to be π, then evaluating Eq. (9.11) at the upper limit results in a zero contribution at this deflection. Therefore, Eq. (9.11) becomes

$$\overline{\Delta g_z} = \left(\frac{q_1 q_2}{2\varepsilon_0 m_r g^2}\right)^2 n_1 f_T(\mathbf{v}) \frac{g^2}{\pi} \ln \sin\left(\frac{\chi_{\min}}{2}\right) \tag{9.12}$$

We shall let $\Lambda = \sin(\chi_{\min}/2)$, so

$$\overline{\Delta g_z} = \left(\frac{q_1 q_2}{2\varepsilon_0 m_r g^2}\right)^2 n_1 f_T(\mathbf{v}) \frac{g^2}{\pi} \ln \Lambda \tag{9.13}$$

which is the final form for the average over deflection angles. Often $\ln \Lambda$ is about -20.

We must now integrate Eq. (9.13) over all values of target velocity to obtain the average over target particle velocities. However, before doing this, we shall have to write Eq. (9.13) on the basis of the individual particle velocity, rather than the relative velocity, as it is presently written. To do this, we first modify Eq. (9.13) to read

$$\overline{\Delta \mathbf{g}} = \frac{1}{\pi}\left(\frac{q_1 q_2}{2\varepsilon_0 m_r}\right)^2 \ln \Lambda \, n_1 \frac{\mathbf{g}_{21}}{|\mathbf{g}_{21}|^3} f(\mathbf{v}_1) \tag{9.14}$$

We may write (9.14) as a vector because $\overline{\Delta g_x}$ and $\overline{\Delta g_y}$ have been shown to be zero, so the only component remaining is that in the z direction. We have called the velocity of the target particles $\mathbf{v}_1$; $\mathbf{v}_2$ will be the velocity of the test particles.

Now we recall from the development in Chapter 2 that

$$\mathbf{v}_2 = \mathbf{G} + M_1 \mathbf{g}_{21} \qquad \text{and} \qquad \mathbf{v}_1 = \mathbf{G} + M_2 \mathbf{g}_{12} \tag{2.13}$$

where $\mathbf{G}$ is the velocity of the center of mass, while

$$M_1 = \frac{m_1}{m_1 + m_2} \qquad \text{and} \qquad M_2 = \frac{m_2}{m_1 + m_2}$$

So therefore the changes in $\mathbf{v}_2$ and $\mathbf{v}_1$ are

$$\Delta \mathbf{v}_2 = M_1 \, \Delta \mathbf{g}_{21} \qquad \text{and} \qquad \Delta \mathbf{v}_1 = M_2 \, \Delta \mathbf{g}_{12} \tag{9.15}$$

since $\mathbf{G}$ remains constant. The foregoing is also true if the angular averages are taken. Making these substitutions into Eq. (9.14) results in

$$\Delta \mathbf{v}_2 = \frac{n_1}{\pi}\left(\frac{q_1 q_2}{2\varepsilon_0 m_r}\right)^2 M_1 \ln \Lambda \frac{\mathbf{g}_{21}}{|\mathbf{g}_{21}|^3} f(\mathbf{v}_1) \tag{9.16}$$

$$\Delta \mathbf{v}_2 = \frac{n_1 q_1^2 q_2^2}{4\pi\varepsilon_0^2 m_r m_2} \ln \Lambda \frac{\mathbf{g}_{21}}{|\mathbf{g}_{21}|^3} f(\mathbf{v}_1) \tag{9.17}$$

Since $\mathbf{g}_{21} = \mathbf{v}_2 - \mathbf{v}_1$, (9.17) may be multiplied by the incremental volume in $\mathbf{v}_1$ space and written as

$$\Delta \mathbf{v}_2 \, dv_{1x} \, dv_{1y} \, dv_{1z} = \frac{n_1 q_1^2 q_2^2}{4\pi\varepsilon_0^2 m_r m_2} \ln \Lambda f(\mathbf{v}_1) \frac{\mathbf{v}_2 - \mathbf{v}_1}{|\mathbf{v}_2 - \mathbf{v}_1|^3} \, dv_{1x} \, dv_{1y} \, dv_{1z} \quad (9.18)$$

Then, the final average necessary for the Fokker–Planck term may be made by integrating over $\mathbf{v}_1$, the velocity space of the target particles. This average is

$$\langle \Delta \mathbf{v}_2 \rangle = \frac{n_1 q_1^2 q_2^2}{4\pi\varepsilon_0^2 m_r m_2} \ln \Lambda \int f(\mathbf{v}_1) \frac{\mathbf{v}_2 - \mathbf{v}_1}{|\mathbf{v}_2 - \mathbf{v}_1|^3} \, dv_{1x} \, dv_{1y} \, dv_{1z} \quad (9.19)$$

The triple integration is understood. Equation (9.19) looks, in form, exactly like the integral for the expression for an electrostatic field from a charge distribution. We can thus make a few comments concerning these similarities and draw some conclusions from them. Let us examine the integral expression for the electrostatic field produced by a distribution of charges. It is

$$\mathbf{E}(\mathbf{r}_2) = \frac{1}{4\pi\varepsilon_0} \int \rho_E(\mathbf{r}_1) \frac{\mathbf{r}_2 - \mathbf{r}_1}{|\mathbf{r}_2 - \mathbf{r}_1|^3} \, dx_1 \, dy_1 \, dz_1 \quad (9.20)$$

The analog is made by letting $\mathbf{r}_1$ be $\mathbf{v}_1$, $\mathbf{r}_2$ be $\mathbf{v}_2$, $\rho_E(\mathbf{r}_1)$ be $\sim f(\mathbf{v}_1)$, and so on. This analog permits us to see that the same types of equations that are utilized to obtain the solutions to electrostatic problems can be used for dynamical friction problems. For example, we know that a *spherically symmetric charge distribution* produces a radial electric field with no effects resulting from charges outside of the radius where the field is computed. A similar effect will appear when a *spherically symmetric velocity distribution* is considered. In other words, a net dynamical frictional force should result only from those particles that are moving *slower* than the test particle. The direction of the dynamical frictional drag force is such as to slow the test particle down (or speed it up if the velocity distribution is centered about a point other than $v = 0$) to the bulk of the velocity distribution; hence the name dynamical friction. (Recall that $\ln \Lambda$ is negative.)

We now consider the second term in the Fokker–Planck equation, which is the velocity space diffusion. In order to compute the quantity $\langle \Delta \mathbf{v} \, \Delta \mathbf{v} \rangle$ the process is exactly the same. We consider the angular and velocity averages of the products of $\Delta g_x \, \Delta g_x$, $\Delta g_x \, \Delta g_y$, $\Delta g_y \, \Delta g_z$, and so on. To find them, we first take the products and average over the deflection angles. The only terms that remain after this averaging are the products $\overline{\Delta g_x \, \Delta g_x}$, $\overline{\Delta g_z \, \Delta g_z}$, and $\overline{\Delta g_y \, \Delta g_y}$.

The integrals for the averages of each of these products are

$$\overline{\Delta g_z \Delta g_z} = \int_{\chi_{\min}}^{\chi_{\max}} \int_0^{2\pi} n_1 \left(\frac{q_1 q_2}{8\pi\varepsilon_0 m_r g^2} \right)^2 4 f_T(\mathbf{v}) g^3 \frac{\sin^4(\chi/2) \sin(\chi)\, d\chi\, d\varphi}{\sin^4(\chi/2)} \tag{9.21}$$

$$\overline{\Delta g_x \Delta g_x} = \int_{\chi_{\min}}^{\chi_{\max}} \int_0^{2\pi} n_1 \left(\frac{q_1 q_2}{8\pi\varepsilon_0 m_r g^2} \right)^2 4 f_T(\mathbf{v}) g^3 \times \frac{\sin^2(\chi/2) \cos^2(\chi/2) \sin(\chi) \cos^2(\varphi)\, d\chi\, d\varphi}{\sin^4(\chi/2)} \tag{9.22}$$

$$\overline{\Delta g_y \Delta g_y} = \int_{\chi_{\min}}^{\chi_{\max}} \int_0^{2\pi} n_1 \left(\frac{q_1 q_2}{8\pi\varepsilon_0 m_r g^2} \right)^2 4 f_T(\mathbf{v}) g^3 \times \frac{\sin^2(\chi/2) \cos^2(\chi/2) \sin(\chi) \sin^2(\varphi)\, d\chi\, d\varphi}{\sin^4(\chi/2)} \tag{9.23}$$

After completing the integrations and determining the dyadic expression $\overline{\Delta \mathbf{g}\, \Delta \mathbf{g}}$, it may be integrated over the velocity distribution of the target particles. It will be found that the components $\overline{\Delta g_x \Delta g_x}$ and $\overline{\Delta g_y \Delta g_y}$ are approximately the same. Their values are, after completing the integration over scattering angles,

$$\overline{\Delta g_x \Delta g_x} \simeq \overline{\Delta g_y \Delta g_y} \simeq \frac{n_1}{\pi} \left(\frac{q_1 q_2}{2\varepsilon_0 m_r} \right)^2 \frac{\ln \Lambda}{g} f_T(\mathbf{v}) \tag{9.24}$$

if we neglect terms which do not contain $\ln \Lambda$. $\overline{\Delta g_z \Delta g_z}$ will turn out to be quite small also, since it does not contain a term of the order of $\ln \Lambda$. Changing variables to $\mathbf{v}_1$ and $\mathbf{v}_2$ and then integrating to obtain the velocity average of Eq. (9.24) results in an expression similar to that for dynamical friction. This expression, however, is a dyadic that has components in both the x and y directions. For example,

$$\langle \Delta v_{2x} \Delta v_{2x} \rangle = \frac{n_1}{\pi} \frac{q_1 q_2}{2\varepsilon_0} \frac{\ln \Lambda}{m_2} \int \frac{f(\mathbf{v}_1)}{|\mathbf{v}_2 - \mathbf{v}_1|}\, dv_{1x}\, dv_{1y}\, dv_{1z} \tag{9.25}$$

is the x component of this dyadic. Again we note, in (9.25), the analog to electrostatics.

The developments up to this point have shown what happens to a given test particle moving with velocity $\mathbf{v}_2$ when it moves through a distribution of target particles having a velocity distribution $f(\mathbf{v}_1)$. In a plasma, there is usually no way to distinguish between target and test particles, since all particles are always in the process of making *some* kind of collision. As a result, the Fokker–Planck equation must now be integrated twice: first over the target particle distribution, and then over the test particle distribution before it can be placed in the Boltzmann equation. If electrons, for example, collide with electrons, ions, neutrals, and so on, then each of these latter distribution functions must be separately considered as the target distribution for the electrons. The results can then usually be summed directly.

In order to investigate this in somewhat greater detail, we shall extend the electrostatic analogy. Since the expression for $\langle \Delta \mathbf{v}_2 \rangle$ may be obtained in the same manner as the electrostatic field, that is, by integration over the target particles ("charge distribution"), then a scalar potential function should also materialize. For $\langle \Delta \mathbf{v}_2 \rangle$ the potential function should be such that the gradient of this function results in $\langle \Delta \mathbf{v}_2 \rangle$. Let us call this potential function H_{21}. Then, in general, we should obtain

$$\langle \Delta \mathbf{v}_2 \rangle = \Gamma_2 \frac{\partial H_{21}}{\partial \mathbf{v}_2} \tag{9.26}$$

From Eq. (9.19) we can define Γ_2 and H_{21} to be

$$\Gamma_2 = \left(\frac{q_1 q_2}{2\varepsilon_0 m_2} \right)^2 \frac{\ln \Lambda}{\pi} \tag{9.27a}$$

and

$$H_{21} = n_1 \frac{m_2}{m_r} \int \frac{f(\mathbf{v}_1)}{|\mathbf{v}_2 - \mathbf{v}_1|} dv_{1x} dv_{1y} dv_{1z} \tag{9.27b}$$

The velocity gradient operator $\partial/\partial \mathbf{v}_2$ should be respecified here. It is defined in Cartesian coordinates as

$$\frac{\partial}{\partial \mathbf{v}_2} = \frac{\partial}{\partial v_{2x}} \hat{a}_x + \frac{\partial}{\partial v_{2y}} \hat{a}_y + \frac{\partial}{\partial v_{2z}} \hat{a}_z = \nabla_{v_2} \tag{9.28}$$

These last two equations show that in Eq. (9.26) the direction of $\langle \Delta \mathbf{v}_2 \rangle$ is in the direction of $\mathbf{v}_2$, just as the electric field is in the direction of its gradient.

We can also introduce a similar potential function for $\langle \Delta \mathbf{v}_2 \, \Delta \mathbf{v}_2 \rangle$. Since, however, a dyadic must appear, the simple "gradient" operator of Eq. (9.26) is not sufficient. Here the gradient of a potential must be taken twice with respect to the velocity coordinates. The operator $\partial^2/\partial \mathbf{v}_2 \, \partial \mathbf{v}_2$ may be used as the appropriate dyadic operator. It is written out in components as

$$\begin{aligned} \frac{\partial^2}{\partial \mathbf{v}_2 \, \partial \mathbf{v}_2} = \frac{\partial}{\partial \mathbf{v}_2} \left[\frac{\partial}{\partial \mathbf{v}_2} \right] &= \hat{a}_x \frac{\partial}{\partial v_{2x}} \frac{\partial}{\partial v_{2x}} \hat{a}_x + \hat{a}_x \frac{\partial}{\partial v_{2x}} \frac{\partial}{\partial v_{2y}} \hat{a}_y + \hat{a}_x \frac{\partial}{\partial v_{2x}} \frac{\partial}{\partial v_{2z}} \hat{a}_z \\ &+ \hat{a}_y \frac{\partial}{\partial v_{2y}} \frac{\partial}{\partial v_{2x}} \hat{a}_x + \hat{a}_y \frac{\partial}{\partial v_{2y}} \frac{\partial}{\partial v_{2y}} \hat{a}_y + \hat{a}_y \frac{\partial}{\partial v_{2y}} \frac{\partial}{\partial v_{2z}} \hat{a}_z \\ &+ \hat{a}_z \frac{\partial}{\partial v_{2z}} \frac{\partial}{\partial v_{2x}} \hat{a}_x + \hat{a}_z \frac{\partial}{\partial v_{2z}} \frac{\partial}{\partial v_{2y}} \hat{a}_y + \hat{a}_z \frac{\partial}{\partial v_{2z}} \frac{\partial}{\partial v_{2z}} \hat{a}_z \end{aligned} \tag{9.29}$$

We can define the potential function F_{21} to be

$$F_{21} = n_1 \int f(\mathbf{v}_1) | \mathbf{v}_2 - \mathbf{v}_1 | \, dv_{1x} \, dv_{1y} \, dv_{1z} \tag{9.30}$$

Then $\langle \Delta \mathbf{v}_2 \, \Delta \mathbf{v}_2 \rangle$ will turn out to be

$$\langle \Delta \mathbf{v}_2 \, \Delta \mathbf{v}_2 \rangle = \Gamma_2 \frac{\partial^2 F_{21}}{\partial \mathbf{v}_2 \, \partial \mathbf{v}_2} \tag{9.31}$$

Γ_2 is the same constant as defined in Eq. (9.27).

Now we can write the Fokker–Planck equation for a plasma in terms of the potentials H_{21} and F_{21}. It becomes

$$\left(\frac{\partial f}{\partial t} \right)_{\text{coll}} = \Gamma_2 \left[- \frac{\partial}{\partial \mathbf{v}_2} \left(f(\mathbf{v}_2) \frac{\partial H_{21}}{\partial \mathbf{v}_2} \right) + \frac{1}{2} \frac{\partial^2}{\partial \mathbf{v}_2 \, \partial \mathbf{v}_2} \left(f(\mathbf{v}_2) \frac{\partial^2 F_{21}}{\partial \mathbf{v}_2 \, \partial \mathbf{v}_2} \right) \right] \tag{9.32}$$

The potentials F_{21} and H_{21}, often called Rosenbluth potentials, are related to each other. The relation between them which can be easily obtained is

$$\frac{m_2}{m_r} \frac{\partial^2}{\partial \mathbf{v}_2 \, \partial \mathbf{v}_2} F_{21} = 2H_{21} \tag{9.33}$$

We shall now evaluate the Fokker–Planck equation for a plasma. Since several types of interactions may be present, the potentials H_{21} and F_{21} must be evaluated for each type of interaction separately. For example, if the Fokker–Planck equation is to be evaluated for electrons, the potentials must be developed for electron interactions with ions, other electrons, neutrals, and so on.

For each interaction, both the velocity distribution and the collision cross section must be known. All of these interactions are coupled. For example, the electron distribution perturbs the ion distribution, which perturbs the neutral distribution, which perturbs the electron distribution, and so forth. The result is a set of coupled nonlinear differential equations. To solve this problem requires at the very least some approximations or a resort to numerical methods.

A simplified problem of this type that can be solved conveniently is that of a plasma composed of positive and negative charges. The charges of each species are assumed to be equal and opposite. Also, the particles will have the same mass and distribution function. The distribution function is to be homogeneous and isotropic, but *not necessarily* Maxwellian. This means that we may consider, for the first time, nonequilibrium distribution functions. If a non-Maxwellian distribution function is assumed, a numerical evaluation of the Fokker–Planck equation will show that the distribution will relax to a Maxwellian as time progresses. As is to be expected, the results show that the "tail" of the non-Maxwellian distribution takes a longer time to "fill out" than the rest of the non-Maxwellian distribution, since only by the velocity space diffusion process can the plasma particles gain enough energy to move at high velocity.

An additional example can be expressed more quantitatively. Suppose that the velocity distribution of the target particles is assumed to be Maxwellian. If in addition the target particles are assumed to be isotropically distributed in velocity space, then $\langle \Delta \mathbf{v}_2 \rangle$ can be seen to always point toward the center of the spherically symmetric velocity distribution. Figure 9.3 shows this arrangement.

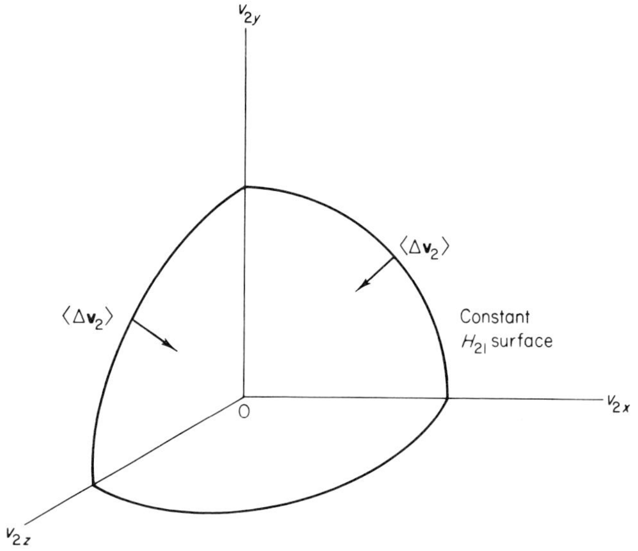

Fig. 9.3. Potential function in velocity space for a spherically symmetric distribution.

This situation means that, as can be seen from the potential definition for $\langle \Delta \mathbf{v}_2 \rangle$, the "equipotentials" for H_{21} are concentric spheres surrounding the origin in velocity space. $\langle \Delta \mathbf{v}_2 \, \Delta \mathbf{v}_2 \rangle$ will turn out to lie along the equipotential surfaces of H_{21}. The equipotential surfaces for F_{21} must also be spherically symmetric, since double differentiation is required to obtain $\langle \Delta \mathbf{v}_2 \, \Delta \mathbf{v}_2 \rangle$.

Let us assume first that the form of the distribution function for the target particles is Maxwellian, namely,

$$f(\mathbf{v}_1) = \left(\frac{m_1}{2\pi k T_1} \right)^{3/2} \exp\left(- \frac{m_1 v_1{}^2}{2 k T_1} \right) \tag{9.34}$$

Note that $f(\mathbf{v}_1)$ is *isotropic*, so the previous discussion applies. We can now develop an expression for $\langle \Delta \mathbf{v}_2 \rangle$ taking the symmetry of the problem into account. The expression for H_{21} for this distribution function is

$$H_{21} = \frac{m_2}{m_r} \int \frac{n_1}{|\mathbf{v}_2 - \mathbf{v}_1|} \left(\frac{m_1}{2\pi k T_1} \right)^{3/2} \exp\left(-\frac{m v_1^2}{2 k T_1} \right) dv_{1x}\, dv_{1y}\, dv_{1z} \quad (9.35)$$

Converting to spherical velocity space coordinates $(v_1, \theta_1, \varphi_1)$ results in

$$H_{21} = \frac{m_2}{m_r} n_1 \left(\frac{m_1}{2\pi k T_1} \right)^{3/2} \int_0^{2\pi} \int_0^{\pi} \int_0^{\infty} \frac{\exp\left(-\dfrac{m_1 v_1^2}{2 k T_1} \right)}{|\mathbf{v}_2 - \mathbf{v}_1|} v_1^2 \sin\theta_1 \, d\theta_1 \, d\varphi_1 \, dv_1 \quad (9.36)$$

We now proceed to evaluate this integral. Figure 9.4 shows the general geometrical configuration for this problem. We now reiterate that $\mathbf{v}_2$ is the velocity

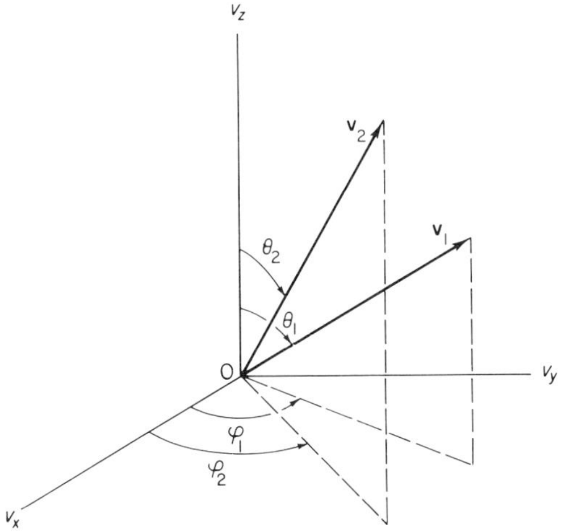

Fig. 9.4. Two sets of spherical velocity space coordinates, one each for $\mathbf{v}_1$ and $\mathbf{v}_2$.

of the *test* particle and $\mathbf{v}_1$ is the velocity of the target particle. We first integrate (9.36) over $\mathbf{v}_1$. Without loss of generality, we may orient the coordinates so that $\mathbf{v}_2$ lies along the v_z axis. Then the coordinates are as shown in Fig. 9.5. We have redefined the origins of the angles θ_1 and φ_1 but the formulation in (9.36) remains the same. The analogy to electrostatics can now be applied. We note that the expression for the electrostatic potential in terms of a charge distribution is

$$\Phi(r_1, \theta_2, \varphi_2) = \frac{1}{4\pi\varepsilon_0} \int_0^{\infty} \int_0^{2\pi} \int_0^{\pi} \frac{\rho_E(\mathbf{r}_1)}{|\mathbf{r}_2 - \mathbf{r}_1|} r_1^2 \sin\theta_1 \, d\theta_1 \, d\varphi_1 \, dr_1 \quad (9.37)$$

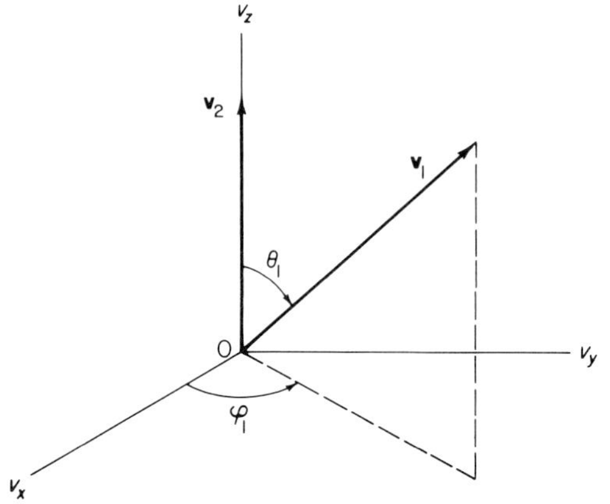

Fig. 9.5. Rotated velocity space coordinates.

expressed in spherical coordinates. The subscript 2 refers to the coordinate where the potential is measured. The subscript 1 is a reference to the coordinates of the charge distribution. Note that the integration is only over the charge distribution coordinates. Equation (9.37) is identical to the integral defining H_{21}. This means that both Φ and H_{21} satisfy Laplace's (or Poisson's) equation.

In spherical coordinates, the functions in the integral of (9.37) can be expanded in spherical harmonics. The general series for these harmonics is

$$\mathscr{F}(r_1, \theta_1, \varphi_1) = \sum_{l=0}^{\infty} \sum_{m=-l}^{+l} \left[A_{lm} r_1{}^l + B_{lm} r_1^{-(l+1)}\right] P_l^m(\cos\theta_1) \exp(jm\varphi_1) \tag{9.38}$$

where $P_l^m(\cos\theta_1)$ are the Legendre polynomials. If we assume that there is no φ_1 variation ($m = 0$), then (9.38) reduces to

$$\mathscr{F}(r_1, \theta_1) = \sum_{l=0}^{\infty} [A_{l0} r_1{}^l + B_{l0} r_1^{-(l+1)}] P_l^0(\cos\theta_1) \tag{9.39}$$

The coefficients A_{l0} and B_{l0} may be evaluated from Eq. (9.37) as follows. First, we note that the portion of the expansion (9.39) in the r_1 direction is independent of θ_1. Since the *constants* are also independent of θ_1, we may fix a value of θ_1 to simplify the expansion without loss of generality.

We will now expand $1/|\mathbf{r}_2 - \mathbf{r}_1|$ in spherical harmonics. For example, if $\mathbf{r}_1$ and $\mathbf{r}_2$ happen to lie along the z axis (that is, $\theta_1 = 0$), the quantity $|\mathbf{r}_2 - \mathbf{r}_1|$ becomes a scalar $|r_2 - r_1|$. We can always do this because r_2 is the point of observation and the z axis can be rotated to pass through the charge point. We may expand $1/|r_2 - r_1|$ from (9.37) into one of two series, depending on whether $r_2 > r_1$ or $r_1 > r_2$.

If $r_2 > r_1$, the series is

$$\frac{1}{|r_2 - r_1|} = \frac{1}{r_2}\left[\frac{1}{1-(r_1/r_2)}\right] = \frac{1}{r_2}\left[1 + \frac{r_1}{r_2} + \left(\frac{r_1}{r_2}\right)^2 + \cdots\right] \tag{9.40}$$

If $r_2 < r_1$, then the series is

$$\frac{1}{|r_2 - r_1|} = \frac{1}{r_1}\left[\frac{1}{1-(r_2/r_1)}\right] = \frac{1}{r_1}\left[1 + \frac{r_2}{r_1} + \left(\frac{r_2}{r_1}\right)^2 + \cdots\right] \tag{9.41}$$

The expansions are made differently so that both series will converge. Remember that in Chapter 8 we used this same form of a series to investigate analytic continuation. We can evaluate the coefficients A_{l0} and B_{l0} by noting that the expansion in spherical harmonics is to be made in powers of r_1. Equation (9.40) then becomes

$$\frac{1}{|r_2 - r_1|} = \frac{1}{r_2} + \left(\frac{1}{r_2^{\,2}}\right)r_1 + \left(\frac{1}{r_2^{\,3}}\right)r_1^{\,2} + \left(\frac{1}{r_2^{\,4}}\right)r_1^{\,3} + \cdots \qquad \text{for} \quad r_2 > r_1 \tag{9.42}$$

and (9.41) will be

$$\frac{1}{|r_2 - r_1|} = 0 + \frac{1}{r_1} + r_2\left(\frac{1}{r_1}\right)^2 + r_2^{\,2}\left(\frac{1}{r_1^{\,3}}\right) + \cdots \qquad \text{for} \quad r_1 > r_2 \tag{9.43}$$

The complete expansions for $1/|\mathbf{r}_2 - \mathbf{r}_1|$ ($\mathbf{r}_2$ and $\mathbf{r}_1$ not aligned together, but $\mathbf{r}_2$ always along the z-axis) are then

$$\frac{1}{|\mathbf{r}_2 - \mathbf{r}_1|} = \sum_{l=0}^{\infty}\left[\left(\frac{1}{r_2}\right)^{l+1} r_1^{\,l}\right] P_l^{\,0}(\cos\theta_1) \qquad \text{for} \quad r_2 > r_1 \tag{9.44}$$

and

$$\frac{1}{|\mathbf{r}_2 - \mathbf{r}_1|} = \sum_{l=0}^{\infty}\left[r_2^{\,l}\, r_1^{-(l+1)}\right] P_l^{\,0}(\cos\theta_1) \qquad \text{for} \quad r_1 > r_2 \tag{9.45}$$

We must now return to velocity space. The corresponding expansion for Eq. (9.36) is to be made with the term $1/|\mathbf{v}_2 - \mathbf{v}_1|$. These variables may be substituted for $\mathbf{r}_2$ and $\mathbf{r}_1$ in the expansions (9.44) and (9.45) to produce the velocity coordinate expansion. The integral in (9.36) with respect to v_1 can be evaluated in two separate parts. First, from 0 to v_2 and then, from v_2 to ∞. We need not expand the distribution function in spherical harmonics, because it is isotropic and a function only of v_1. The first part of the integral (from 0 to v_2) in (9.36) is

$$\frac{m_2}{m_r} n_1 \left(\frac{m_1}{2\pi k T_1}\right)^{3/2} \int_0^{\pi}\int_0^{2\pi}\int_0^{v_2} \exp\left(-\frac{m_1 v_1^{\,2}}{2kT_1}\right) v_1^{\,2}$$
$$\times \sum_{l=0}^{\infty}\left[\left(\frac{1}{v_2}\right)^{l+1} v_1^{\,l}\right] P_l^{\,0}(\cos\theta_1)\sin\theta_1\, d\theta_1\, d\varphi_1\, dv_1 \tag{9.46}$$

The integration with respect to φ_1 results only in a contribution of 2π, since the integral is defined to be independent of φ_1. The θ_1 integration of each term in the series (9.46) is of the form

$$\int_0^{\pi} P_l{}^0(\cos\theta_1)\sin\theta_1\, d\theta_1 \tag{9.47}$$

Since $\sin\theta_1$ is $P_1{}^1$, and Legendre polynomials are orthogonal in the index l, expression (9.47) is identically equal to zero, since $P_1{}^1$ never can appear, except for the term where $l = 0$. The only term in the series that remains must be the one where $l = 0$, because term $P_0{}^0 = 1$, so the integral of (9.47) under this condition has exactly the value of 2. The integral (9.46) is then

$$H_{21} = \frac{m_2}{m_r} n_1 \left(\frac{m_1}{2\pi k T_1}\right)^{3/2} \frac{4\pi}{v_2} \int_0^{v_2} \exp\left(-\frac{m_1 v_1{}^2}{2kT_1}\right) v_1{}^2\, dv_1 \tag{9.48}$$

The integration from v_2 to ∞ need not be performed, since, by application of Gauss's Law, the contribution to the potential from a spherical distribution outside the point in question is zero. We recall that there is only one term remaining in the series for the integral from 0 to v_2 because of the orthogonality of the Legendre functions. The velocity integral in Eq. (9.48) can be evaluated by integration by parts, if the integral is first multiplied and divided by -2α, where $\alpha = m_1/2kT_1$. The integral in Eq. (9.48) now becomes

$$\int_0^{v_2} \exp(-\alpha v_1{}^2) v_1{}^2\, dv_1 = \int_0^{v_2} \frac{v_1[-2\alpha \exp(-\alpha v_1{}^2)]}{-2\alpha}\, dv_1 \tag{9.49a}$$

We will evaluate the right-hand side of (9.49a) by integrating it by parts, which results in

$$-\frac{1}{2\alpha}\left\{[v_1 \exp(-\alpha v_1{}^2)]_0^{v_2} - \int_0^{v_2} \exp(-\alpha v_1{}^2)\, dv_1\right\} \tag{9.49b}$$

We can use the foregoing integral and return to the complete expression for potential H_{21}. This is now

$$H_{21} = -\frac{m_2}{m_1} n_1 \left(\frac{m_1}{2\pi k T_1}\right)^{1/2} \frac{2}{v_2} \times \left[v_1 \exp\left(-\frac{m_1 v_1{}^2}{2kT_1}\right)\Bigg|_0^{v_2} - \int_0^{v_2} \exp\left(-\frac{m_1 v_1{}^2}{2kT_1}\right) dv_1 \right]$$

or

$$H_{21} = -\frac{m_2}{m_1} n_1 \left(\frac{m_1}{2\pi k T_1}\right)^{1/2} \frac{2}{v_2} \times \left\{ v_2 \exp\left(-\frac{m v_2{}^2}{2kT_1}\right) - \left(\frac{\pi k T_1}{2m_1}\right)^{1/2} \operatorname{erf}\left[\left(\frac{m_1}{2kT_1}\right)^{1/2} v_2\right]\right\} \tag{9.50}$$

where $\mathrm{erf}[(m_1/2kT_1)^{1/2}v_z]$ is the *error function* defined as

$$\mathrm{erf}(x) = \frac{2}{\sqrt{\pi}} \int_0^x \exp(-y^2)\, dy \tag{9.51}$$

The expression (9.50) may now be placed in Eq. (9.26) and differentiated with respect to $\mathbf{v}_2$ to obtain $\langle \Delta\mathbf{v}_2 \rangle$. Equation (9.33) will then yield F_{21}, which may be placed in Eq. (9.31) to obtain $\langle \Delta\mathbf{v}_2\, \Delta\mathbf{v}_2 \rangle$. These values of $\langle \Delta\mathbf{v}_2 \rangle$ and $\langle \Delta\mathbf{v}_2\, \Delta\mathbf{v}_2 \rangle$ are the terms needed for the Fokker–Planck equation and we should now be able to use it to produce a term for the right-hand side of the Boltzmann equation. Note that if an isotropic Maxwellian is used, we know equilibrium is maintained. Hence, the net value of the Fokker–Planck collision term $(\partial f/\partial t)_{\mathrm{coll}}$ must be zero. However, if the distribution is non-Maxwellian, the integration methods used to obtain the Fokker–Planck term are basically the same as for the isotropic Maxwellian case.

Similar calculations may be performed for various other forms of distribution functions and, as can be imagined, the integration often has to be performed numerically.

B. VALIDITY OF COLLISIONAL MODELS

The question of the validity of the collisional models assumed will now be considered. A general statement is often made that the collision "time" should be short compared to the "time of interest" or the "times looked at" in an experimental apparatus. However, a somewhat more definitive remark can be made.

There are several time parameters previously developed that are of a fundamental nature in a plasma device: (1) ω, the frequency to be measured in the experiment; (2) ω_p, the plasma frequency; (3) ω_c, the cyclotron frequency. The last two quantities are, of course, easily defined for both ions and electrons.

We must now make some sort of definition of a collision or interaction time between two particles. This is an approximation to the interaction, since, especially for Coulomb interactions, the collision time is infinite. However, if we consider Fig. 9.6, the following definition for collision time can be applied.

If the velocity of the colliding particle relative to the target particle is g and the impact parameter is b, we shall assume that *most* of the interaction occurs when the particles are within a distance $b/2$ of the point of closest approach to each other. The collision time can then be defined as the time

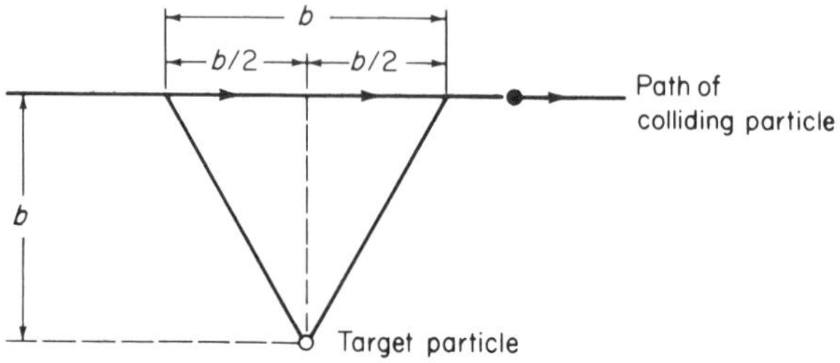

Fig. 9.6. Straight-line approximation to particle interactions.

over which the particles are within a distance $b/2$ of their closest approach or, in this example,

$$\tau_{\text{coll}} = \frac{b}{g} \tag{9.52}$$

If we define the collision time this way, the maximum collision time is directly related to the maximum impact parameter. The minimum collision time is also set by the minimum impact parameter, which is, in turn, set by the *maximum* scattering angle. This angle is normally set somewhere between 90 and 180°.

We are more concerned, however, with the maximum scattering time, since the corresponding small angle collisions are the most common in a plasma.

First, let us consider the situation when the frequency ω and the cyclotron frequency ω_c are both less than the plasma frequency ω_p. Here the plasma should be considered highly dense and a good "conductor." This means that the Debye shielding distance will be a reasonably good value to use for the maximum value of the impact parameter, and the maximum collision time for a given particle species would therefore be

$$\tau_{\max} = \frac{\lambda_D}{g} \tag{9.53}$$

If the cyclotron frequency were still kept low but the frequency of the measurement were raised so that it was greater than the plasma frequency, then the plasma would not react quite so quickly to fluctuations as before. Therefore, the maximum collision time ought to be made shorter. We may assume it to be

$$\tau_{\max} = \frac{1}{\omega} \tag{9.54}$$

Equation (9.54) may now be placed in Eq. (9.52) to obtain the maximum impact parameter for this case, which is

$$b_{\max} = \frac{g}{\omega} \tag{9.55}$$

If we now also raise ω_c to be larger than ω_p *and* ω, the maximum collision time can be defined as

$$\tau_{\max} = \frac{1}{\omega_c} \tag{9.56}$$

and the maximum impact parameter would then be

$$b_{\max} = \frac{g}{\omega_c} \tag{9.57}$$

Table 9.1 gives a summary of these cases.

Table 9.1 Maximum Impact Parameter and Minimum Scattering Time in Three Different Frequency Ranges

	$\omega < \omega_p$ $\omega < \omega_c$	$\omega > \omega_p$ $\omega_c < \omega_p$	$\omega_c > \omega > \omega_p$
$b_{\max}$	λ_D	$\frac{g}{\omega}$	$\frac{g}{\omega_c}$
$\tau_{\max}$	$\frac{\lambda_D}{g}$	$\frac{1}{\omega}$	$\frac{1}{\omega_c}$

C. DEVELOPMENT OF THE BOGOLIUBOV, BORN, GREEN, KIRKWOOD, YVON (BBGKY) HIERARCHY

The last section presented a criterion developed to ascertain when the Fokker–Planck methods may be applied to the Boltzmann equation. That is, the collisional terms should be used in the Boltzmann equation whenever it can be determined whether the values of $(\partial f/\partial t)_{\text{coll}}$ from the Fokker–Planck equation are significant. Table 9.1 gives some approximate times when the Fokker–Planck term *may* be important.

As can be seen, the lower the plasma frequency relative to the frequency of the experiment, the shorter the maximum collision time and the less information we can obtain about the interactions from the Fokker–Planck approach.

We now proceed to the different approach mentioned at the beginning of this chapter. It attacks the interactions between particles from a more fundamental viewpoint, that is, directly from the Liouville equation of Chapter 4. We recall that this equation was written in terms of a distribution function f_N for all N particles making up the plasma. However, the interaction restrictions necessary for the Boltzmann equation are not binding on Liouville's theorem.

In this treatment we will normalize f_N to 1 as

$$\int f_N(dq_1 \cdots dp_r)_1(dq_1 \cdots dp_r)_2 \cdots\cdot = 1 \tag{9.58}$$

rather than to N, so that f_N is the *probability* of finding N particles in an incremental volume in phase space (γ-space). Each particle has r degrees of freedom. The q's are the position coordinates and the p's are the momentum coordinates. Liouville's theorem is rewritten as

$$\frac{df_N}{dt} = \frac{\partial f_N}{\partial t} + [f_N, H_N] = 0 \tag{9.59}$$

where H_N is the Hamiltonian for the system of N particles. The brackets are those of the Poisson bracket, which was also defined previously in Chapter 4. Remember that since there are N particles [N separate sets of coordinates $(q_1 \cdots p_r)$], the Poisson bracket is

$$[f_N, H_N] = \sum_{j=1}^{N} \sum_{i=1}^{r} \left[\left(\frac{\partial f_N}{\partial q_i}\frac{\partial H_N}{\partial p_i}\right)_j - \left(\frac{\partial f_N}{\partial p_i}\frac{\partial H_N}{\partial q_i}\right)_j\right] \tag{9.60}$$

The *reduced* distribution functions will now be of interest since we will be interested in single-particle distributions in particular and multiple-particle distributions, other than that of the entire system, in general. For example, remember that the Boltzmann equation itself is written on a per particle basis. The general form of a reduced probability function for s particles can be written as

$$\frac{f_s}{V_s} = \int f_N(dq_1 \cdots dp_r)_{s+1} \cdots (dq_1 \cdots dp_r)_N \tag{9.61}$$

Note that we have "integrated out" the unwanted particle coordinates ($s + 1$ to N) so f_s/V_s is only a function of s sets of position and momentum coordinates.

We shall adopt the abbreviated notation dX_i to mean $(dq_1 \cdots dp_r)_i$. V_s is the total volume of s sets of *position* space coordinates, namely,

$\int (dq_1 \cdots dq_r)_{1 \cdots s}$. Equation (9.61) now becomes, in terms of the new notation,

$$\frac{f_s}{V_s} = \int f_N \, dX_{s+1} \cdots dX_N \tag{9.62}$$

Again, since f_N is a function of $X_1 \cdots X_N$, f_s/V_s will be a function only of $X_1 \cdots X_s$, hence the result of the integration is an s-particle distribution function where $s \le N$.

We shall perform the foregoing type of integration with respect to the coordinates of Eq. (9.62) on Liouville's theorem itself to obtain an equation in f_s. Liouville's theorem, when integrated in this way, becomes

$$\int \frac{\partial f_N}{\partial t} \, dX_{s+1} \cdots dX_N + \int [f_N, H_N] \, dX_{s+1} \cdots dX_N = 0 \tag{9.63}$$

The coordinates are independent of time, so we may remove the time differentiation to outside of the integral in the first term of Eq. (9.63). The integrated result for this term is

$$\int \frac{\partial f_N}{\partial t} \, dX_{s+1} \cdots dX_N = \frac{1}{V_s} \frac{\partial f_s}{\partial t} \tag{9.64}$$

We now turn our attention to the integration over the Poisson bracket, which is the second term in (9.63). Equation (9.60), which defines the Poisson bracket, may be integrated term by term in the series. In order to proceed further, we must specify something about the nature of the Hamiltonian H_N. For this development, since we are considering a plasma, we will define the Hamiltonian for N particles as

$$H_N = \sum_{i=1}^{N} \frac{\mathbf{p}_i^2}{2m_i} + \frac{1}{2} \sum_{i=1}^{N} q_i \sum_{\substack{j=1 \\ i \neq j}}^{N} \phi_{ij} \tag{9.65}$$

We use vector notation for the $\mathbf{p}_i$'s and $\mathbf{q}_i$'s to eliminate the summation $1, \ldots, r$ as in (9.60). This notation is such that $\mathbf{p}_i = (p_1, p_2, \ldots, p_r)_i$, that is, the vector $\mathbf{p}_i$ is a representation of all of the coordinates $p_1, \ldots, p_r$, and $\mathbf{q}_i$ is taken similarly for the q's. Each component must be used separately, as is the case for all vectors. q_i is the electric charge of particle i.

Let ϕ_{ij} be a Coulomb potential, dependent only upon the particle separations; ϕ_{ij} is then defined as

$$\phi_{ij} = \frac{q_j}{4\pi\varepsilon_0} \frac{1}{|\mathbf{q}_i - \mathbf{q}_j|} \tag{9.66}$$

where q_j is the electric charge of particle j. Note that the factor of $\frac{1}{2}$ is required in Eq. (9.65), so that the potentials are not counted twice, that is,

$\phi_{ij} = \phi_{ji}$. If the plasma is to be assumed to be of infinite extent, and magnetic field forces are neglected, then Eq. (9.65) is a proper representation of the Hamiltonian H_N.

A typical term in the Poisson bracket integral for (9.63) is then

$$\int \left(\frac{\partial f_N}{\partial \mathbf{q}_i} \cdot \frac{\partial H_N}{\partial \mathbf{p}_i} - \frac{\partial f_N}{\partial \mathbf{p}_i} \cdot \frac{\partial H_N}{\partial \mathbf{q}_i} \right) dX_{s+1} \cdots dX_N \tag{9.67}$$

We will now look at the terms $\partial H_N / \partial \mathbf{p}_i$ and $\partial H_N / \partial \mathbf{q}_i$ in (9.67). H_N, except for the term involving $\mathbf{p}_i^2 / 2m_i$, is independent of $\mathbf{p}_i$. Therefore

$$\frac{\partial H_N}{\partial \mathbf{p}_i} = \frac{\mathbf{p}_i}{m_i} \tag{9.68}$$

In addition, the only term in the Hamiltonian that is a function of the $\mathbf{q}_i$'s is ϕ_{ij}. Therefore we can obtain the relation

$$\frac{\partial H_N}{\partial \mathbf{q}_i} = \mathscr{q}_i \sum_{\substack{j=1 \\ i \neq j}}^{N} \frac{\partial \phi_{ij}}{\partial \mathbf{q}_i} = -\mathscr{q}_i \mathbf{E}_i \tag{9.69}$$

$\mathbf{E}_i$, the net electric field at charge i, is the vector sum of the individual electrostatic fields of *each* of the charges. The complete integral for the Poisson bracket in Liouville's theorem then becomes

$$\sum_{i=1}^{N} \int \left[\frac{\partial f_N}{\partial \mathbf{q}_i} \cdot \frac{\mathbf{p}_i}{m_i} - \mathscr{q}_i \sum_{\substack{j=1 \\ i \neq j}}^{N} \frac{\partial f_N}{\partial \mathbf{p}_i} \cdot \frac{\partial \phi_{ij}}{\partial \mathbf{q}_i} \right] dX_{s+1} \cdots dX_N \tag{9.70}$$

There are two cases to consider in evaluating this integral, since i may be less than or greater than s as the summation is made. To take this into account we will first split the integral into its two parts, so Eq. (9.70) is now the difference of two integrals, which are

$$\sum_{i=1}^{N} \int \left(\frac{\partial f_N}{\partial \mathbf{q}_i} \cdot \frac{\mathbf{p}_i}{m_i} \right) dX_{s+1} \cdots dX_N \tag{9.71}$$

and

$$\sum_{i=1}^{N} \mathscr{q}_i \int \sum_{\substack{j=1 \\ i \neq j}}^{N} \frac{\partial f_N}{\partial \mathbf{p}_i} \cdot \frac{\partial \phi_{ij}}{\partial \mathbf{q}_i} \, dX_{s+1} \cdots dX_N \tag{9.72}$$

Note that we are still using the complete summation. Let us first consider (9.71). It should be emphasized that *each* term in this series gets integrated over *all* coordinates from X_{S+1} to X_N. There must therefore be an integration in each term in the series (9.71) when $i > s$, that is,

$$\int \left(\frac{\partial f_N}{\partial \mathbf{q}_i} \cdot \frac{\mathbf{p}_i}{m_i} \right) d\mathbf{q}_i \tag{9.73}$$

since $dX_i = (dq_1 \cdots dp_r)_i$. Note that this term may only appear for values of $i > s$. Since the $\mathbf{p}$'s and $\mathbf{q}$'s are orthogonal, the integrand can be written as a differential of a product. Thus (9.73) becomes

$$\int \frac{\partial}{\partial \mathbf{q}_i} \cdot \left[\frac{f_N \mathbf{p}_i}{m_i}\right] d\mathbf{q}_i = 0 \tag{9.74}$$

To show that (9.74) is equal to (9.73), note that the product in (9.74) may be differentiated with respect to $\mathbf{q}_i$ and (9.73) will result, since $\partial \mathbf{p}_i / \partial \mathbf{q}_i \equiv 0$. When Eq. (9.74) is integrated by parts, the result shows only the term $f_N \mathbf{p}_i / m_i$ to be evaluated at the extremes of $\mathbf{p}_i = \pm \infty$. We will assume that f_N approaches zero very rapidly at these extremes, and thus the value of the quantity $f_N \mathbf{p}_i$ in (9.74) is zero. Hence, (9.73) is zero. Further integration over all the remaining coordinates in expression (9.71) for the given value i results only in additional multiplicative terms to the value of (9.73) of zero. Therefore, this means that the integration of the series in (9.71) will have no contributions from any terms for which $i > s$. The integrated series (9.71) is thus only composed of the following shortened series

$$\sum_{i=1}^{s} \int \left(\frac{\partial f_N}{\partial \mathbf{q}_i} \cdot \frac{\mathbf{p}_i}{m_i}\right) dX_{s+1} \cdots dX_N \tag{9.75}$$

Restating this, we have just shown every term for which $i > s$ in the integrated series of Eq. (9.71) is now identically zero. We can finish the integration in (9.75), by noting that the subscript i is *never* one of the values of the X's, so the values of $\mathbf{p}_i$ and $\mathbf{q}_i$ are *never* found in the X's. Therefore the integration *only* affects f_N. The final value of the integrated series that can be used for the integral in (9.71) is

$$\frac{1}{V_s} \sum_{i=1}^{s} \left(\frac{\partial f_s}{\partial \mathbf{q}_i} \cdot \frac{\mathbf{p}_i}{m_i}\right) \tag{9.76}$$

We will return to Eq. (9.76) subsequently. A similar process will be followed to integrate the series (9.72). We first break it up into two terms as shown in Eq. (9.77).

$$\begin{aligned}
&\sum_{i=1}^{N} \mathcal{G}_i \int \sum_{\substack{j=1 \\ i \neq j}}^{N} \left(\frac{\partial f_N}{\partial \mathbf{p}_i} \cdot \frac{\partial \phi_{ij}}{\partial \mathbf{q}_i}\right) dX_{s+1} \cdots dX_N \\
&\quad = \sum_{i=1}^{s} \mathcal{G}_i \int \sum_{\substack{j=1 \\ i \neq j}}^{N} \left(\frac{\partial f_N}{\partial \mathbf{p}_i} \cdot \frac{\partial \phi_{ij}}{\partial \mathbf{q}_i}\right) dX_{s+1} \cdots dX_N \\
&\qquad + \sum_{i=s+1}^{N} \mathcal{G}_i \int \sum_{\substack{j=1 \\ i \neq j}}^{N} \left(\frac{\partial f_N}{\partial \mathbf{p}_i} \cdot \frac{\partial \phi_{ij}}{\partial \mathbf{q}_i}\right) dX_{s+1} \cdots dX_N
\end{aligned} \tag{9.77}$$

Both series on the right-hand side of (9.77) will be integrated. The second term on the right-hand side of (9.77) can be shown to be zero. This is true because there will be, for every term in the series whenever $i > s$, an integration of the form

$$\int \mathscr{G}_i \sum_{\substack{j=1 \\ i \neq j}}^{N} \left(\frac{\partial f_N}{\partial \mathbf{p}_i} \cdot \frac{\partial \phi_{ij}}{\partial \mathbf{q}_i} \right) d\mathbf{p}_i \tag{9.78}$$

Then, by virtue of the orthogonality of the **p**'s and **q**'s (9.78) becomes

$$\int \mathscr{G}_i \frac{\partial}{\partial \mathbf{p}_i} \cdot \left(f_N \sum_{\substack{j=1 \\ i \neq j}}^{N} \frac{\partial \phi_{ij}}{\partial \mathbf{q}_i} \right) d\mathbf{p}_i \equiv 0 \tag{9.79}$$

Again this is so because of the zero value of f_N at the extremes of $\mathbf{p}_i$. This is exactly the same reasoning that was used to arrive at expressions (9.75) and (9.76). The only term left in (9.77) is the first one on the right-hand side. This term may, in turn, be broken into two parts involving the summation over j. The two parts are

$$\sum_{i=1}^{s} \mathscr{G}_i \int \sum_{\substack{j=1 \\ i \neq j}}^{s} \left(\frac{\partial f_N}{\partial \mathbf{p}_i} \cdot \frac{\partial \phi_{ij}}{\partial \mathbf{q}_i} \right) dX_{s+1} \cdots dX_N \tag{9.80}$$

and

$$\sum_{i=1}^{s} \mathscr{G}_i \int \sum_{j=s+1}^{N} \left(\frac{\partial f_N}{\partial \mathbf{p}_i} \cdot \frac{\partial \phi_{ij}}{\partial \mathbf{q}_i} \right) dX_{s+1} \cdots dX_N \tag{9.81}$$

The integrated result for (9.80) is relatively straightforward. Since none of the factors, except f_N, in the integrand involve the coordinates $X_{s+1} \cdots X_N$, we may perform the integration first, which results in the following series for (9.80):

$$\frac{1}{V_s} \sum_{i=1}^{s} \mathscr{G}_i \sum_{\substack{j=1 \\ i \neq j}}^{s} \frac{\partial f_s}{\partial \mathbf{p}_i} \cdot \frac{\partial \phi_{ij}}{\partial \mathbf{q}_i} \tag{9.82}$$

Again, we will retain this series for future use. We now examine expression (9.81). It may be integrated term by term in the series for j. The first term to be integrated in this series ($j = s + 1$) is

$$\sum_{i=1}^{s} \mathscr{G}_i \int \left(\frac{\partial f_N}{\partial \mathbf{p}_i} \cdot \frac{\partial \phi_{i,s+1}}{\partial \mathbf{q}_i} \right) dX_{s+1} \cdots dX_N \tag{9.83}$$

We may easily integrate over all coordinates from $X_{s+2} \cdots X_N$ since again none of the terms in the integrand, except for f_N, depend upon *these* coordinates. The integration over X_{s+1} remains.

Now since, from the definition of the reduced probability distribution

$$\int f_N \, dX_{s+2} \cdots dX_N = \frac{f_{s+1}}{V_{s+1}} \tag{9.84}$$

expression (9.83) may be changed to read

$$\frac{1}{V_{s+1}} \sum_{i=1}^{s} \varphi_i \int \left(\frac{\partial f_{s+1}}{\partial \mathbf{p}_i} \cdot \frac{\partial \phi_{i,\, s+1}}{\partial \mathbf{q}_i} \right) dX_{s+1} \tag{9.85}$$

Note that expression (9.85) has an integration operation remaining and that this integration involves f_{s+1}, which is the reduced probability distribution of one order greater than f_s.

There are now a total of $(N - s)$ such terms as shown in (9.85) in the series for j of expression (9.81). Each one of them may be integrated in exactly the same way to arrive at a term like expression (9.85). Therefore, all $(N - s)$ terms look *exactly* like (9.85) in that they all involve f_{s+1}. Now since the actual identities of the particles are not required, it would appear that the expression (9.85) multiplied by $(N - s)$ would give the complete required integrated series expression for (9.81). This series is then written as

$$\frac{N-s}{V_{s+1}} \sum_{i=1}^{s} \varphi_i \int \left(\frac{\partial f_{s+1}}{\partial \mathbf{p}_i} \cdot \frac{\partial \phi_{i,\, s+1}}{\partial \mathbf{q}_i} \right) dX_{s+1} \tag{9.86}$$

This completes the integrations necessary for Liouville's theorem.

And now, after this painstaking venture, we may examine the integrated equation in detail. The expression that is to be integrated is shown in (9.59). By using (9.64), (9.76), (9.82), and (9.86), we may write out the integrated form of Liouville's equation as

$$\frac{1}{V_s} \frac{\partial f_s}{\partial t} + \frac{1}{V_s} \sum_{i=1}^{s} \left(\frac{\partial f_s}{\partial \mathbf{q}_i} \cdot \frac{\mathbf{p}_i}{m_i} \right) - \frac{1}{V_s} \sum_{i=1}^{s} \varphi_i \sum_{\substack{j=1 \\ i \neq j}}^{s} \frac{\partial f_s}{\partial \mathbf{p}_i} \cdot \frac{\partial \phi_{ij}}{\partial \mathbf{q}_i}$$

$$- \frac{(N-s)}{V_{s+1}} \sum_{i=1}^{s} \varphi_i \int \left(\frac{\partial f_{s+1}}{\partial \mathbf{p}_i} \cdot \frac{\partial \phi_{i,\, s+1}}{\partial \mathbf{q}_i} \right) dX_{s+1} = 0 \tag{9.87}$$

Each term in (9.87) involves f_s *except* the last one, which involves f_{s+1}. We will discuss this situation shortly.

If we now multiply Eq. (9.87) through by V_s and rearrange terms, we may obtain

$$\frac{\partial f_s}{\partial t} + \sum_{i=1}^{s} \frac{\partial f_s}{\partial \mathbf{q}_i} \cdot \frac{\mathbf{p}_i}{m_i} - \sum_{i=1}^{s} \varphi_i \sum_{\substack{j=1 \\ i \neq j}}^{s} \frac{\partial f_s}{\partial \mathbf{p}_i} \cdot \frac{\partial \phi_{ij}}{\partial \mathbf{q}_i}$$

$$= \frac{N-s}{V} \sum_{i=1}^{s} \varphi_i \int \frac{\partial f_{s+1}}{\partial \mathbf{p}_i} \cdot \frac{\partial \phi_{i,\, s+1}}{\partial \mathbf{q}_i} \, dX_{s+1} \tag{9.88}$$

where $V = V_{s+1}/V_s$, the total volume of one set of position space coordinates. If we now let N and V approach infinity, but require their quotient to remain constant, then $(N - s)/V \to N/V \to n_0$ which is the mean particle density per unit volume of position space.

This last equation, (9.88), is very interesting because the left-hand side of (9.88) is entirely in terms of f_s, the distribution function for s particles. The right-hand side, however, is in terms of f_{s+1}. Let us consider an example to see what this result means.

Suppose that we wish to know f_1. This is, except perhaps for a normalization constant, the f distribution used previously. We then can write Eq. (9.88) for f_1. This is done by setting $s = 1$:

$$\frac{\partial f_1}{\partial t} + \frac{\mathbf{p}_1}{m_1} \cdot \frac{\partial f_1}{\partial \mathbf{q}_1} = n_0 \varphi_1 \int \frac{\partial \phi_{12}}{\partial \mathbf{q}_1} \cdot \frac{\partial f_2}{\partial \mathbf{p}_1}\, dX_2 \tag{9.89}$$

Note that the left-hand side of (9.89) only involves the one-particle distribution function, while the right-hand side of (9.89) involves f_2 (the two-particle function), the next higher-order function in this *hierarchy* of distributions. [The third term on the left-hand side of Eq. (9.88) does not appear in Eq. (9.89) because particle number 1 does not interact with itself.] Since we now must know f_2 to determine f_1, the equation for f_2 must be solved. This equation is

$$\begin{aligned}\frac{\partial f_2}{\partial t} + \frac{\mathbf{p}_1}{m_1} \cdot \frac{\partial f_2}{\partial \mathbf{q}_1} + \frac{\mathbf{p}_2}{m_2} \cdot \frac{\partial f_2}{\partial \mathbf{q}_2} - \varphi_1 \frac{\partial \phi_{12}}{\partial \mathbf{q}_1} \cdot \frac{\partial f_2}{\partial \mathbf{p}_1} - \varphi_2 \frac{\partial \phi_{21}}{\partial \mathbf{q}_2} \cdot \frac{\partial f_2}{\partial \mathbf{p}_2} \\ = n_0 \varphi_1 \int \frac{\partial \phi_{13}}{\partial \mathbf{q}_1} \cdot \frac{\partial f_3}{\partial \mathbf{p}_1}\, dX_3 + n_0 \varphi_2 \int \frac{\partial \phi_{23}}{\partial \mathbf{q}_2} \cdot \frac{\partial f_3}{\partial \mathbf{p}_2}\, dX_3\end{aligned} \tag{9.90}$$

Here the left-hand side of (9.90) only depends on f_2, while the right side involves f_3. To solve for f_3 then, requires a further equation, and so on. If the hierarchy of equations is not truncated somewhere, the result will be that at the very end of the hierarchy, the knowledge of f_N is required and no additional knowledge is gained from Liouville's theorem.

We can, however, proceed with a solution by investigating an expansion for the distribution functions in terms of a characteristic parameter. The parameter will enable us to decide how to truncate the hierarchy. To obtain this parameter, we first make Eq. (9.89) and (9.90) dimensionless by substituting the following three dimensionless quantities:

$$T = t\omega_\mathrm{p} \tag{9.91}$$

$$\mathbf{Q}_i = \frac{\mathbf{q}_i}{\lambda_\mathrm{D}} \quad \text{and} \quad \mathbf{P}_i = \frac{\mathbf{p}_i}{\omega_\mathrm{p} \lambda_\mathrm{D} m} \tag{9.92}$$

Here we have assumed that each particle has mass m. λ_D and ω_p have been defined previously as the Debye length and plasma frequency, respectively.

Equation (9.89) then becomes, after substitution of the quantities above,

$$\omega_p \frac{\partial f_1}{\partial T} + \omega_p \mathbf{P}_1 \cdot \frac{\partial f_1}{\partial \mathbf{Q}_1} = \frac{n_0 q_1}{\omega_p \lambda_D{}^2 m} \int \frac{\partial \phi_{12}}{\partial \mathbf{Q}_1} \cdot \frac{\partial f_2}{\partial \mathbf{P}_1} \, dX_2 \tag{9.93}$$

Then, dividing (9.93) through by ω_p, we get

$$\frac{\partial f_1}{\partial T} + \mathbf{P}_1 \cdot \frac{\partial f_1}{\partial \mathbf{Q}_1} = \frac{n_0 q_1}{\omega_p{}^2 \lambda_D m} \int \frac{\partial \phi_{12}}{\partial \mathbf{Q}_1} \cdot \frac{\partial f_2}{\partial \mathbf{P}_1} \, dX_2 \tag{9.94}$$

We note that the functions f_1, and so on, must now be written as functions of $Q_1 \cdots P_r$, the dimensionless parameters. dX_1, dX_2, and so on, will also now be dimensionless. Now, since

$$\phi_{12} = \frac{q_2}{4\pi\varepsilon_0 |\mathbf{q}_2 - \mathbf{q}_1|}$$

for Coulomb fields, we may apply the dimensionless parameters to this function to obtain

$$\phi_{12} = \frac{q_2}{4\pi\varepsilon_0 \lambda_D |\mathbf{Q}_2 - \mathbf{Q}_1|} = \Phi_{12} \frac{q_2}{4\pi\varepsilon_0 \lambda_D} \tag{9.95}$$

where Φ_{12} is a *dimensionless* potential function. Equation (9.94) now becomes

$$\frac{\partial f_1}{\partial T} + \mathbf{P}_1 \cdot \frac{\partial f_1}{\partial \mathbf{Q}_1} = \frac{n_0 q_1 q_2 \lambda_D{}^3}{\omega_p{}^2 4\pi\varepsilon_0 \lambda_D{}^3 m} \int \frac{\partial \Phi_{12}}{\partial \mathbf{Q}_1} \cdot \frac{\partial f_2}{\partial \mathbf{P}_1} \, d\mathbf{X}_2 \tag{9.96}$$

The factor of $\lambda_D{}^3$ in the numerator of the right-hand side of (9.96) appears as a result of making the volume, which appears after the integration with respect to X_2, dimensionless.

From the definition of the sth-order distribution function, and remembering that n_0 is the density per unit volume in position space, we can simplify the constant term on the right-hand side of Eq. (9.96). If $q_1 = q_2 = q$, then the multiplicative constant in (9.96) becomes simply $1/4\pi$, since $\omega_p{}^2 = q^2 n_0 / m\varepsilon_0$ where n_0 is the density per unit volume. Equation (9.89) is, in final dimensionless form

$$\frac{\partial f_1}{\partial T} + \mathbf{P}_1 \cdot \frac{\partial f_1}{\partial \mathbf{Q}_1} = \frac{1}{4\pi} \int \frac{\partial \Phi_{12}}{\partial \mathbf{Q}_1} \cdot \frac{\partial f_2}{\partial \mathbf{P}_1} \, d\mathbf{X}_2 \tag{9.97}$$

A similar development may be followed for Eq. (9.90). Written in terms of the same dimensionless quantities it becomes

$$\frac{\partial f_2}{\partial T} + \mathbf{P}_1 \cdot \frac{\partial f_2}{\partial \mathbf{Q}_1} + \mathbf{P}_2 \cdot \frac{\partial f_2}{\partial \mathbf{Q}_2} - \frac{\xi}{4\pi} \frac{\partial \Phi_{12}}{\partial \mathbf{Q}_1} \cdot \frac{\partial f_2}{\partial \mathbf{P}_1} - \frac{\xi}{4\pi} \frac{\partial \Phi_{21}}{\partial \mathbf{Q}_2} \cdot \frac{\partial f_2}{\partial \mathbf{P}_2}$$

$$= \frac{1}{4\pi} \int \frac{\partial \Phi_{13}}{\partial \mathbf{Q}_1} \cdot \frac{\partial f_3}{\partial \mathbf{P}_1} \, d\mathbf{X}_3 + \frac{1}{4\pi} \int \frac{\partial \Phi_{23}}{\partial \mathbf{Q}_2} \cdot \frac{\partial f_3}{\partial \mathbf{P}_2} \, d\mathbf{X}_3 \tag{9.98}$$

The parameter $\xi = 1/n_0 \lambda_D{}^3$, which is inversely proportional to the number of particles in a Debye sphere, is typically a small quantity for plasmas. To solve (9.97), (9.98), and any of the higher-order equations, we may expand them in terms of the parameter ξ. This method will permit a truncation of the hierarchy. The expansion is developed as follows.

D. CLUSTER EXPANSION

If a very large number of particles ($N \to \infty$) are considered, we may make the following nonequilibrium *Mayer cluster* expansion of the distribution functions. We first recall that the forms of the distribution functions f_s depend only upon the coordinates of the number of particles up to the order of the function, that is, $f_1 = f_1(X_1), f_2 = f_2(X_1, X_2), f_3 = f_3(X_1, X_2, X_3)$, and so on. We then make the following expansions for these distribution functions.

$$
\begin{aligned}
f_1(X_1) &= f_1(X_1) \\
f_2(X_1, X_2) &= f_1(X_1)f_1(X_2) + \mathscr{P}(X_1, X_2) \\
f_3(X_1, X_2, X_3) &= f_1(X_1)f_1(X_2)f_1(X_3) + f_1(X_1)\mathscr{P}(X_2, X_3) \\
&\quad + f_1(X_2)\mathscr{P}(X_1, X_3) + f_1(X_3)\mathscr{P}(X_1, X_2) \\
&\quad + T(X_1, X_2, X_3)
\end{aligned}
\tag{9.99}
$$

This last set of equations states that, for example, f_2 is to be expressed as the sum of two terms. The first term is called the *factorable part*. In other words, this part of the expansion for f_2 is a product of two f_1 distributions, with no correlations between them. The second term $\mathscr{P}(X_1, X_2)$ (the *pair correlation function*) gives the correlated part of the distribution function f_2, if any. In this case, correlation means that state X_1 depends on the values of X_2 and vice versa. The expansion for f_3 is much the same, with both factorable and correlated terms involving $\mathscr{P}$. We may write $\mathscr{P}(X_1, X_2)$, $\mathscr{P}(X_1, X_3)$, and $\mathscr{P}(X_2, X_3)$, since these are all pair correlation functions. There can also be a third interaction $T(X_1, X_2, X_3)$ (the *triplet* correlation function), which gives the correlation of three-body interactions. Further expansions of the higher-order distribution functions will have correspondingly higher correlating functions.

We shall now assume that the $\mathscr{P}$ and T terms are always small compared to the factorable parts of the distribution functions. If this assumption is justified, then it may be possible for a *nonequilibrium* perturbation expansion to be written.

We shall assume that $\mathscr{P}/f_1 f_1$ is of order ξ. We will then retain only those terms in Eq. (9.97) and (9.98) that are of order ξ and below. We will neglect $T(X_1, X_2, X_3)$. This situation implies that we are expanding in ξ. Written in terms of the cluster expansion, we may return to the *dimensional* hierarchy equations and substitute expansions like (9.99) for the distribution functions into them. The first-order *dimensional* equation is now

$$\frac{\partial f_1}{\partial t} + \frac{\mathbf{p}_1}{m_1} \cdot \frac{\partial f_1}{\partial \mathbf{q}_1} = n_0 \mathscr{g}_1 \int \frac{\partial \phi_{12}}{\partial \mathbf{q}_1} \cdot \frac{\partial f_1(X_1)}{\partial \mathbf{p}_1} f_1(X_2)\, dX_2 + n_0 \mathscr{g}_1 \int \frac{\partial \phi_{12}}{\partial \mathbf{q}_1} \cdot \frac{\partial \mathscr{P}(X_1, X_2)}{\partial \mathbf{p}_1}\, dX_2 \tag{9.100}$$

The second-order equation becomes

$$\begin{aligned}
&\left(\frac{\partial}{\partial t} + \frac{\mathbf{p}_1}{m_1} \cdot \frac{\partial}{\partial \mathbf{q}_1} + \frac{\mathbf{p}_2}{m_2} \cdot \frac{\partial}{\partial \mathbf{q}_2}\right) \mathscr{P}(X_1, X_2) \\
&\quad - \left[\mathscr{g}_1 \frac{\partial \phi_{12}}{\partial \mathbf{q}_1} \cdot \frac{\partial}{\partial \mathbf{p}_1} + \mathscr{g}_2 \frac{\partial \phi_{21}}{\partial \mathbf{q}_2} \cdot \frac{\partial}{\partial \mathbf{p}_2}\right] [f_1(X_1) f_1(X_2) + \mathscr{P}(X_1, X_2)] \\
&= n_0 \mathscr{g}_1 \int f_1(X_3) \frac{\partial \phi_{13}}{\partial \mathbf{q}_1} \cdot \frac{\partial \mathscr{P}(X_1, X_2)}{\partial \mathbf{p}_1}\, dX_3 \\
&\quad + n_0 \mathscr{g}_2 \int f_1(X_3) \frac{\partial \phi_{23}}{\partial \mathbf{q}_2} \cdot \frac{\partial \mathscr{P}(X_1, X_2)}{\partial \mathbf{p}_2}\, dX_3 \\
&\quad + n_0 \mathscr{g}_1 \frac{\partial f_1(X_1)}{\partial \mathbf{p}_1} \cdot \int \frac{\partial \phi_{13}}{\partial \mathbf{q}_1} \mathscr{P}(X_2, X_3)\, dX_3 \\
&\quad + n_0 \mathscr{g}_2 \frac{\partial f_1(X_2)}{\partial \mathbf{p}_2} \cdot \int \frac{\partial \phi_{23}}{\partial \mathbf{q}_2} \mathscr{P}(X_1, X_3)\, dX_3
\end{aligned} \tag{9.101}$$

We may now convert our notation from the X's, $\mathbf{q}$'s, and $\mathbf{p}$'s back to position and velocity by the following substitutions, neglecting relativity

$$\mathbf{q} \to \mathbf{x}$$

$$\mathbf{p} \to m\mathbf{v}$$

$$d\mathbf{p}\, f_1(\mathbf{q}, \mathbf{p}) \to f_1(\mathbf{x}, \mathbf{v})\, d\mathbf{v} \tag{9.102}$$

$$\mathscr{P}(X_1, X_2)\, d\mathbf{p}_1\, d\mathbf{p}_2 \to \mathscr{P}(\mathbf{x}_1, \mathbf{v}_1, \mathbf{x}_2, \mathbf{v}_2)\, d\mathbf{v}_1\, d\mathbf{v}_2 \tag{9.103}$$

The mass m is assumed to be made part of the functions $f_1(\mathbf{x}, \mathbf{v})$ and $\mathscr{P}(\mathbf{x}_1, \mathbf{v}_1, \mathbf{x}_2, \mathbf{v}_2)$.

Equations (9.100) and (9.101) may then be rewritten with the foregoing substitutions as

$$\frac{\partial f_1(\mathbf{x}_1, \mathbf{v}_1)}{\partial t} + \mathbf{v}_1 \cdot \frac{\partial f_1(\mathbf{x}_1, \mathbf{v}_1)}{\partial \mathbf{x}_1} = \frac{n_0 q_1}{m} \int \frac{\partial \phi_{12}}{\partial \mathbf{x}_1} \cdot \frac{\partial f_1(\mathbf{x}_1, \mathbf{v}_1)}{\partial \mathbf{v}_1} f_1(\mathbf{x}_2, \mathbf{v}_2)\, d\mathbf{x}_2\, d\mathbf{v}_2 + \frac{n_0 q_1}{m} \int \frac{\partial \phi_{12}}{\partial \mathbf{x}_1} \cdot \frac{\partial \mathscr{P}(\mathbf{x}_1, \mathbf{v}_1, \mathbf{x}_2, \mathbf{v}_2)}{\partial \mathbf{v}_1} d\mathbf{x}_2\, d\mathbf{v}_2 \tag{9.104}$$

and

$$\begin{aligned}
&\left(\frac{\partial}{\partial t} + \mathbf{v}_1 \cdot \frac{\partial}{\partial \mathbf{x}_1} + \mathbf{v}_2 \cdot \frac{\partial}{\partial \mathbf{x}_2}\right) \mathscr{P}(\mathbf{x}_1, \mathbf{v}_1, \mathbf{x}_2, \mathbf{v}_2) \\
&= \left[\frac{q_1}{m} \frac{\partial \phi_{12}}{\partial \mathbf{x}_1} \cdot \frac{\partial}{\partial \mathbf{v}_1} + \frac{q_2}{m} \frac{\partial \phi_{21}}{\partial \mathbf{x}_2} \cdot \frac{\partial}{\partial \mathbf{v}_2}\right] [f_1(\mathbf{x}_1, \mathbf{v}_1) f_1(\mathbf{x}_2, \mathbf{v}_2)] \\
&+ \frac{n_0 q_1}{m} \int f_1(\mathbf{x}_3, \mathbf{v}_3) \frac{\partial \phi_{13}}{\partial \mathbf{x}_1} \cdot \frac{\partial \mathscr{P}(\mathbf{x}_1, \mathbf{v}_1, \mathbf{x}_2, \mathbf{v}_2)}{\partial \mathbf{v}_1} d\mathbf{x}_3\, d\mathbf{v}_3 \\
&+ \frac{n_0 q_2}{m} \int f_1(\mathbf{x}_3, \mathbf{v}_3) \frac{\partial \phi_{23}}{\partial \mathbf{x}_2} \cdot \frac{\partial \mathscr{P}(\mathbf{x}_1, \mathbf{v}_1, \mathbf{x}_2, \mathbf{v}_2)}{\partial \mathbf{v}_2} d\mathbf{x}_3\, d\mathbf{v}_3 \\
&+ \frac{n_0 q_1}{m} \frac{\partial f_1(\mathbf{x}_1, \mathbf{v}_1)}{\partial \mathbf{v}_1} \cdot \int \frac{\partial \phi_{13}}{\partial \mathbf{x}_1} \mathscr{P}(\mathbf{x}_2, \mathbf{v}_2, \mathbf{x}_3, \mathbf{v}_3)\, d\mathbf{x}_3\, d\mathbf{v}_3 \\
&+ \frac{n_0 q_2}{m} \frac{\partial f_1(\mathbf{x}_2, \mathbf{v}_2)}{\partial \mathbf{v}_2} \cdot \int \frac{\partial \phi_{23}}{\partial \mathbf{x}_2} \mathscr{P}(\mathbf{x}_1, \mathbf{v}_1, \mathbf{x}_3, \mathbf{v}_3)\, d\mathbf{x}_3\, d\mathbf{v}_3
\end{aligned} \tag{9.105}$$

Note that, for convenience, $m_1 = m_2 = m_3 = m$, and so on.

We now must attempt to solve these equations. They are coupled in $\mathscr{P}$ and f_1. Note that we have neglected $\mathscr{P}(\mathbf{x}_1, \mathbf{v}_1, \mathbf{x}_2, \mathbf{v}_2)$ when it appeared with $f_1(\mathbf{x}_1, \mathbf{v}_1) f_1(\mathbf{x}_2, \mathbf{v}_2)$ in Eq. (9.101) in order to obtain Eq. (9.105).

If we first consider the quantity ξ to be zero, then $\mathscr{P} \to 0$. That is, no correlations between particles will then exist. Equation (9.104) then becomes

$$\frac{\partial f_1(\mathbf{x}_1, \mathbf{v}_1)}{\partial t} + \mathbf{v}_1 \cdot \frac{\partial f_1(\mathbf{x}_1, \mathbf{v}_1)}{\partial \mathbf{x}_1} - \frac{n_0 q_1}{m} \frac{\partial f_1(\mathbf{x}_1, \mathbf{v}_1)}{\partial \mathbf{v}_1} \cdot \int \frac{\partial \phi_{12}}{\partial \mathbf{x}_1} f_1(\mathbf{x}_2, \mathbf{v}_2)\, d\mathbf{x}_2\, d\mathbf{v}_2 = 0 \tag{9.106}$$

We may simplify (9.106) to

$$\frac{\partial f}{\partial t} + \mathbf{v} \cdot \frac{\partial f}{\partial \mathbf{x}} + \frac{q \mathbf{E}}{m} \cdot \frac{\partial f}{\partial \mathbf{v}} = 0 \tag{9.107}$$

We have removed the subscripts 1, since there is no correlation to consider. The net electric field $\mathbf{E}$ is defined as

$$\mathbf{E} = -n_0 \int \frac{\partial \phi_{12}}{\partial \mathbf{x}_1} f_1(\mathbf{x}_2, \mathbf{v}_2)\, d\mathbf{x}_2\, d\mathbf{v}_2 \tag{9.108}$$

Equation (9.107) is the collisionless Boltzmann equation! It appears from this and Eq. (9.104) and (9.105) that we have now succeeded in truncating the BBGKY hierarchy by means of the Mayer cluster expansion. We have expanded in terms of the parameter ξ, and have neglected the triplet (T) and higher-order interactions.

We have ended up with two equations in f_1 and $\mathscr{P}$: (9.104) and (9.105). These, in principle, may be solved for f_1 and $\mathscr{P}$ but, as can be envisaged from the equations, they are very complicated. Since f_1 is usually what is desired, then if $\mathscr{P}$ can be obtained in terms of f_1, a "simple" kinetic equation for f_1 can be written, at least in principle.

The initial conditions on $\mathscr{P}$ and f_1 will enter into the solution. However, it is to be hoped that for times long enough, these transient terms may disappear and a mostly Markovian formulation for f_1 may be obtained. (Markovian implies that f_1 only depends on its value at the instant of time immediately before the time of interest, and not on the initial conditions. See the discussion of the Fokker–Planck equation.)

There are thus three regimes of time, beginning at $t = 0$, in which we can look for solutions. The first regime would be those times near $t = 0$. Here f_1 and $\mathscr{P}$ depend primarily upon their initial values. Then, at somewhat later times, the Markovian processes become dominant. Finally, the fluid-type situation appears at extremely long times when "equilibrium" is achieved. The single-particle approach attempts to cover the first time period, kinetic theory covers the second period, and fluid mechanics covers the third period.

The question might now be asked as to the necessity for the BBGKY hierarchy when the previously developed Fokker–Planck methods seem to cover the effects of collisions if the cross section is known. The answer is that the Fokker–Planck equation does *not* consider correlations like $\mathscr{P}$ and T. It covers only the interactions in which one test particle interacts with each target particle separately. The correlation effects, however, may be extended to cover all particles interacting with each other at the same time. This may be important, especially because of the presence of the long-range Coulomb forces in a plasma. Therefore these correlations should be considered, and hence the need for the BBGKY hierarchy.

Dupree has developed a method that simplifies the solution of the two equations (9.104) and (9.109). The solution is obtained using this method by first assuming f_1 constant and then calculating $\mathscr{P}$ as an initial value problem.

In other words, we hold f_1 constant and allow $\mathscr{P}$ to change over the time of interest. In addition, for simplicity, the plasma is assumed to be uniform. These assumptions mean that Eq. (9.105) contains no terms in which f_1 varies with time or space. Note that $\mathscr{P}$ does vary with time. Under these conditions, Eq. (9.105) becomes

$$\left[\frac{\partial}{\partial t} + \mathbf{v}_1 \cdot \frac{\partial}{\partial \mathbf{x}_1} + \mathbf{v}_2 \cdot \frac{\partial}{\partial \mathbf{x}_2}\right]\mathscr{P}(\mathbf{x}_1, \mathbf{v}_1, \mathbf{x}_2, \mathbf{v}_2)$$

$$= \frac{q_1}{m}\frac{\partial \phi_{12}}{\partial \mathbf{x}_1} \cdot \frac{\partial f_1(\mathbf{x}_1, \mathbf{v}_1)}{\partial \mathbf{v}_1} f_1(\mathbf{x}_2, \mathbf{v}_2) + \frac{q_2}{m}\frac{\partial \phi_{21}}{\partial \mathbf{x}_2} \cdot \frac{\partial f_1(\mathbf{x}_2, \mathbf{v}_2)}{\partial \mathbf{v}_2} f_1(\mathbf{x}_1, \mathbf{v}_1)$$

$$+ \frac{n_0 q_1}{m}\frac{\partial f_1(\mathbf{x}_1, \mathbf{v}_1)}{\partial \mathbf{v}_1} \cdot \int \frac{\partial \phi_{13}}{\partial \mathbf{x}_1} \mathscr{P}(\mathbf{x}_2, \mathbf{v}_2, \mathbf{x}_3, \mathbf{v}_3)\, d\mathbf{x}_3\, d\mathbf{v}_3$$

$$+ \frac{n_0 q_2}{m}\frac{\partial f_1(\mathbf{x}_2, \mathbf{v}_2)}{\partial \mathbf{v}_2} \cdot \int \frac{\partial \phi_{23}}{\partial \mathbf{x}_2} \mathscr{P}(\mathbf{x}_1, \mathbf{v}_1, \mathbf{x}_3, \mathbf{v}_3)\, d\mathbf{x}_3\, d\mathbf{v}_3 \qquad (9.109)$$

We write the symbol $\mathscr{P}$ for $\mathscr{P}(\mathbf{x}_1, \mathbf{v}_1, \mathbf{x}_2, \mathbf{v}_2)$. Also note that $f_1(\mathbf{x}_1, \mathbf{v}_1) = f_1(\mathbf{v}_1)$, since no spatial variation has been assumed. The explicit time dependence of (9.109) is of first order and we may write it in a simple terminology as

$$\frac{\partial \mathscr{P}}{\partial t} + (H_1 + H_2)\mathscr{P} = Y \qquad (9.110)$$

where H_1 and H_2 are operators that can be defined by showing how they operate on an arbitrary function $h(\mathbf{x}_1, \mathbf{v}_1, \mathbf{x}_2, \mathbf{v}_2)$, as shown in the following two equations. The two operators are

$$H_1 h(\mathbf{x}_1, \mathbf{v}_1, \mathbf{x}_2, \mathbf{v}_2) = \mathbf{v}_1 \cdot \frac{\partial h(\mathbf{x}_1, \mathbf{v}_1, \mathbf{x}_2, \mathbf{v}_2)}{\partial \mathbf{x}_1}$$

$$- \frac{n_0 q_1}{m}\frac{\partial f_1(\mathbf{v}_1)}{\partial \mathbf{v}_1} \cdot \int \frac{\partial \phi_{13}}{\partial \mathbf{x}_1} h(\mathbf{x}_3, \mathbf{v}_3, \mathbf{x}_2, \mathbf{v}_2)\, d\mathbf{x}_3\, d\mathbf{v}_3 \qquad (9.111)$$

for the operator H_1 and

$$H_2 h(\mathbf{x}_1, \mathbf{v}_1, \mathbf{x}_2, \mathbf{v}_2) = \mathbf{v}_2 \cdot \frac{\partial h(\mathbf{x}_1, \mathbf{v}_1, \mathbf{x}_2, \mathbf{v}_2)}{\partial \mathbf{x}_2}$$

$$- \frac{n_0 q_2}{m}\frac{\partial f_1(\mathbf{v}_2)}{\partial \mathbf{v}_2} \cdot \int \frac{\partial \phi_{23}}{\partial \mathbf{x}_2} h(\mathbf{x}_3, \mathbf{v}_3, \mathbf{x}_1, \mathbf{v}_1)\, d\mathbf{x}_3\, d\mathbf{v}_3 \qquad (9.112)$$

for the operator H_2.

The operators H_1 and H_2 commute since they are operating on different orthogonal coordinates. That is, we can operate on a function first with H_1 and then with H_2 or vice versa, and obtain the same result.

The source function Y (which is not a function of $\mathscr{P}$) can then be written after comparing (9.109) to (9.110) as

$$Y = \frac{q_1}{m}\frac{\partial \phi_{12}}{\partial \mathbf{x}_1} \cdot \frac{\partial f_1(\mathbf{v}_1)}{\partial \mathbf{v}_1} f_1(\mathbf{v}_2) + \frac{q_2}{m}\frac{\partial \phi_{21}}{\partial \mathbf{x}_2} \cdot \frac{\partial f_1(\mathbf{v}_2)}{\partial \mathbf{v}_2} f_1(\mathbf{v}_1) \qquad (9.113)$$

Equation (9.110) is to be valid over a range from $t = 0$ to some later time t. The solution of (9.110) may be obtained by assuming a product solution for $\mathscr{P}$. To obtain this, we first will define a new operator $H = H_1 + H_2$, which is the *sum* of the operators defined in (9.111) and (9.112). Then the solution of (9.110) for $\mathscr{P}$ can be defined as the product of two functions, namely,

$$\mathscr{P} = Q\mathscr{R} \qquad (9.114)$$

Q is the operator solution to the homogeneous equation

$$\frac{\partial Q}{\partial t} + HQ = 0 \qquad (9.115)$$

The solution of (9.115) shows that

$$Q(t) = Ae^{-Ht} \qquad (9.116)$$

We shall let $Q(0) = 1$, so $A = 1$. The inverse operator Q^{-1} is defined so that

$$Q^{-1}Q = 1 \qquad (9.117)$$

so, solving for Q^{-1}, we obtain

$$Q^{-1} = e^{Ht} \qquad (9.118)$$

If we substitute the product $Q\mathscr{R}$ into Eq. (9.110), we can obtain, using (9.116)

$$-He^{-Ht}\mathscr{R} + e^{-Ht}\frac{\partial \mathscr{R}}{\partial t} + He^{-Ht}\mathscr{R} = Y \qquad (9.119)$$

The first and third terms on the left-hand side of (9.119) cancel and we may then solve (9.119) for $\mathscr{R}$. The solution for $\mathscr{R}$ is

$$\mathscr{R} = \int_0^t e^{H\tau} Y(\tau)\, d\tau + \mathscr{R}(t = 0) \qquad (9.120)$$

The complete solution for $\mathscr{P}$ is then obtained, since we know both Q and $\mathscr{R}$. It is

$$\mathscr{P} = Q\mathscr{R} = e^{-Ht}\int_0^t e^{H\tau} Y(\tau)\, d\tau + e^{-Ht}\mathscr{R}(t = 0) \qquad (9.121)$$

Now, we recall that $\mathscr{P}$ is a function of $\mathbf{x}_1$, $\mathbf{v}_1$, $\mathbf{x}_2$, and $\mathbf{v}_2$. Since H_1 and H_2 commute, we can assume a new product solution for Q to be

$$Q(\mathbf{x}_1, \mathbf{v}_1, \mathbf{x}_2, \mathbf{v}_2) = Q_1(\mathbf{x}_1, \mathbf{v}_1)Q_2(\mathbf{x}_2, \mathbf{v}_2) \tag{9.122}$$

Equation (9.115) may now be split into two separate equations in Q_1 and Q_2 by first substituting (9.122) into (9.115) as shown in Eq. (9.123).

$$Q_1\left[\frac{\partial Q_2}{\partial t} + H_2 Q_2\right] + Q_2\left[\frac{\partial Q_1}{\partial t} + H_1 Q_1\right] = 0 \tag{9.123}$$

We note that terms with $H_2 Q_1$ or $H_1 Q_2$ do not appear because $H_1 Q_2$ and $H_2 Q_1$ commute and they have been removed from Eq. (9.123).

Since the quantities in the brackets of (9.123) are functions of different variables, they should both be equal to zero in order to satisfy (9.123) under all conditions. We therefore obtain two separated equations. They are

$$\frac{\partial Q_2}{\partial t} + H_2 Q_2 = 0$$

and

$$\frac{\partial Q_1}{\partial t} + H_1 Q_1 = 0 \tag{9.124}$$

The form of the solutions for Q_1 and Q_2 are of the same form as obtained for Q previously, that is,

$$Q_1 = e^{-H_1 t} \qquad \text{and} \qquad Q_2 = e^{-H_2 t} \tag{9.125}$$

This is true if the single-particle distribution functions making up H_1 and H_2 do not change with time over the time of interest. This condition is called the "adiabatic" assumption of Bogoliubov.

Therefore, we can write the solution for (9.121) as

$$\mathscr{P} = e^{-(H_1+H_2)t}\int_0^t e^{(H_1+H_2)\tau}\, Y(\tau)\, d\tau + e^{-(H_1+H_2)t}\mathscr{R}(t=0) \tag{9.126}$$

where we use (9.120) as the basic definition of $\mathscr{R}$. It should be noted that we can find two solutions for $\mathscr{R}$, one for Q_1 and one for Q_2 that are of the same form as (9.120) using H_1 and H_2. We can combine them into a single function as shown in (9.126).

We will carry the explicit time dependence of Y along, but later we will remove it assuming this problem to be time independent. We can put the quantity $\exp[-(H_1 + H_2)t]$ inside the integral of (9.126), so the solution for $\mathscr{P}$ can be written as

$$\mathscr{P} = \int_0^t e^{-(H_1+H_2)(t-\tau)}\, Y(\tau)\, d\tau + e^{-(H_1+H_2)t}\mathscr{R}(t=0) \tag{9.127}$$

Note that Eq. (9.127) is of the form

$$\int_0^t g_1(t-\tau)g_2(\tau)\,d\tau \tag{9.128}$$

where g_1 and g_2 are two different functions of time. Equation (9.128) is also equal to the following integral.

$$\int_0^t g_1(\tau)g_2(t-\tau)\,d\tau \tag{9.129}$$

We note that (9.128) and (9.129) are equal because both integrals are *convolution* integrals of the same functions. This is a result taken from Laplace transform theory in which the Laplace transform of both integrals can be shown to be the product of the Laplace transforms of g_1 and g_2. Hence, the integrals (9.128) and (9.129) must be equal to each other.

We may use this convolution theorem to shift the variable of integration in Eq. (9.127). The integral with the variables shifted is

$$\mathscr{P} = \int_0^t e^{-(H_1+H_2)\tau}Y(t-\tau)\,d\tau + e^{-(H_1+H_2)t}\mathscr{R}(t=0) \tag{9.130}$$

We can now substitute the operator definitions for H_1 and H_2 found in Eq. (9.125) into (9.130). The result is that (9.130) becomes

$$\mathscr{P} = \int_0^t Q_1(\tau)Q_2(\tau)\,Y(t-\tau)\,d\tau + Q_1(t)Q_2(t)\mathscr{R}(t=0) \tag{9.131}$$

If the adiabatic hypothesis is made, then Y is not an explicit function of time and the operators $Q_1(\tau)$ and $Q_2(\tau)$ exhibit the only explicit time dependence. Therefore, these operators will produce the complete function of time when they operate on an arbitrary vector function. That is, if h is an arbitrary vector function of $(\mathbf{x}_1, \mathbf{v}_1)$, then

$$h(\mathbf{x}_1, \mathbf{v}_1, t) = Q_1(t)h(\mathbf{x}_1, \mathbf{v}_1) \tag{9.132}$$

If h is a function of $(\mathbf{x}_2, \mathbf{v}_2)$, then

$$h(\mathbf{x}_2, \mathbf{v}_2, t) = Q_2(t)h(\mathbf{x}_2, \mathbf{v}_2)$$

Note that Q_1 is a function of the $\mathbf{x}_1$, $\mathbf{v}_1$ coordinates and Q_2 is a function of the $\mathbf{x}_2$, $\mathbf{v}_2$ coordinates. If the arbitrary vector function h is a function of both $\mathbf{x}_1$, $\mathbf{v}_1$ and $\mathbf{x}_2$, $\mathbf{v}_2$ then we must operate with both $Q_1(t)$ and $Q_2(t)$ on h to obtain the complete time dependence. That is,

$$h(\mathbf{x}_1, \mathbf{v}_1, \mathbf{x}_2, \mathbf{v}_2, t) = Q_1(t)Q_2(t)h(\mathbf{x}_1, \mathbf{v}_1, \mathbf{x}_2, \mathbf{v}_2) \tag{9.133}$$

Now we can utilize these last developments to solve Eq. (9.109). Due to the nature of the operators H_1 and H_2, h must obey the two equations

$$\frac{\partial h(\mathbf{x}_1, \mathbf{v}_1, t)}{\partial t} + H_1 h(\mathbf{x}_1, \mathbf{v}_1, t) = 0 \tag{9.134}$$

and

$$\frac{\partial h(\mathbf{x}_2, \mathbf{v}_2, t)}{\partial t} + H_2 h(\mathbf{x}_2, \mathbf{v}_2, t) = 0 \tag{9.135}$$

These equations may be explicitly written out, since we know the definitions of H_1 and H_2. Note that since H_1 does not depend on $\mathbf{x}_2$, $\mathbf{v}_2$ and vice versa, it is permissible to make h a function of $\mathbf{x}_1$, $\mathbf{v}_1$ and $\mathbf{x}_2$, $\mathbf{v}_2$ and still have it satisfy both (9.134) and (9.135).

Let us investigate (9.134). Written out explicitly, it becomes

$$\frac{\partial h(\mathbf{x}_1, \mathbf{v}_1, \mathbf{x}_2, \mathbf{v}_2, t)}{\partial t} + \mathbf{v}_1 \cdot \frac{\partial}{\partial \mathbf{x}_1} h(\mathbf{x}_1, \mathbf{v}_1, \mathbf{x}_2, \mathbf{v}_2, t)$$
$$- \frac{n_0 \varphi_1}{m} \frac{\partial f_1(\mathbf{v}_1)}{\partial \mathbf{v}_1} \cdot \int \frac{\partial \phi_{13}}{\partial \mathbf{x}_1} h(\mathbf{x}_2, \mathbf{v}_2, \mathbf{x}_3, \mathbf{v}_3, t)\, d\mathbf{x}_3\, d\mathbf{v}_3 = 0 \tag{9.136}$$

This equation *is* the linearized Boltzmann equation and may be solved by Laplace transformation in time and Fourier transformation in the spatial coordinate $\mathbf{x}$, alone, as we have done in Chapter 8. The solution for h can be expressed as

$$h(\mathbf{x}_1, \mathbf{v}_1, \mathbf{x}_2, \mathbf{v}_2, t)$$
$$= \frac{1}{(2\pi)^2} \int_{-\infty + j\sigma_1}^{+\infty + j\sigma_1} \int_{-\infty}^{+\infty} h(\mathbf{k}_1, \mathbf{v}_1, \omega_1, \mathbf{x}_2, \mathbf{v}_2) \exp[j(\mathbf{k}_1 \cdot \mathbf{x}_1 - \omega_1 t)]\, d\mathbf{k}_1\, d\omega_1 \tag{9.137}$$

where $h(\mathbf{k}_1, \mathbf{v}_1, \omega_1, \mathbf{x}_2, \mathbf{v}_2)$ is the solution to the Laplace transform in time and the Fourier transform in variable $\mathbf{x}_1$ of Eq. (9.136), which again is Eq. (9.134) written out explicitly. We may now operate in the same manner with state $\mathbf{x}_2$ and Eq. (9.135) to obtain a similar solution. Doing both of these things twice permits us to obtain the complete solution for h when *both* operators are utilized. This solution is

$$h(\mathbf{x}_1, \mathbf{v}_1, \mathbf{x}_2, \mathbf{v}_2, t)$$
$$= Q_1(t) Q_2(t) h(\mathbf{x}_1, \mathbf{v}_1, \mathbf{x}_2, \mathbf{v}_2)$$
$$= \frac{1}{(2\pi)^4} \int_{-\infty + j\sigma_1}^{+\infty + j\sigma_1} \int_{-\infty}^{+\infty} \int_{-\infty + j\sigma_2}^{+\infty + j\sigma_2} \int_{-\infty}^{+\infty} h(\mathbf{k}_1, \mathbf{k}_2, \mathbf{v}_1, \mathbf{v}_2, \omega_1, \omega_2)$$
$$\times \exp[j(\mathbf{k}_1 \cdot \mathbf{x}_1 + \mathbf{k}_2 \cdot \mathbf{x}_2 - (\omega_1 + \omega_2) t)]\, d\mathbf{k}_1\, d\mathbf{k}_2\, d\omega_1\, d\omega_2 \tag{9.138}$$

Note that Eq. (9.138) shows that we have Laplace and Fourier transformed twice. That is, we have transformed to obtain the variables $\mathbf{k}_1$, ω_1, $\mathbf{k}_2$, and ω_2.

The question now becomes that of deciding what sort of h will be utilized. We shall use for h, the Fourier transform of Y, the source function in Eq. (9.109) operated on by the transformed operators $Q_1(\omega_1)$ and $Q_2(\omega_2)$. The Laplace transform is not required for Y because Y does not now actually have a time dependence, since we have made the adiabatic assumption.

By noting again that Eq. (9.110) is the form of Eq. (9.109), we can now write the explicit form of Y. It is

$$Y = \frac{q_1}{m}\frac{\partial \phi_{12}}{\partial \mathbf{x}_1} \cdot \frac{\partial f_1(\mathbf{v}_1)}{\partial \mathbf{v}_1} f_1(\mathbf{v}_2) + \frac{q_2}{m}\frac{\partial \phi_{21}}{\partial \mathbf{x}_2} \cdot \frac{\partial f_1(\mathbf{v}_2)}{\partial \mathbf{v}_2} f_1(\mathbf{v}_1) \qquad (9.139)$$

We must take the Fourier transform of Y twice, once with respect to each of the coordinates $\mathbf{x}_1$ and $\mathbf{x}_2$. Once these transforms of (9.139) are taken, we can use this result in the expression for the h required in Eq. (9.138).

Now, upon looking at (9.138), we note that it looks very much like the *integral* in Eq. (9.131). In fact, we could now complete the integration in (9.138) and obtain $\mathscr{P}$, the pair correlation, directly, which is the desired quantity. Knowing $\mathscr{P}$ permits us to find the first-order correction to the single-particle distribution function, which will now include these correlations. Naturally, ϕ_{12} and ϕ_{21} [in (9.139)] must be known functions in order to complete the integration.

The method, to summarize, requires successive substitution of the Fourier-transformed function Y into Eqs. (9.132) and (9.133) to obtain the transformed function necessary for (9.138). The result of the inverse integration in (9.138) is $\mathscr{P}$ itself.

The process, although conceptually straightforward, is a most difficult problem, and many of the integrations involved really cannot often be done, not even numerically. However, certain simplifications may be made that render the integration possible, especially for long times, which, in many cases, is the response that is desired.

We can take the expression for $\mathscr{P}$ and place it in Eq. (9.100), which is the first equation in the BBGKY hierarchy. Since we have assumed a homogeneous problem in space, the only term on the left-hand side of Eq. (9.100) that remains is $\partial f_1/\partial t$. The right-hand side remains only a function of $\mathbf{v}_1$ after all of the integrations have been performed. We can rewrite Eq. (9.100) in the form

$$\frac{\partial f_1}{\partial t} = -\frac{\partial}{\partial \mathbf{v}_1} \cdot \mathbf{J}(\mathbf{v}_1) \qquad (9.140)$$

This equation is the *Lenard–Balescu* equation. It is another kinetic equation for plasmas. The method for obtaining this equation, proposed by Dupree, was to assume that f_1 was constant over the time of integration.

To obtain the time response, we first obtain $\mathcal{P}$ from (9.138) and then use Eq. (9.140) to obtain the "change" in f_1. A higher-order approach, developed by Guernsey, assumes perturbations for *both* $\mathcal{P}$ and f_1, which are of the form

$$\mathcal{P} = \mathcal{P}_0{}^0 + \mathcal{P}_0{}^1 \tag{9.141}$$

and

$$f_1 = f_1{}^0 + f_1{}^1 \tag{9.142}$$

where $f_1{}^0$ and $\mathcal{P}_0{}^0$ are the equilibrium distribution and correlation functions.

These expressions can be utilized to obtain from Eq. (9.110), the following perturbation equation.

$$\frac{\partial \mathcal{P}_0{}^1}{\partial t} + (H_1 + H_2)\mathcal{P}_0{}^1 = Y^1 \tag{9.143}$$

We can then, in principle, develop a kinetic equation from this equation that is of a higher order than (9.138).

The preceding developments have shown what is necessary to produce results that lead to kinetic and nonequilibrium equations for plasmas. These equations can be used to describe the behavior of the single- or multiple-particle distribution functions in plasmas. In turn, since the distribution function is related to other parameters in the plasma, we can then compute the plasma density, electric field strength, temperature, and so on.

The difficulty with these methods is twofold. First, the development of the BBGKY hierarchy, while rigorous, involves assumptions about the coupling forces between particles and a truncation of the hierarchy. The second difficulty is that there is, even under the conditions described above, a great complexity in the integration processes necessary to determine the solutions. It should also be stated that the models described in this chapter for collisions and expansions of Liouville's theorem are not the only ones used. A large number of different formulations have been made.

However, with all of this now said, the kinetic model, by which most plasmas studied on earth are described, will yield information of a very important nature.

Suggested Reading

Bogoliubov, N. N., Problems of a Dynamical Theory in Statistical Physics, Part A *in* "Studies in Statistical Mechanics" (J. D. Boer and G. E. Uhlenbeck, eds.), North Holland Publ., Amsterdam, 1962.

Dupree, T. H., *Phys. Fluids* **4**, 696 (1961).

Montgomery, D. C., and Tidman, D. A., "Plasma Kinetic Theory," McGraw-Hill, New York, 1964. A very good treatment of basic concepts in kinetic theory of plasmas, along with applications to experiments, is given.

Schmidt, G., "Physics of High Temperature Plasmas—An Introduction," Academic Press, New York, 1966.
Shkarofsky, I. P., Bachynsky, T. W., and Watson, M. P., "The Particle Kinetics of Plasmas," Addison-Wesley, Reading, Massachusetts, 1967. This is a complete reference work on many aspects of kinetic theory and radiation.
Stix, T. H., "The Theory of Plasma Waves," McGraw-Hill, New York, 1963.

Problems

1 Prove the relation (9.33) and then find the potential F_{21}, knowing the potential H_{21} as given in Eq. (9.50).

2 Show, in Eq. (9.26), that $\langle \Delta \mathbf{v}_2 \rangle$ does point in the direction of $\mathbf{v}_2$.

3 Solve for $\langle \Delta \mathbf{v}_2 \rangle$ and $\langle \Delta \mathbf{v}_2 \, \Delta \mathbf{v}_2 \rangle$, given the results of Problems 1 and 2, for an isotropic Maxwellian. Show that these terms, when placed in the Fokker–Planck equation, cancel.

4 Suppose that we wish to find the pair correlation function for a plasma, characterized by a collection of single-particle one-dimensional Maxwellian distributions. If Coulomb interactions are assumed, compute the function Y for two particles only, as developed in Eq. (9.139). Then, following the discussion in the chapter, take the Fourier and Laplace transforms as required, assuming that the two particle coordinates x_1 and x_2 are each a distance L on either side of the origin of the position coordinates at $t = 0$. Finally, use this result in Eq. (9.131) or (9.138) to obtain $\mathscr{P}$. Note: You will not be able to take the Fourier transform of the Coulomb potential since it has a singularity when the charges are placed at the same point. To remove this, try truncating the Coulomb potential for charge separations greater than, perhaps, the minimum impact parameter.

5 Show that the angular averages $\overline{\Delta g_x \, \Delta g_y}$, $\overline{\Delta g_y \, \Delta g_z}$, ..., are zero.

6 Obtain Eq. (9.24) from (9.21), (9.22), and (9.23). Then find $\overline{\Delta g_z \, \Delta g_z}$.

7 Obtain Eq. (9.50) without using spherical harmonics. You may assume Gauss's Law applies.

8 For a spherically symmetric velocity distribution, determine the expression for the equipotentials of F_{21} and H_{21}. Find the direction of $\langle \Delta \mathbf{v}_2 \rangle$ and $\langle \Delta \mathbf{v}_2 \, \Delta \mathbf{v}_2 \rangle$ with respect to these equipotentials.

9 Do the operators Q_1 and Q_2 commute? Prove your answer.

10 Show that Eq. (9.136) is the linearized Boltzmann equation.

10 Radiation Processes and Correlation Functions

Up to this point we have not considered the emission of radiation during the interactions between particles in a plasma. If the particles are charged, then any accelerating force on the particles can and usually does result in radiation. Since there are many different processes by which a charged particle can be accelerated, there will likewise be many different types of radiation processes occurring.

Some of these different acceleration processes that produce radiation are Coulomb collisions, electron or ion–neutral collisions, cyclotron motion in a magnetic field, collective or wave phenomena, and instabilities. Each of these processes may result in some form of acceleration of charged particles that produces emission of radiation. Such radiations can often be measured and their spectra plotted. Very often each process produces its own characteristic identifying spectral pattern. The pattern is usually in a broad band, or continuum, but may have various peaks at various characteristic frequencies, such as the cyclotron frequency or those of radiative transitions between energy levels in ions or atoms. Measurements of the spectra often yield much information about the nature of the plasma producing the radiation.

A. GENERALIZED EMISSION AND ABSORPTION PROCESSES

Let us first consider some of the general quantities needed for a description of all of these radiation processes. In the various types of radiation, plasma particles may emit photons or absorb them. The nature of the emitting processes, however, is the same. These processes can be lumped together into three groups:

(1) *Spontaneous Emission.* A photon of radiation may be emitted *spontaneously*, that is, without any known "triggering" mechanism.

(2) *Stimulated Emission.* In this process, a photon that is already present "causes" a particle to emit another photon of the *same* energy.

(3) *Absorption.* Here a photon that is already present "causes" a particle to absorb it. Due to the similarity with process (2), it is sometimes called "stimulated" absorption.

These three types of radiation phenomena can be grouped together in various ways to describe the nature of plasma radiation processes. After considering the specific nature of some of the radiation processes, we will study the behavior of fluctuations and correlations of radiation as well as those of other quantities in a plasma.

We introduce *emission* and *absorption coefficients*, which show the rates of emission of radiation from and absorption of radiation by a plasma. They are defined as follows.

Let us consider an electron with a momentum between $\mathbf{p}'$ and $\mathbf{p}' + d\mathbf{p}'$ in a unit volume located at $\mathbf{q}$. Then, the quantity $\eta_\omega(\mathbf{p}', \mathbf{q}, \xi)$ will be defined for a given process as the rate of spontaneous emission of radiation energy per unit time per unit solid angle, in direction ξ with frequency between ω and $\omega + d\omega$. The *total* rate of spontaneous emission of energy for this process per unit volume per unit solid angle in direction ξ from a plasma is

$$N\eta_\omega(\mathbf{p}', \mathbf{q}, \xi) f(\mathbf{p}', \mathbf{q})\, d\mathbf{p}' \tag{10.1}$$

where $f(\mathbf{p}', \mathbf{q})$ is the velocity distribution function of the particles that emit the given type of radiation. The total number of particles of the given species is N.

A similar definition may be developed for stimulated emission. The differential energy emission rate for *stimulated* emission will be written as $\eta_{\omega S}(\mathbf{p}, \mathbf{q}, \xi)$, but the rate of emission per unit volume for this process must be proportional to the product of the distribution function *and* the intensity of radiation at the desired frequency. This is because both a particle and a photon must be present for stimulated emission to occur.

The net energy emission rate for stimulated emission is therefore written as

$$N\eta_{\omega S}(\mathbf{p}', \mathbf{q}, \xi) f(\mathbf{p}', \mathbf{q}) I_\omega(\mathbf{q}, \xi)\, d\mathbf{p}' \tag{10.2}$$

where $I_\omega(\mathbf{q}, \xi)$ is the specific intensity of radiation (watts per square meter) per unit solid angle at position $\mathbf{q}$ in the frequency band between ω and $\omega + d\omega$ in direction ξ. This result means that $\eta_{\omega S}$ does not have the same dimensions as η_ω.

The absorption process (remember it is sometimes called *stimulated absorption*) has a differential absorption rate $\eta_{\omega A}(\mathbf{p}, \mathbf{q}, \xi)$ and a net absorption rate of

$$N\eta_{\omega A}(\mathbf{p}, \mathbf{q}, \xi) f(\mathbf{p}, \mathbf{q}) I_\omega(\mathbf{q}, \xi)\, d\mathbf{p} \tag{10.3}$$

We again note that both particles *and* photons must be present for absorption to occur, so the same quantities must appear in (10.3) as were in (10.2). Note that the momentum in (10.3) is written without the prime superscript to keep the two processes separate. It will be assumed that a particle of momentum $\mathbf{p}'$ decays to a momentum of $\mathbf{p}$ when emission occurs, and a particle with momentum $\mathbf{p}$ changes to $\mathbf{p}'$ for absorption.

To eliminate the momentum dependence, we can integrate over all momentum space in Eqs. (10.1)–(10.3). Since $I_\omega(\mathbf{q}, \xi)$ does not depend upon momentum, it does not have to be included in the integral. This integration enables us to define the following coefficients.

(*a*) *Spontaneous Emission Coefficient*

$$j_\omega(\mathbf{q}, \xi) = N \int \eta_\omega(\mathbf{p}', \mathbf{q}, \xi) f(\mathbf{p}', \mathbf{q})\, d\mathbf{p}' \tag{10.4}$$

(*b*) *Stimulated Emission Coefficient*

$$\alpha_{\omega S}(\mathbf{q}, \xi) = N \int \eta_{\omega S}(\mathbf{p}', \mathbf{q}, \xi) f(\mathbf{p}', \mathbf{q})\, d\mathbf{p}' \tag{10.5}$$

(*c*) *Absorption Coefficient*

$$\alpha_{\omega A}(\mathbf{q}, \xi) = N \int \eta_{\omega A}(\mathbf{p}, \mathbf{q}, \xi) f(\mathbf{p}, \mathbf{q})\, d\mathbf{p} \tag{10.6}$$

Note that the variable of integration in (10.6) is $\mathbf{p}$ and not $\mathbf{p}'$. Equations (10.5) and (10.6) must be multiplied by the intensity of radiation to get the net appropriate emission and absorption rates.

A word should now be said regarding the absorption that radiation itself encounters while traversing a plasma. Imagine that a signal source of a given frequency sends signals through a plasma. If, in a plasma, the three processes mentioned above are simultaneously taking place, then the stimulated emission process will tend to increase the level of radiation, and the absorption process will tend to decrease the level. Spontaneous emission will be always present, usually in a broad band, whether or not the incident radiation is there. *Normally*, the absorption rate is greater than the stimulated emission rate, so there will result a net absorption of radiation. The experimentally measured absorption coefficient is therefore the difference between Eqs. (10.6) and (10.5); expressed in equation form it is

$$\alpha_\omega = \alpha_{\omega A} - \alpha_{\omega S} \tag{10.7}$$

In equilibrium, if a plasma could be isolated from its surroundings, the rate of emitting transitions would be exactly balanced by the rate of absorbing transitions, so that energy is conserved. That is, the number of emitting transitions where electron momentum changes from $\mathbf{p}'$ to $\mathbf{p}$ must be exactly balanced by the number of absorbing transititions, which change the momentum from $\mathbf{p}$ to $\mathbf{p}'$. This statement is the principle of *detailed balance.*

Under these conditions, it is possible to obtain a relationship between the emission and absorption coefficients if collective effects of the plasma can be neglected. In order to do this, however, we must first consider the photon distribution. A plasma will have, at any instant, not only a distribution of particle momenta and positions, but also a distribution of energies of radiation photons.

The treatment of the distribution of radiation photons in a volume of plasma may be considered statistically. However, instead of obeying Maxwell–Boltzmann statistics, as we have previously assumed for plasma particles, the photons obey Bose–Einstein statistics, which were originally defined in Eq. (4.66). We assume that the following energy frequency relation holds for each photon.

$$\mathscr{E}_i = \hbar\omega = \frac{h}{2\pi}\omega \tag{10.8}$$

where h is Planck's constant, $\hbar = h/2\pi$, and ω is the angular radian frequency of the radiation. $\mathscr{E}_i$ is the energy of the photon. The blackbody radiation function for such a collection of photons has been found to be

$$B(\omega, T) = n_r^{\,2}\,\frac{\hbar\omega^3}{8\pi^3c^2}\,\frac{1}{\exp(\hbar\omega/kT) - 1} \tag{10.9}$$

The units of $B(\omega, T)$ are in watts per steradian per unit area per unit frequency interval. n_r is the "ray refractive index" of the material and has the value of 1 in vacuum. This means that the blackbody intensity in a medium is straightforwardly related to the vacuum blackbody intensity by this factor, $n_r^{\,2}$. In an *isotropic* material $n_r = n$, the index of refraction of the material. Note that in the expression for the index of refraction for a cold isotropic plasma in which only the electrons move, for example, n is the square root of the dielectric constant, namely,

$$n = \left(1 - \frac{\omega_{\mathrm{p_e}}^2}{\omega^2}\right)^{1/2} \tag{10.10}$$

If this formulation is applied to Eq. (10.9), the radiation function becomes

$$B(\omega, T) = \left(1 - \frac{\omega_{\mathrm{p_e}}^2}{\omega^2}\right)B_0(\omega, T) \tag{10.11}$$

For a *blackbody*, the intensity of radiation is equal to (10.9), that is, $I_\omega = B(\omega, T)$. The expression $B_0(\omega, T)$ is the vacuum blackbody function, which is the radiation from a collection of photons in vacuum. For an anisotropic medium, n_r is a function of the angular dependence of the index of refraction.

To develop the relationship between the emission coefficients, we will assume that the electron energies are described by a nonrelativistic Maxwellian distribution for velocities that is independent of position, such as

$$Nf(v) = N\left(\frac{m_e}{2\pi \ell T_e}\right)^{3/2} \exp\left(-\frac{m_e v^2}{2\ell T_e}\right) = A \exp\left(-\frac{\mathscr{E}}{kT_e}\right) \tag{10.12}$$

For a blackbody, using detailed balance, we recall that the rate of emission from state $\mathbf{p}'$ to $\mathbf{p}$, must be balanced by the absorption from $\mathbf{p}$ to $\mathbf{p}'$, which is written in Eq. (10.13),

$$\eta_\omega(\mathbf{p}')\exp\left(-\frac{\mathscr{E}'}{\ell T_e}\right) d\mathbf{p}' = I_\omega\left[\eta_{\omega A}(\mathbf{p})\exp\left(-\frac{\mathscr{E}}{\ell T_e}\right)d\mathbf{p} - \left(\eta_{\omega S}(\mathbf{p}')\exp-\frac{\mathscr{E}'}{\ell T_e}\right) d\mathbf{p}'\right] \tag{10.13}$$

The constant A is common to both sides and is removed from (10.13). The relation between $\mathscr{E}$ and $\mathscr{E}'$ which allows (10.13) to be satisfied must be

$$\mathscr{E}' - \mathscr{E} = \mathscr{E}_{\text{photon}} = \text{constant} = \hbar\omega \tag{10.14}$$

The constant has the value $\hbar\omega$, which is the energy of the emitted photon. That is, the net change in energy of the electron is equal to the energy of the photon. We can now solve Eq. (10.13) for I_ω, which is then

$$I_\omega = \frac{\eta_\omega(\mathbf{p}')/\eta_{\omega S}(\mathbf{p}')}{[\eta_{\omega A}(\mathbf{p})\, d\mathbf{p}/\eta_{\omega S}(\mathbf{p}')\, d\mathbf{p}'][\exp(\hbar\omega/\ell T)] - 1} \tag{10.15}$$

Because we have been considering a blackbody we know that I_ω is equal to $B(\omega, T)$. This now permits us to set (10.15) equal to Eq. (10.9), and from this relation, the following two expressions for the relationship between the different rates may be obtained.

$$\eta_\omega(\mathbf{p}') = n_r^{\,2}\frac{\hbar\omega^3}{8\pi^3 c^2}\eta_{\omega S}(\mathbf{p}') \tag{10.16}$$

and

$$\eta_{\omega A}(\mathbf{p})\, d\mathbf{p} = \eta_{\omega S}(\mathbf{p}')\, d\mathbf{p}' \tag{10.17}$$

These two equations relate the three emission rates, so that if one is known, the others may be found. Again, this is only strictly the case if the plasma is a blackbody. A plasma may be considered to be a blackbody over a given range of frequencies if its continuum emission versus energy curve looks like the blackbody emission function of Eq. (10.9). Typically, a laboratory plasma looks like a blackbody at low frequencies (perhaps microwaves) and its emission falls off from the blackbody function as the frequency is raised.

We will now investigate some of the specific types of processes that produce radiation. The following processes are quite important in plasma radiation phenomena.

(a) Bremsstrahlung: radiation by acceleration of charges due to collisions, either with other plasma particles or with the walls of the confining vessel, if any;

(b) cyclotron and synchrotron emission: radiation by orbiting particles in a magnetic field;

(c) other radiation processes due to external acceleration; and

(d) spectral line emission.

The method of production of radiation for all but the last of these processes is practically the same; acceleration of charged particles. The important items to consider in each process are the spectrum and angular distribution of the intensity of the radiation. We shall not, however, consider the spectral line emission resulting from energy level transitions in excited atoms or ions in this chapter. This topic alone is broad enough to be the subject of many texts.

B. RADIATION FIELDS FROM A CHARGED PARTICLE

We now turn to a calculation of the emission coefficients for accelerated charges. To do this, we will make use of the Lienard–Wiechart potentials for a point charge in order to calculate the radiation fields. They will be developed and expanded in the following discussion.

Let the velocity of the charge be $\mathbf{u}$. Let $\mathbf{r}$ be the radius vector from the point where the charge is to the point of observation as shown in Fig. 10.1. This figure shows a set of vectors to be used in the following development. The Lienard–Wiechart potentials, both scalar and vector, are, in terms of these vectors, respectively, assuming a charge q_e,

$$\phi = \frac{1}{4\pi\varepsilon_0}\left[\frac{q_e}{|\mathbf{r}| - (\mathbf{r}\cdot\mathbf{u})/c}\right] \tag{10.18}$$

$$\mathbf{A} = \frac{\mu_0}{4\pi}\left[\frac{q_e\mathbf{u}}{|\mathbf{r}| - (\mathbf{r}\cdot\mathbf{u})/c}\right] \tag{10.19}$$

These potentials are those observed at point C. Note that the potentials are computed from the position and velocity of a particle at point A. Once the radiation leaves the particle, it is independent of the particle's further motion.

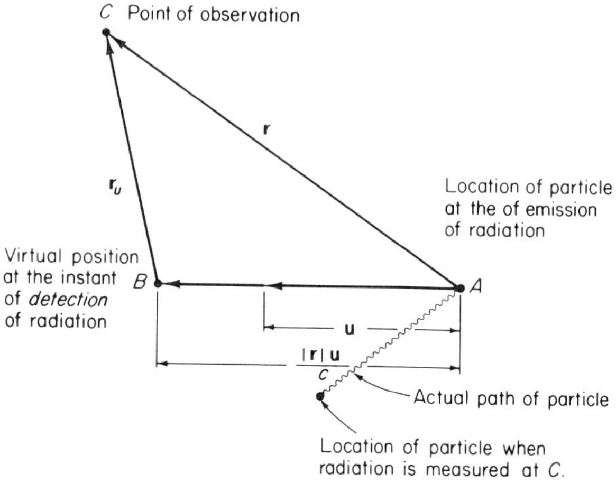

Fig. 10.1. Vector diagram used to formulate radiation fields.

Since it takes a finite time for the radiation to travel from A to C, the particle may move away from point A during this time. When we observe the potentials at C we must always recall that they were produced at point A some time previously. In this case, the time difference is $|\mathbf{r}|/c$. We may find $\mathbf{B}$ and $\mathbf{E}$ from the Lienard–Wiechart potentials by noting that

$$\mathbf{B} = \nabla_r \times \mathbf{A} \qquad \text{and} \qquad \mathbf{E} = -\nabla_r \phi - \frac{\partial \mathbf{A}}{\partial t}$$

$\mathbf{r}$ is called the *retarded* radius vector. We shall also define a *virtual present radius* vector to be

$$\mathbf{r}_u \equiv \mathbf{r} - \frac{|\mathbf{r}|\mathbf{u}}{c} \tag{10.20}$$

The origin of $\mathbf{r}_u$ is the position that the particle would occupy if it had continued with velocity $\mathbf{u}$ during the time it takes for the radiation produced at point A by the charged particle to travel from A to C. Note that if the particle is *not* accelerated or decelerated it *will*, in fact, be at point B when the radiation arrives at point C. The time difference between production and measurement of the radiation is, again, $|\mathbf{r}|/c$. Since it is necessary for the charge to be accelerated in order that radiation be produced, it probably will *not* be at the virtual position location (point B) after time $|\mathbf{r}|/c$ has elapsed. This is why $\mathbf{r}_u$ is usually called a *virtual* present radius vector.

It can be shown from Maxwell's equations that if the radiation part of the electric field $\mathbf{E}$ is known, then the magnetic field $\mathbf{B}$ due to the radiation is

$$\mathbf{B}_{\text{rad}} = \frac{\mathbf{r} \times \mathbf{E}_{\text{rad}}}{|\mathbf{r}|c} \tag{10.21}$$

Thus, **B** is perpendicular to both the retarded radius vector **r**, and the electric field **E**. The radiation part of the electric field can be shown to be

$$\mathbf{E}_{\mathrm{rad}} = \frac{q_e}{4\pi\varepsilon_0 s^3 c^2} [\mathbf{r} \times (\mathbf{r}_u \times \dot{\mathbf{u}})] \tag{10.22}$$

Normally, radiation parts of electric and magnetic fields are those components, or parts of components, of the fields that vary as $1/r$. The quantity

$$s = |\mathbf{r}| - \frac{\mathbf{u} \cdot \mathbf{r}}{c}$$

is the denominator of the scalar and vector potentials ϕ and **A**, respectively, which were defined in Eq. (10.18) and (10.19). $\dot{\mathbf{u}}$ is the acceleration $d\mathbf{u}/dt$.

It should again be restated at this point that the quantities $\mathbf{r}$, $\mathbf{r}_u$, $\mathbf{u}$, and $\dot{\mathbf{u}}$ are all *evaluated at the time of emission* of the radiation (retarded time).

When the velocity is small enough to that $u/c \ll 1$, then $\mathbf{r}_u \approx \mathbf{r}$ and $s \approx r$, so that

$$\mathbf{E}_{\mathrm{rad}} \cong \frac{q_e}{4\pi\varepsilon_0 c^2 r^3} [\mathbf{r} \times (\mathbf{r} \times \dot{\mathbf{u}})] \tag{10.23}$$

and

$$\mathbf{H}_{\mathrm{rad}} = \frac{\mathbf{B}_{\mathrm{rad}}}{\mu_0} \cong \frac{q_e}{4\pi c r^2} [\dot{\mathbf{u}} \times \mathbf{r}] \tag{10.24}$$

The Poynting vector for this set of fields is

$$(\mathbf{E} \times \mathbf{H})_{\mathrm{rad}} = \frac{q_e^2}{16\pi^2\varepsilon_0 c^2 r^5} \{[\mathbf{r} \times (\mathbf{r} \times \dot{\mathbf{u}})] \times (\dot{\mathbf{u}} \times \mathbf{r})\} \tag{10.25}$$

Now the radiated power per unit solid angle emitted from this charge may be determined by taking the outward normal component of the Poynting vector in the **r** direction and multiplying it by r^2, which is the incremental surface area in spherical coordinates $r^2 \sin\theta \, d\theta \, d\varphi$ divided by the incremental solid angle $\sin\theta \, d\theta \, d\varphi$. Note that the radiation is independent of **u** and therefore only $\dot{\mathbf{u}}$ and **r** need be known. The outward radiation flux per unit solid angle is then

$$r^2(\mathbf{E} \times \mathbf{H}) \cdot \hat{a}_r = \frac{q_e^2}{16\pi^2\varepsilon_0 c^3} |\dot{\mathbf{u}}|^2 \sin^2\theta \tag{10.26}$$

The angular dependence of the radiation is independent of φ and is plotted in Fig. 10.2. The polar angle θ is measured with respect to the $\dot{\mathbf{u}}$ direction for this figure. The figure may be rotated about the $\dot{\mathbf{u}}$ axis without change.

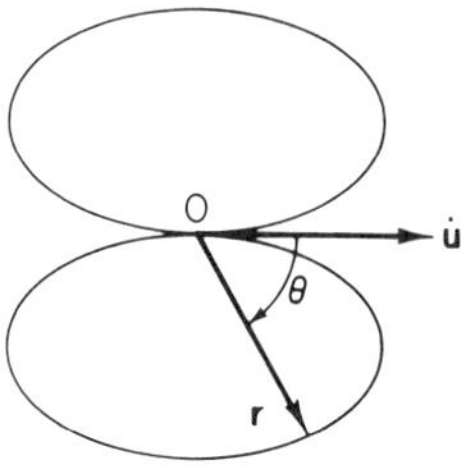

Fig. 10.2. Radiation pattern of a charged particle in the nonrelativistic approximation.

Equation (10.26) has the same field configuration as that of an oscillating electric dipole, and the assumption $u/c \ll 1$ is therefore called the *dipole approximation*. Note that the dipole radiation is dominant only when $u/c \ll 1$. As u/c gets larger, higher-order terms appear, that is, quadrupole, hexupole, and so on. These refer to terms in the expansion of Eq. (10.18) and (10.19) in powers of u/c.

If Eq. (10.26) is integrated over the entire spherical angular space as follows

$$\int_0^{2\pi}\int_0^{\pi} r^2(\mathbf{E}\times\mathbf{H})\cdot\hat{a}_r \sin\theta\, d\theta\, d\varphi = \frac{q_e^{\,2}|\dot{\mathbf{u}}|^2}{6\pi\varepsilon_0 c^3} \tag{10.27}$$

then Eq. (10.27) must be the total energy lost per second (watts) from the accelerated charge, in the dipole approximation.

If we make the restriction that the acceleration $\dot{\mathbf{u}}$ is parallel to $\mathbf{u}$, the radiation fields are

$$\mathbf{E}_{\text{rad}} = \frac{q_e}{4\pi\varepsilon_0 c^2 s^3}[\mathbf{r}\times(\mathbf{r}\times\dot{\mathbf{u}})] \tag{10.28}$$

and

$$\mathbf{H}_{\text{rad}} = \frac{q_e|\mathbf{r}|}{4\pi c s^3}[\dot{\mathbf{u}}\times\mathbf{r}] \tag{10.29}$$

These equations have exactly the same form as (10.23) and (10.24) regardless of the value of u/c, even at relativistic velocities. They may be related directly to Eqs. (10.23) and (10.24) by multiplying (10.28) by a factor s^3/r^3. This produces (10.23), which can, with the use of (10.21), produce (10.24). This factor may be written out as

$$\frac{s^3}{r^3} = \left[1 - \frac{u}{c}\cos\theta\right]^3 \tag{10.30}$$

Note that $\mathbf{u}$ and $\dot{\mathbf{u}}$ are in the same direction for this case, hence θ is still defined as in Fig. 10.2.

We may now form the outward radiation flux density for the radiation described in (10.28) and (10.29) to be

$$r^2(\mathbf{E} \times \mathbf{H}) \cdot \hat{a}_r = \frac{q_e}{16\pi\varepsilon_0 c^3} |\dot{\mathbf{u}}|^2 \sin^2\theta \left[1 - \frac{u}{c}\cos\theta\right]^{-6} \tag{10.31}$$

We note that the radiation fields have been evaluated at the observation point. The time at which they are evaluated is called t, but **E** and **H** are functions of the time of emission of the radiation and not t. Yet, the Poynting vector states the energy loss per unit t. If we wish to know the angular distribution of radiation flux centered about point A so we may use coordinates in Fig. 10.1, we must convert the value of (10.31) so that it measures energy loss in units of the time t' at which the radiation has been emitted. Note that in the previous case, the dipole approximation, this does not have to be done, because $t' \cong t$. The superscript prime for the time is used to denote the time as measured *at the emitting charge* during the emission of the signal.

The Poynting vector itself is a function of position since **E** and **H** vary in space. Its divergence measures the rate of energy loss or accumulation in a incremental volume per unit time at whatever position it is evaluated. Again, the time at the point where the Poynting vector is measured is called t (without the prime). That is, the Poynting vector's value is in terms of the time where it is evaluated even though the fields given here are evaluated in terms of the time at the radiator. There is a relation between t and t', which can be ascertained by examination of Fig. 10.1. It is

$$t = t' + \frac{|\mathbf{r}|}{c} \tag{10.32}$$

since it takes the radiation $|\mathbf{r}|/c$ seconds to travel to the point of observation. The rate of energy emitted into an incremental solid angle $d\Omega$ centered about the electron is, in terms of the Poynting vector at time t',

$$\frac{dW}{dt'}\,d\Omega = |\mathbf{E} \times \mathbf{H}|\,\frac{dt}{dt'}\,r^2\,d\Omega \tag{10.33}$$

The absolute value signs may be used instead of $(\mathbf{E} \times \mathbf{H}) \cdot \hat{a}_r$, since the radiation fields always point in the **r** direction as measured from point A of Fig. 10.1. The factor dt/dt' is necessary to convert the time scale over which the Poynting vector is evaluated to that at the point of emission of the radiation.

We can obtain dt/dt' from Eq. (10.32). This relation is, first, for the differential of t

$$dt = dt' + \frac{1}{c}\frac{\partial|\mathbf{r}|}{\partial t'}\,dt' \tag{10.34}$$

Then, this may be used to obtain

$$\frac{dt}{dt'} = 1 + \frac{1}{c}\frac{\partial |\mathbf{r}|}{\partial t'} \tag{10.35}$$

Also, note that

$$\frac{\partial |\mathbf{r}|}{\partial t'} = -\frac{\mathbf{r} \cdot \mathbf{u}}{|\mathbf{r}|} \tag{10.36}$$

which can be seen from Fig. 10.1, by obtaining the shift in $|\mathbf{r}|$ for a "virtual" movement of the source point A over a distance $\mathbf{u}\,dt'$.

Equation (10.35) now becomes

$$\frac{dt}{dt'} = 1 - \frac{\mathbf{r} \cdot \mathbf{u}}{|\mathbf{r}|c} = \frac{s}{|\mathbf{r}|} \tag{10.37}$$

We can insert this result into Eq. (10.33) so that it removes the direct dependence on dt/dt'. This is, then

$$\frac{dW}{dt'}\,d\Omega = |\mathbf{E} \times \mathbf{H}|\,\frac{s}{|\mathbf{r}|}\,r^2\,d\Omega \tag{10.38}$$

Using Eq. (10.38) we can change Eq. (10.31) to show the outward flux centered around the electron at point A of Fig. 10.1.

$$\frac{dW}{dt'}\,d\Omega = \frac{q_e^{\,2}}{16\pi\varepsilon_0 c^3}\,|\dot{\mathbf{u}}|^2 \sin^2\theta \left[1 - \frac{u}{c}\cos\theta\right]^{-5} d\Omega \tag{10.39}$$

A plot of this function is shown in Fig. 10.3.

We now may take the general expressions for the radiation fields, Eqs. (10.22) and (10.21), and obtain the general Poynting vector, regardless of the

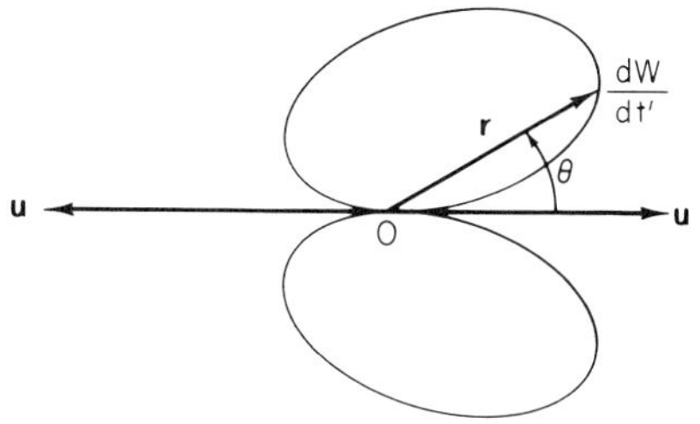

Fig. 10.3. Radiation pattern of a relativistic particle when the acceleration is in the same direction as its velocity.

acceleration. The magnitude of the general Poynting vector with no restrictions on the acceleration or velocity, in terms of the **E** field, is

$$|\mathbf{E} \times \mathbf{H}| = \left| \mathbf{E} \times \frac{\mathbf{r} \times \mathbf{E}}{|\mathbf{r}| c\mu_0} \right| = \left| \frac{\mathbf{r}(E^2)}{|\mathbf{r}| c\mu_0} + \frac{\mathbf{E}(\mathbf{E} \cdot \mathbf{r})}{|\mathbf{r}| c\mu_0} \right| \tag{10.40}$$

Now, from Eq. (10.22) we note that the radiation **E** field is perpendicular to **r** so Eq. (10.40) can be written as

$$|\mathbf{E} \times \mathbf{H}| = \frac{E^2}{c\mu_0} = \frac{q_e^{\,2}}{\mu_0 16\pi^2 \varepsilon_0^{\,2} s^6 c^5} |\mathbf{r} \times (\mathbf{r}_u \times \dot{\mathbf{u}})|^2 \tag{10.41}$$

We may now compute the outward radiation flux by returning to Eq. (10.38) and substituting in (10.41). The flux is then

$$\frac{dW}{dt'} d\Omega = \frac{q_e^{\,2}}{16\pi^2 \varepsilon_0 c^3 s^5} |\mathbf{r} \times (\mathbf{r}_u \times \dot{\mathbf{u}})|^2 |\mathbf{r}| \, d\Omega \tag{10.42}$$

C. THE RADIATION SPECTRA

At this point we have two paths that may be followed, provided that we know **r**, $\mathbf{r}_u$, **u**, and $\dot{\mathbf{u}}$. We wish to know both the spectrum and angular distribution of the emitted radiation flux. We can take the Fourier transform in time of (10.42), which will yield the spectrum and the angular distribution of the radiation. If the spectrum alone is desired, (10.42) may first be integrated over all solid angles and then the Fourier transform may be taken of this result. Note that (10.42) is the differential energy emission rate η_ω.

The definition of the Fourier transform requires an integration with respect to time. The choice of the time integration variable is important. Since the time at the point of evaluation of the Poynting vectoı is t, the Fourier integration must be made with respect to t (not t') if we wish to use the time at the point of observation. This can be found by using Eq. (10.41), multiplying by $r^2 \, d\Omega$, and taking the Fourier transform as shown below.

$$W(\omega, \Omega) \, d\Omega = d\Omega \int_{-\infty}^{+\infty} \frac{dW}{dt'} e^{-j\omega t} \, dt = \left[\int_{-\infty}^{+\infty} r^2 |\mathbf{E} \times \mathbf{H}| e^{-j\omega t} \, dt \right] d\Omega \tag{10.43}$$

Note, however, that dW/dt' is evaluated at t'. We can convert the integral in (10.43) to an integration with respect to t' by using Eq. (10.33) or (10.38). In doing this, (10.43) becomes

$$W(\omega, \Omega) \, d\Omega = \int_{-\infty}^{+\infty} \frac{s}{|\mathbf{r}|} \frac{dW}{dt'} \exp\left[-j\omega\left(t' + \frac{|\mathbf{r}|}{c} \right) \right] dt' \, d\Omega \tag{10.44}$$

The expression dW/dt' defined in Eq. (10.42) is written as the square of the absolute value of a time function. We will now represent the entire integrand in (10.44) as the square of a function of time. We then must examine this kind of a time function to determine just what kind of spectrum may be obtained.

The normalized energy of such a time function is defined as

$$\mathscr{E} = \int_{-\infty}^{+\infty} f^2(t)\, dt \tag{10.45}$$

If $\mathscr{E}$ is finite, the function $f(t)$ contains only a specified amount of energy and is known as an *energy function* (energy signal). $\mathscr{E}$ may exhibit a frequency spectrum. If the value of $\mathscr{E}$ in infinite, which is the case, for example, for a strictly periodic (monochromatic) function, the concept of energy is undefined, but the *time-average power* of this type of function may be noninfinite and have a spectrum. The time-average power is defined to be

$$P_{\mathrm{AV}} = \overline{f^2(t)} = \lim_{T\to\infty} \frac{1}{2T} \int_{-T}^{+T} f^2(t)\, dt \tag{10.46}$$

where T is a time interval.

The interactions previously described in this chapter are often of the finite energy type. This can be made clear by examining the kinds of interactions possible. In many cases, the interactions are such that a quantum of energy is emitted, as for example, light photons or x-ray photons. Thus the notion of finite energy signals readily appears. We proceed to determine the spectra of energy functions.

Now, if $f(t)$ has a Fourier transform defined as

$$F(\omega) = \int_{-\infty}^{+\infty} f(t) e^{-j\omega t}\, dt \tag{10.47}$$

and an inverse transformation

$$f(t) = \frac{1}{2\pi} \int_{-\infty}^{+\infty} F(\omega) e^{j\omega t}\, d\omega \tag{10.48}$$

then we may modify Eq. (10.45) for the energy to include the Fourier transform as shown below.

$$\mathscr{E} = \int_{-\infty}^{+\infty} f^2(t)\, dt = \int_{-\infty}^{+\infty} f(t) \left[\frac{1}{2\pi} \int_{-\infty}^{+\infty} F(\omega) e^{j\omega t}\, d\omega \right] dt \tag{10.49}$$

We may now interchange the order of integration on the right-hand side of Eq. (10.49) and obtain the following expression.

$$\mathscr{E} = \frac{1}{2\pi}\int_{-\infty}^{+\infty} F(\omega)\left[\int_{-\infty}^{+\infty} f(t)e^{j\omega t}\,dt\right] d\omega \tag{10.50}$$

The quantity in the brackets of Eq. (10.50) is the Fourier transform of $f(t)$ when the substitution $\omega = -\omega$ is made. That is

$$F(-\omega) = \int_{-\infty}^{+\infty} f(t)e^{j\omega t}\,dt \tag{10.51}$$

Therefore

$$\mathscr{E} = \frac{1}{2\pi}\int_{-\infty}^{+\infty} F(\omega)F(-\omega)\,d\omega = \int_{-\infty}^{+\infty} f^2(t)\,dt \tag{10.52}$$

Now for a real $f(t)$, $F(\omega)F(-\omega) = F(\omega)F^*(\omega) = |F(\omega)|^2$ and $f^2(t) = |f(t)|^2$. $F^*(\omega)$ is the complex conjugate of $F(\omega)$. Equation (10.52) now becomes

$$\int_{-\infty}^{+\infty} |f^2(t)|\,dt = \frac{1}{2\pi}\int_{-\infty}^{+\infty} |F(\omega)|^2\,d\omega \tag{10.53}$$

Equation (10.53) is known as *Parseval's theorem.*

We define $|F(\omega)|^2$ to be the energy spectrum of $f(t)$. The energy spectrum of Eq. (10.44) is therefore

$$W(\omega, \Omega) = \frac{q_e^2}{16\pi^2\varepsilon_0 c^3}\left|\int_{-\infty}^{+\infty} \frac{\exp[-j\omega(t' + |\mathbf{r}|/c)][\mathbf{r} \times (\mathbf{r}_u \times \dot{\mathbf{u}})]\,dt'}{s^2}\right|^2 \tag{10.54}$$

This can be obtained by placing (10.44) into the form required by (10.53). We may integrate before taking the absolute value if desired. In the nonrelativistic limit, $t' \to t$, $\mathbf{r}_u \to \mathbf{r}$, and $s \to r$, so Eq. (10.54) becomes, under this condition,

$$W(\omega, \Omega) = \frac{q_e^2}{16\pi^2\varepsilon_0 c^3}\left|\int_{-\infty}^{+\infty} \frac{e^{-j\omega t}[\mathbf{r} \times (\mathbf{r} \times \dot{\mathbf{u}})]\,dt}{r^2}\right|^2 \tag{10.55}$$

This last equation gives the spectrum of the *dipole* radiation component of an electron of charge q_e. Remember that it is only valid for low energies. If Eq. (10.55) is integrated over all solid angles, it becomes

$$W(\omega) = \frac{q_e^2}{6\pi^2\varepsilon_0 c^3}\left|\int_{-\infty}^{+\infty} \dot{\mathbf{u}}e^{-j\omega t}\,dt\right|^2 \tag{10.56}$$

for the energy spectrum, regardless of angle.

These last expressions determine the energy spectrum and angular distribution of the radiation if the acceleration and velocity of the interacting particles are known. The dipole approximation may be used as a first approximation, and then higher-order terms in the expansion may be included as the energies increase.

For completeness, we will mention another approach to obtain the emission of radiation by charged particles. This approach begins *at* the relativistic limit and expands back toward lower energies. The method used to arrive at the radiation intensity is to utilize Dirac's relativistic wave equation for a colliding electron interacting with an electromagnetic field. The interaction consists of two parts. These are (1) the interaction of the electron with the radiation field and (2) the interaction of the electron with the field of the scattering center. These interactions are considered as perturbations of the total energy of the system in an expansion. In general, higher-order terms are utilized when the energies of the electrons are *decreased* from very high values.

We now turn to applications of the previous results. In particular, we will consider two of the processes in a plasma that produce radiation: (a) bremsstrahlung (literally, braking radiation); and (b) cyclotron radiation.

D. BREMSSTRAHLUNG

The first process, bremsstrahlung, occurs whenever a charged particle is *decelerated* by making some kind of collisional process. If the charged particle remains unbound both before and after the collision, the process is called free–free bremsstrahlung. Otherwise, it is known as free–bound radiation, in that the charged particle has been captured by another particle as it has emitted the radiation. In both cases the energy of the emitted photon is the difference between the energy of the charged particle before and after the collision. That is,

$$\mathscr{E}_{\text{photon}} = \mathscr{E}_{\substack{\text{particle before}\\ \text{coll}}} - \mathscr{E}_{\substack{\text{particle after}\\ \text{coll}}} \tag{10.57}$$

or

$$\hbar\omega = \tfrac{1}{2}mv'^2 - \tfrac{1}{2}mv^2$$

Free–free or free–bound (remember that the particle is trapped after the latter interaction) radiation occurs either when particles strike the walls of the vacuum chamber containing them or when they make Coulomb interactions between particles of the same or other species. This type of radiation process may result from any of the following interactions:

(a) electron–electron;
(b) electron–ion;
(c) electron–neutral (atom);
(d) ion–ion;
(e) ion–neutral.

Note that neutral–neutral collisions do not normally produce bremsstrahlung radiation.

In the nonrelativistic limit, collisions between like species of particles do not give rise to radiation, if we consider only the dipole approximation. This can be qualitatively understood by realizing that the motions of the particles will produce radiation fields that exactly cancel under the dipole approximation. The quadrupole (next term in the expansion) component of radiation does begin to be significant when the kinetic energies of the charged particles approach their rest energy ($m_0 c^2$) where m_0 is the rest mass of the particle. For electrons, this is an energy of about 511 keV.

We shall instead examine electron–ion bremsstrahlung in more detail. This normally produces more radiation than electron–electron or ion–ion bremsstrahlung. A nonquantum-mechanical nonrelativistic treatment is considered first. This means that the energy of the electron before and after the collision is now not only always nonrelativistic, but is assumed to remain approximately constant. The geometry of this type of collision is that described in Chapter 2 in Figs. 2.2 and 2.3.

Now, if the assumption that a *photon* (wave packet or pulse) is emitted while the particles are colliding is accepted, a truncation of the interaction time should be made, since otherwise the radiation would be emitted continually, and we could only obtain a power spectrum, not an energy spectrum. Note that Coulomb forces *are* long range, but their range of effectiveness has often been mathematically limited at a distance of about one Debye length, so there is some justification for the truncation.

In this approximation, the interaction time will be set to be about

$$\tau_{\text{int}} = \frac{b}{v} \tag{10.58}$$

where b is the impact parameter and v is the velocity of the particle. In addition, the particle will be assumed to follow a straight-line trajectory as shown in Fig. 10.4. The magnitude of the force that the electron sees is expressed as a function of time in Eq. (10.59).

$$F = m\frac{dv}{dt} = \frac{q_e^{\,2}}{4\pi\varepsilon_0(b^2 + v^2t^2)} \tag{10.59}$$

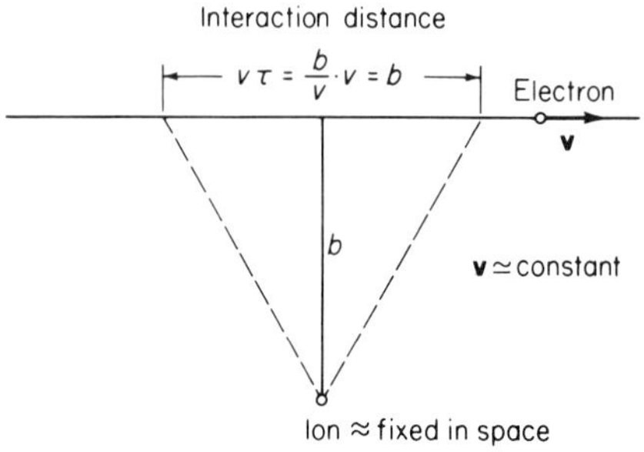

Fig. 10.4. Straight-line approximation.

Here $t = 0$ is assumed to be at the point of closest approach, and the ion has the same magnitude of charge as the electron. The force in the transverse direction (perpendicular to the direction of the trajectory) is determined by trigonometry to be

$$F_{\text{tran}} = \frac{q_e^2 b}{4\pi\varepsilon_0 (b^2 + v^2t^2)^{3/2}} \tag{10.60}$$

The *transverse* acceleration is then

$$\dot{v}_{\text{tran}} = \frac{q_e^2 b}{4\pi\varepsilon_0\, m(b^2 + v^2t^2)^{3/2}} \tag{10.61}$$

The *parallel* acceleration will be neglected. This is justified as a consequence of the assumptions of the straight-line trajectory *and* the finite interaction time, which is proportional to b/v. Equation (10.61) may be substituted into any of the nonrelativistic spectral equations given previously, such as (10.55) or (10.56), to obtain the energy spectrum.

Equation (10.61) is defined for all time and, since the Fourier transform is required to obtain the spectrum, we can easily do the Fourier integration. We can show $\dot{v}_{\text{tran}}(t)$ to be an energy function if Eq. (10.61) is used in (10.45), so we can safely compute the energy spectrum.

The Fourier transform of $\dot{v}_{\text{tran}}$ is

$$\dot{v}_{\text{tran}}(\omega) = \int_{-\infty}^{+\infty} \dot{v}_{\text{tran}}(t) e^{-j\omega t}\, dt = \frac{q_e^2 b}{4\pi\varepsilon_0\, m} \int_{-\infty}^{+\infty} \frac{e^{-j\omega t}\, dt}{(b^2 + v^2t^2)^{3/2}} \tag{10.62}$$

Since the quantity $1/(b^2 + v^2t^2)^{3/2}$ is an even function of time, the exponential factor may be replaced by $\cos \omega t$. Then, twice the integral in (10.62) from 0 to ∞ may be used as a replacement, so that Eq. (10.62) is changed into the following form

$$\dot{v}_{\text{tran}}(\omega) = \frac{2q_e^2 b}{4\pi\varepsilon_0\, m} \int_0^{\infty} \frac{\cos \omega t\, dt}{(b^2 + v^2t^2)^{3/2}} \tag{10.63}$$

This integral is related to the defining integral for the modified Bessel function of the second kind, which is

$$K_1\left(\frac{\omega b}{v}\right) = \frac{b}{v\omega}\int_0^\infty \frac{\cos \omega t \, dt}{[(b/v)^2 + t^2]^{3/2}} \tag{10.64}$$

Therefore, the Fourier transform of the acceleration becomes, in final form,

$$\dot{v}_{\text{tran}}(\omega) = \frac{2q_e^{\,2}}{4\pi\varepsilon_0\, m\, v^2}\,\frac{\omega}{} K_1\left(\frac{\omega b}{v}\right) \tag{10.65}$$

This result may now be squared and utilized in Eq. (10.56) to obtain the energy spectrum directly, or used in (10.55) and/or (10.54) to obtain the angular distribution of the radiation spectrum.

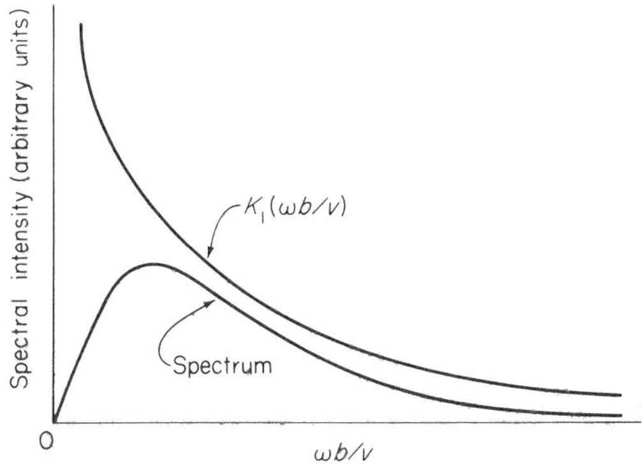

Fig. 10.5. Bremsstrahlung spectrum of a particle.

Figure 10.5 shows the angularly independent energy spectrum as a function of $(\omega b/v)$. This graph shows that the electron–ion bremsstrahlung spectral intensity drops off monotonically as the frequency increases. The emission tends to be low when $\omega b/v > 1$. This spectrum is, again, only the *dipole* radiation from the interaction. Relativity and quantum-mechanical effects (the electron was assumed to have essentially the same energy before and after the collision) have been neglected.

The angular distribution of the nonrelativistic radiation at a given frequency may be obtained by utilizing Eq. (10.55). Note that **r** is not a quantity that will affect the Fourier transformation, and $\dot{v}$ is pointing only in one direction (for example, along the polar axis of a spherical coordinate system) as shown in Fig. 10.6. The magnitude of the triple cross product in Eq. (10.55) is $(r^2 \sin\theta)\dot{v}$. This result may be applied to Eq. (10.55) to obtain the pattern.

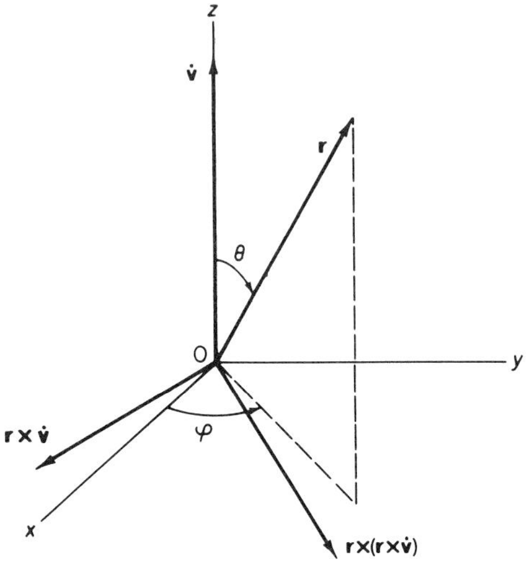

Fig. 10.6. Spherical coordinates used to obtain the angular distribution of the bremsstrahlung spectrum.

This pattern, which is for nonrelativistic, nonquantum-mechanical bremsstrahlung, is shown in Fig. 10.7. The mathematical expression for the pattern is

$$W(\omega, \Omega) = \frac{q_e^2}{16\pi^2 \varepsilon_0 c^3} \left(\frac{2q_e^2 \omega}{4\pi\varepsilon_0 mv^2}\right)^2 \left[K_1\left(\frac{\omega b}{v}\right)\right]^2 \sin^2\theta \tag{10.66}$$

The radiation pattern is still the usual $\sin^2\theta$ curve found from dipole antennas. Note that the *shape* of the radiation pattern is the same, *regardless* of frequency. Only the magnitude of the pattern is frequency dependent.

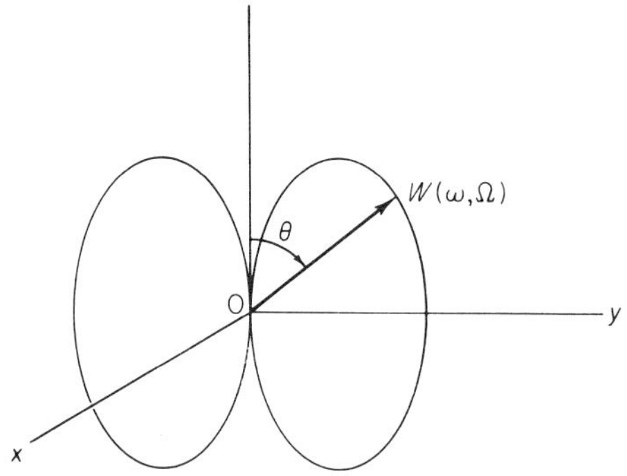

Fig. 10.7. Radiation pattern of nonrelativistic, nonquantum-mechanical bremsstrahlung.

In order to find the spectrum and pattern of the radiation produced by a complete plasma, and not just that from one radiating particle, several items must be considered. First, the spectrum must be integrated over the range of allowed impact parameters. Then, an integration over the appropriate velocity distribution must be considered.

The first item, the integration over the range of impact parameters, may be made straightforwardly. The collisional impact parameter integration formulation from Chapter 2 is utilized to form an integral that results in a spectrum to be integrated over the allowed range of impact parameters as shown.

$$\langle W(\omega, \Omega)\rangle_b = \frac{q_e^2}{16\pi^2\varepsilon_0 c^3}\left(\frac{2q_e^2\omega}{4\pi\varepsilon_0 m v^2}\right)^2 \sin^2\theta \int_{b_{\min}}^{b_{\max}} \left[K_1\left(\frac{\omega b}{v}\right)\right]^2 2\pi b\, db \quad (10.67)$$

The integral may be evaluated without much difficulty.

The integration of Eq. (10.67) over the velocity distribution must now be made. The result will yield a spectrum that is independent of velocity and is a representation of the actual emitted spectrum produced by a plasma under the conditions of validity of the approximations used. The required velocity integral is of the form

$$\langle W(\omega, \Omega)\rangle_{b,v} = N\int_0^{2\pi}\int_0^{\pi}\int_{(2\hbar\omega/m)^{1/2}}^{\infty} \langle W(\omega, \Omega)\rangle_b f(v) v^2 \sin\theta\, dv\, d\theta\, d\varphi \quad (10.68)$$

The form of the distribution function used may often affect the magnitude and shape of the radiation pattern. This is especially so if the distribution function is anisotropic in velocity space. Note that the velocity space integration begins at the lowest allowed velocity for the photon energy desired, since those particles whose energies are below the photon energy obviously cannot emit this photon. We have used spherical velocity coordinates in (10.68), so the integral may be integrated with respect to the azimuthal coordinates if required.

A further refinement for the bremsstrahlung calculations is to make the collisional transitions quantum-mechanical but nonrelativistic. Sommerfeld has considered this problem, and calculations of the form of Eq. (10.68) for the radiation pattern, including an anisotropic velocity distribution, show good agreement with experimental measurements of bremsstrahlung radiation.

The combination of relativistic and quantum-mechanical calculations is more complicated, but may be done in other ways, one of which is discussed in general in the following paragraphs.

These last two corrections, quantum-mechanical and relativistic, have often been calculated by developing a factor (Gaunt factor) that, when multiplied by the nonrelativistic, nonquantum-mechanical spectrum, yields, in the appropriate energy range over which the particular calculation is valid, approximate results. The Gaunt factor may be obtained from either theory or experiment. Such calculations, however, are often cumbersome and sometimes have limited range of application, although in many cases, this is the only method possible.

E. CYCLOTRON EMISSION

Another major radiation process that we will consider is cyclotron emission. When a plasma is placed in a magnetic field, the orbiting ions and electrons radiate, due to the constant application of the magnetic acceleration used to drive the particles in their circular orbits.

The acceleration, being always perpendicular to the velocity, results in some simplification of the general relation for the spectrum in Eq. (10.54), (10.55), and (10.56). Again, the first approach will be nonrelativistic and nonquantum mechanical. Normally, nonrelativistic particles do not radiate a significant fraction of their energy in the time of one orbit anyway, so the nonquantum-mechanical approach is often sufficient. It should be noted in passing that, since the particles are constantly experiencing an acceleration, the energy loss by radiation would cause them to "run down" and stop, but this is usually a condition never reached due to quantum-mechanical and/or collisional effects. The charged particle will be assumed to orbit in the xy plane and the magnetic field direction will be the z axis.

Figure 10.8 shows the appropriate vector positions. We will use the center of the orbit as the origin. Note that for this problem $\varphi = \omega_c t'$, where ω_c is the cyclotron frequency. Recall that t' is the time measured at the location of the emitting charge. There is a trigonometric relation between the three angles θ, α, and φ shown in Fig. 10.8. This relation is

$$\cos\theta = \sin\alpha\cos\varphi \tag{10.69}$$

From this relation and Fig. 10.8 it can be seen that θ as well as φ changes with time, but α remains constant. This approximation is justified if we are a long distance from the particle compared to its radius of gyration. The desired quantity needed for utilization in Eq. (10.42) or related equations to find the spectrum is $[\mathbf{r} \times (\mathbf{r}_u \times \dot{\mathbf{u}})]^2$, and this can now be obtained from Fig. 10.8.

The terms necessary to evaluate the triple vector product are defined for the orbiting particle as

$$u = r_g\omega_c \qquad \dot{\mathbf{u}}\cdot\mathbf{r} = |\dot{\mathbf{u}}|\,|\mathbf{r}|\cos\theta\,\frac{\sin\varphi}{\cos\varphi}$$

$$\dot{u} = r_g\omega_c^{\,2} \qquad \mathbf{r}_u = \mathbf{r} - \frac{\mathbf{u}\,|\mathbf{r}|}{c}$$

$$\mathbf{u}\cdot\mathbf{r} = |\mathbf{u}|\,|\mathbf{r}|\cos\theta$$

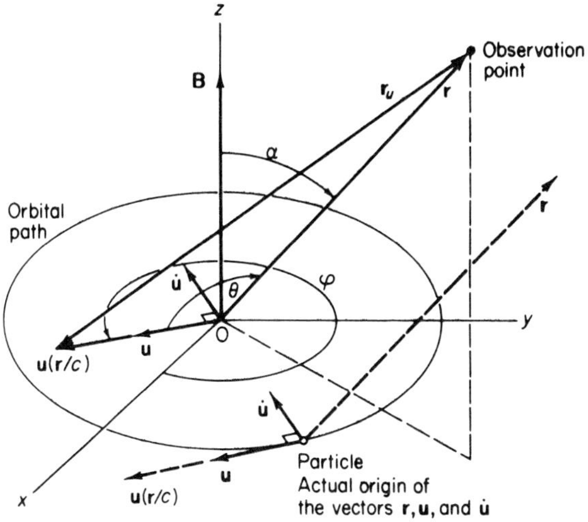

Fig. 10.8. Vector diagram for orbiting particle in a dc magnetic field.

Recall that r_g is the radius of gyration. These terms may be used to form the bracketed triple vector cross product:

$$[\mathbf{r} \times (\mathbf{r}_u \times \dot{\mathbf{u}})]^2 = -(\dot{\mathbf{u}} \cdot \mathbf{r})^2 r^2 \left(1 - \frac{u^2}{c^2}\right) + \dot{u}^2 r^4 \left(1 - \frac{\mathbf{r} \cdot \mathbf{u}}{rc}\right)^2$$

$$= \dot{u}^2 r^4 \left[\left(1 - \frac{u}{c} \cos \theta\right)^2 - \left(1 - \frac{u^2}{c^2}\right) \tan^2 \varphi \cos^2 \theta\right] \qquad (10.70)$$

Therefore, the integrand in Eq. (10.42) for cyclotron radiation becomes

$$\frac{dW}{dt'}\, d\Omega = \frac{q_e^{\,2} \dot{u}^2}{16\pi^2 \varepsilon_0 c^3} \frac{[1 - (u/c) \cos \theta]^2 - [1 - (u^2/c^2)] \tan^2 \varphi \cos^2 \theta}{[1 - (u/c) \cos \theta]^5}\, d\Omega \qquad (10.71)$$

This pattern is zero for all values of t' at an angle θ_0, such that $\cos \theta_0 = u/c$. Note, however, that since θ itself changes with time, so must the position of this "zero" change position with time. We can utilize Eq. (10.69) to eliminate θ from Eq. (10.71) and we obtain

$$\frac{dW}{dt'}\, d\Omega = \frac{q_e^{\,2} \dot{u}^2}{16\pi^2 \varepsilon_0 c^3} \frac{[1 - (u/c) \sin \alpha \cos \varphi]^2 - [1 - (u/c)^2] \sin^2 \varphi \sin^2 \alpha}{[1 - (u/c) \sin \alpha \cos \varphi]^5}\, d\Omega \qquad (10.72)$$

for the spectral pattern. The main reason for substituting α in the expression for the energy flux (10.72) is that angle α is fixed. It is the angle between the axis of rotation and the direction of observation. It does not change as the particle rotates. The *pattern itself does move* as angle φ changes due to the rotation of the particle.

The total radiation flux can be obtained by integrating Eq. (10.72) with respect to a solid angle $d\Omega$. This solid angle is defined to be

$$d\Omega = \sin\alpha \, d\alpha \, d\varphi \tag{10.73}$$

α is the polar angle defined in Fig. 10.8 and φ is the azimuthal angle measured from the x axis. We may consider the particle fixed on the x axis at a given instant of time and so use φ as an azimuthal variable to sweep out the solid angle $d\Omega$. The integrated spectrum is

$$\frac{dW}{dt'} = \frac{q_e^2 \dot{u}^2}{16\pi^2 \varepsilon_0 c^3} \int_0^{2\pi} \int_0^{\pi} \frac{[1 - (u/c)\sin\alpha\cos\varphi]^2 - [1 - (u/c)^2]\sin^2\varphi\sin^2\alpha}{[1 - (u/c)\sin\alpha\cos\varphi]^5} \times \sin\alpha \, d\alpha \, d\varphi \tag{10.74}$$

The results of the integration are

$$\frac{dW}{dt'} = \frac{q_e^2 \dot{u}^2}{6\pi\varepsilon_0 c^3} \frac{1}{[1 - (u^2/c^2)]^2} \tag{10.75}$$

for the total radiated power. This value is also the emission rate for spontaneous emission η_ω for this process. It may then be used in Eq. (10.1) to determine the total rate of emission from the plasma.

If the result is specified for the conditions of cyclotron radiation, then $\dot{u} = r_g \omega_c^2$, $u = r_g \omega_c$, and $\varphi = \omega_c t$ so Eq. (10.75) becomes

$$\frac{dW}{dt'} = \frac{q_e^2 r_g^2 \omega_c^4}{6\pi\varepsilon_0 c^3} \frac{1}{[1 - (r_g \omega_c^2/c^2)]} \tag{10.76}$$

Equations (10.75) and (10.76) imply that electrons will usually radiate more energy than ions, since they usually have a faster orbiting velocity $(r_g \omega)$. Determination of the frequency spectrum of the emitted cyclotron radiation is a more difficult problem. If it is still assumed that a nonquantum-mechanical condition exists, that is, the particle does not lose a significant fraction of its energy per orbit, the problem is simplified greatly. Equation (10.55) is the required one to use to obtain the spectrum as a function of angle. In particular, we must evaluate the exponent, which is $\exp\{-j\omega[t' + (|\mathbf{r}|/c)]\}$ and the quantity s. This latter quantity

$$s = |\mathbf{r}| - \frac{\dot{\mathbf{u}} \cdot \mathbf{r}}{c} = r\left[1 - \frac{u}{c}\sin\alpha\cos\omega_c t'\right] \tag{10.77}$$

is easily obtained from Fig. 10.8.

The exponential in (10.55) is a bit more difficult to handle. It is a function of **r**. Now, since **r** is really measured from the charge to the observer and the charge is constantly changing position, a conversion to a radial coordinate that is measured from the center of the orbital motion will simplify matters and permit us to compute the spectrum easily. This coordinate will be called R. The relation between r and R is

$$r = R - r_g \sin \alpha \sin \omega_c t' \tag{10.78}$$

if R and r are assumed to be parallel, which is valid for large values of r and R. The exponential factor then becomes

$$\exp\left(-\frac{j\omega R}{c} + \frac{j\omega r_g \sin \alpha \sin \omega_c t' - j\omega t'}{c}\right) \tag{10.79}$$

or

$$\exp\left(\frac{j\omega r_g \sin \omega_c t' \sin \alpha}{c}\right) \exp\left(-\frac{j\omega R}{c} - j\omega t'\right) \tag{10.80}$$

Now the first factor in Eq. (10.80) is of the form $\exp(jX \sin \xi)$, where $\xi = \omega_c t'$ and $X = \omega r_g \sin \alpha$. This may be broken up into a series of Bessel functions by means of the identity

$$\exp(jX \sin \xi) = \sum_{n=-\infty}^{+\infty} \exp(jn\xi)\, J_n(X) \tag{10.81}$$

where J_n is a Bessel function of the first kind of order n. This last result shows that the spectrum of radiation will appear as a set of harmonics of the cyclotron frequency. The entire spectrum equation integral (10.55) is now known and the energy spectrum can be obtained. The vector cross product terms are obtained from Eq. (10.69), and Eqs. (10.80) and (10.81) give the appropriate terms for the exponential.

To evaluate the desired integral for the spectrum (Eq. (10.55)) we note that it is a vector integral and has three components. We may expand the triple cross product to be

$$\mathbf{r} \times (\mathbf{r}_u \times \dot{\mathbf{u}}) = (\mathbf{r} \cdot \dot{\mathbf{u}})\mathbf{r}_u - (\mathbf{r} \cdot \mathbf{r}_u)\dot{\mathbf{u}} \tag{10.82}$$

The previous definitions show that $\mathbf{r} \cdot \mathbf{u} = \dot{u} r \cos \theta \tan \varphi$ and $\mathbf{r} \cdot \mathbf{r}_u = r^2[1 - (u/c) \cos \theta]$, so that the cross product is, in three *nonorthogonal* components,

$$\mathbf{r} \times (\mathbf{r}_u \times \dot{\mathbf{u}}) = (\dot{u} r \cos \theta \tan \varphi)\mathbf{r} - \frac{\dot{u}}{c} r^2 \cos \theta \tan \varphi \mathbf{u} - r^2\left(1 - \frac{u}{c} \cos \theta\right)\dot{\mathbf{u}} \tag{10.83}$$

Now it is true for orbital motion that **u** and **u̇** *are* orthogonal. We may break up the **r** vector into perpendicular components as follows, in order to produce three orthogonal terms in (10.83). The three orthogonal components of **r** are

$$\mathbf{r} = r \cos \alpha \, \hat{a}_z + \frac{r \cos \theta}{u} \mathbf{u} + \frac{r \cos \delta}{\dot{u}} \dot{\mathbf{u}} \tag{10.84}$$

where

$$\cos^2 \delta = 1 - \cos^2 \alpha - \cos^2 \theta \tag{10.85}$$

Note that $\hat{a}_z$ is orthogonal to **u** and **u̇** so that the final result for the triple cross product is now stated in terms of three orthogonal components

$$\begin{aligned}[\mathbf{r} \times (\mathbf{r}_u \times \dot{\mathbf{u}})] &= \dot{u} r^2 \cos \alpha \cos \theta \tan \varphi \, \hat{a}_z \\ &+ \left[\frac{\dot{u} r^2 \cos^2 \theta \tan \varphi}{u} - \frac{u r^2}{c} \cos \theta \tan \varphi\right] \mathbf{u} \\ &+ \left[r^2 \cos \delta \cos \theta \tan \varphi - r^2\left(1 - \frac{u}{c} \cos \theta\right)\right] \dot{\mathbf{u}}\end{aligned} \tag{10.86}$$

We may eliminate θ by using Eq. (10.68) and δ by using Eq. (10.85). We may further convert to fixed axes, if desired, by utilizing the transformation

$$\mathbf{u} = u \cos \varphi \hat{a}_x + u \sin \varphi \hat{a}_y$$

and

$$\dot{\mathbf{u}} = -\dot{u} \sin \varphi \hat{a}_x + \dot{u} \cos \varphi \hat{a}_y$$

The integration to obtain the Fourier transform to be used in Eq. (10.54) for the spectra is done via the integrals of the component of the square of the vector cross product in the desired direction of observation *at the point of observation*. For example, the z-direction integral for the spectrum at the point of observation is

$$\begin{aligned}W(\omega, \Omega)_z = \frac{q_e^2}{16\pi\varepsilon_0 c^3} \Bigg| \int_{-\infty}^{+\infty} & \frac{\dot{u} \cos \alpha \cos \theta \tan \varphi}{[1 - (u/c) \sin \alpha \cos \varphi]^2} \exp\left(-\frac{j\omega R}{c} - j\omega t'\right) \\ &\times \exp\left(\frac{j\omega r_g \sin \varphi \sin \alpha}{c}\right) dt' \Bigg|^2\end{aligned} \tag{10.87}$$

Eliminating θ and substituting for φ results in the following.

$$\begin{aligned}W(\omega, \Omega)_z = \frac{q_e^2}{16\pi\varepsilon_0 c^3} \Bigg| \int_{-\infty}^{+\infty} & \frac{\dot{u} \cos \alpha \sin \omega_c t'}{[1 - (u/c) \sin \alpha \cos \omega_c t']^2} \\ &\times \exp\left(-\frac{j\omega R}{c} - j\omega t'\right) \exp\left(\frac{j\omega r_g \sin \omega_c t' \sin \alpha}{c}\right) dt' \Bigg|^2\end{aligned} \tag{10.88}$$

for the spectral components. We note that this component of the spectrum has a term in it that is like Eq. (10.81). That is, we may break it up into harmonic components as

$$\exp\left(\frac{j\omega r_g \sin \omega_c t' \sin \alpha}{c}\right) = \sum_{n=-\infty}^{+\infty} \exp(jn\omega_c t') J_n\left(\frac{\omega r_g \sin \alpha}{c}\right) \tag{10.89}$$

If the relativistic change of mass is included, ω_c becomes

$$\frac{qB}{m_0}\left(1-\frac{u^2}{c^2}\right)^{1/2}$$

the reduced cyclotron frequency, where m_0 is the rest mass. The inclusion of this term shows that the spectral components, when computed for a plasma, with the effects of relativity and a velocity distribution considered, will appear at the "rest" cyclotron frequency and harmonics, but will also result in "broadening" of the spectrum due to these relativistic effects. If the radiation from the orbiting particle is observed directly above the center of its orbit, then angle $\alpha \equiv 0$. The result is that, for the z component of the radiation, Eq. (10.89) has the value of $e^{j\omega_c t'}$ and only the cyclotron frequency itself is present.

We have assumed for this, as in all the other processes previously discussed, that each orbiting particle radiates independently of the others. A conversion to the time frame as seen by the observer may also be made, if desired.

F. FLUCTUATIONS AND CORRELATIONS IN PLASMAS

The previous calculations, both for bremsstrahlung and cyclotron radiation, have been made by considering only isolated particles, that is, the collective effects of the plasma have been neglected.

While it is true that the spectrum may be averaged over a range of velocities obtainable in the plasma itself, by integration as in Eq. (10.68), the collective effects of nearby sources of radiation (other particles) have not been considered.

Since several types of radiation and absorption processes, as well as motions of the particles themselves, may be occurring at the same time in a plasma, the result is likely to be a series of fluctuations in the local values of electric and magnetic fields. Also, since the fields are produced by collisional processes, which are usually inelastic, additional dissipation of energy will occur.

It is sometimes possible to determine the spectrum of these fluctuation processes by utilizing correlation techniques. The method is quite similar to the Markoff process utilized in the discussions surrounding the Fokker-Planck equation. We shall also use the work on Fourier spectra done previously in this chapter.

Since fluctuating signals of the type produced in a plasma may have nonfinite energy (i.e., they cannot be considered as individual photons), we can only obtain a *power* density spectrum rather than an *energy* density spectrum for these signals, as was done for the isolated source case.

The signals produced by the fluctuations are considered to be present for all time. The time-average power of such signals is defined as

$$P_{\text{av}} = \overline{f^2(t)} = \lim_{T\to\infty} \frac{1}{2T} \int_{-T}^{+T} f^2(t)\, dt \tag{10.90}$$

Note that if we do not utilize the division by T, a time interval, the result of Eq. (10.90) would be in the same form as Eq. (10.45) but would be infinite. Equation (10.90) has essentially the units of Eq. (10.45) divided by time, which yields power. To obtain this power density spectrum, we proceed as follows.

We first define a new function $f_T(t)$ by truncating $f(t)$ outside the interval $-T < t < T$; that is, $f_T(t) = 0$ outside the interval, and has the value $f(t)$ inside the interval. This means that as long as T is finite, $f_T(t)$ can have finite *energy*.

We may take the Fourier transform of $f_T(t)$. This is

$$F_T(\omega) = \int_{-\infty}^{+\infty} f_T(t) e^{-j\omega t}\, dt = \int_{-T}^{+T} f(t) e^{-j\omega t}\, dt \tag{10.91}$$

The energy of $f_T(t)$, which is not necessarily infinite, is written as the following integral.

$$\mathscr{E}_T = \int_{-\infty}^{+\infty} f_T{}^2(t)\, dt = \frac{1}{2\pi} \int_{-\infty}^{+\infty} |F_T(\omega)|^2\, d\omega \tag{10.92}$$

utilizing Eq. (10.53). However, since $f_T(t)$ is zero outside the interval, it must be true that

$$\int_{-\infty}^{+\infty} f_T{}^2(t)\, dt = \int_{-T}^{+T} f^2(t)\, dt \tag{10.93}$$

Note that the right-hand sides of both Eqs. (10.91) and (10.93) are in terms of $f(t)$, not $f_T(t)$. The average *power*, from Eq. (10.90), can then be expressed in terms of the Fourier transform as shown in Eq. (10.94).

$$P_{\mathrm{av}} = \overline{f^2(t)} = \lim_{T \to \infty} \frac{1}{2T} \int_{-T}^{+T} f^2(t)\, dt$$

$$= \frac{1}{2\pi} \int_{-\infty}^{+\infty} \lim_{T \to \infty} \frac{|F_T(\omega)|^2}{2T}\, d\omega$$

$$= \int_{-\infty}^{+\infty} \mathscr{S}(\omega)\, d\omega \tag{10.94}$$

where $\mathscr{S}(\omega)$, the *power spectral density*, is defined as

$$\mathscr{S}(\omega) = \frac{1}{2\pi} \left[\lim_{T \to \infty} \frac{|F_T(\omega)|^2}{2T} \right] \tag{10.95}$$

We may also find $\overline{f^2(t)}$ if we know that $f(t)$ has a Fourier transform and an inverse. Its value can be found by taking the *ensemble* average of the product of the Fourier transform and the conjugate Fourier transform of $f(t)$, and then taking the inverse transform. That is

$$\overline{f^2(t)} = \langle f^2(t) \rangle = \frac{1}{(2\pi)^2} \int_{-\infty}^{+\infty} \int_{-\infty}^{+\infty} \langle F(\omega) F^*(\omega) \rangle e^{j(\omega - \omega')t}\, d\omega\, d\omega' \tag{10.96}$$

where

$$F^*(\omega') = \int_{-\infty}^{+\infty} f(t) e^{j\omega' t}\, dt \tag{10.97}$$

Equation (10.96) is true provided that we may consider the fluctuations to be *time stationary*. This assumption means that the average total power is assumed to remain constant with time. Under these conditions, the time and ensemble averages are equivalent.

So we then may say

$$\overline{f^2(t)} = \langle f^2(t) \rangle$$

Recall that the same assumption was made in Chapter 4 in our discussion of statistical mechanics. Since Eq. (10.96) may now be set equal to Eq. (10.94), we may solve for the quantity $\langle F(\omega) F^*(\omega') \rangle$ in terms of the power density function. Equation (10.94), however, has only one variable of integration, ω. We may add the additional variable of integration ω' to Eq. (10.94) by use of the method shown in what follows:

$$\langle f^2(t) \rangle = \int_{-\infty}^{+\infty} \int_{-\infty}^{+\infty} \mathscr{S}(\omega)\, \delta(\omega - \omega')\, d\omega\, d\omega' \tag{10.98}$$

$\delta(\omega - \omega')$ is the Dirac delta function. We may now set the integrands of (10.98) and (10.96) equal, and obtain

$$\langle F(\omega)F^*(\omega')\rangle = (2\pi)^2 \mathscr{S}(\omega)\, \delta(\omega - \omega')e^{-j(\omega-\omega')t} \tag{10.99}$$

However, the function defined in (10.99) has no value unless $\omega = \omega'$, due to the presence of the delta function, so (10.99) is simply

$$\langle F(\omega)F^*(\omega')\rangle = (2\pi)^2 \mathscr{S}(\omega)\, \delta(\omega - \omega') \tag{10.100}$$

The left-hand side of Eq. (10.100) is called the *transformed time correlation function.* The quantity F may be the transform of an electric field, charge density, and so forth.

In addition, a spatially transformed correlation function is possible. In this case the Fourier transformation is taken with respect to the three spatial coordinates. The space-transformed variable is usually written as a vector $\mathbf{k}$. Equation (10.94) now is written in terms of spatial coordinates as

$$\begin{aligned}\langle f^2(\mathbf{r})\rangle = \overline{f^2(\mathbf{r})} &= \lim_{V\to\infty} \frac{1}{8V}\int_{-\Delta x}^{+\Delta x}\int_{-\Delta y}^{+\Delta y}\int_{-\Delta z}^{+\Delta z} f^2(\mathbf{r})\, d\mathbf{r}\\ &= \frac{1}{(2\pi)^3}\int_{-\infty}^{+\infty} \lim_{V\to\infty}\frac{|F(\mathbf{k})|^2\, d\mathbf{k}}{8V} = \int_{-\infty}^{+\infty} \mathscr{S}(\mathbf{k})\, d\mathbf{k}\end{aligned} \tag{10.101}$$

V is the product $\Delta x\, \Delta y\, \Delta z$. The factor of $1/(2\pi)^3$ comes from the vector inverse Fourier transformation for $\mathbf{k}$, since it is in three dimensions. The *spatial* spectral density function is then

$$\mathscr{S}(\mathbf{k}) = \frac{1}{(2\pi)^3}\lim_{V\to\infty}\frac{|F(\mathbf{k})|^2}{8V} \tag{10.102}$$

We have assumed that the spatial Fourier transform of $f(\mathbf{r})$ exists as defined in the following equation.

$$F(\mathbf{k}) = \lim_{V\to\infty}\frac{1}{8V}\int_{-\Delta x}^{+\Delta x}\int_{-\Delta y}^{+\Delta y}\int_{-\Delta z}^{+\Delta z} f(\mathbf{r})e^{-j(\mathbf{k}\cdot\mathbf{r})}\, dx\, dy\, dz$$

We have used the same limiting features as in the time transform. The spatial form for Eq. (10.96) is then

$$\langle f^2(\mathbf{r})\rangle = \frac{1}{(2\pi)^6}\int_{-\infty}^{+\infty}\int_{-\infty}^{+\infty}\langle F(\mathbf{k})F^*(\mathbf{k}')\rangle e^{j(\mathbf{k}-\mathbf{k}')\cdot\mathbf{r}}\, d\mathbf{k}\, d\mathbf{k}' \tag{10.103}$$

We may add the additional variable of integration by using a delta function in Eq. (10.101). The result is

$$\langle f^2(\mathbf{r})\rangle = \int_{-\infty}^{+\infty}\int_{-\infty}^{+\infty}\mathscr{S}(\mathbf{k})\, \delta(\mathbf{k} - \mathbf{k}')\, d\mathbf{k}\, d\mathbf{k}' \tag{10.104}$$

Setting the two integrands equal in Eq. (10.103) and (10.104) yields

$$\langle F(\mathbf{k})F^*(\mathbf{k}')\rangle = (2\pi)^6 \mathscr{S}(\mathbf{k})\,\delta(\mathbf{k}-\mathbf{k}') \tag{10.105}$$

for the transformed spatial correlation function.

It is also possible to produce both spatial and temporal correlations together. We can just combine the results. That is, the average power of a function $f^2(\mathbf{r}, t)$ is

$$\begin{aligned}\langle f^2(\mathbf{r}, t)\rangle &= \frac{1}{(2\pi)^4}\int_{-\infty}^{+\infty}\int_{-\infty}^{+\infty} \lim_{T\to\infty}\lim_{V\to\infty} \frac{|F(\mathbf{k},\omega)|^2}{16TV}\,d\omega\,d\mathbf{k} \\ &= \int_{-\infty}^{+\infty}\int_{-\infty}^{+\infty} \mathscr{S}(\mathbf{k},\omega)\,d\omega\,d\mathbf{k}\end{aligned} \tag{10.106}$$

Note that Eq. (10.106) is a double integral. The spectral power density is

$$\mathscr{S}(\mathbf{k},\omega) = \frac{1}{(2\pi)^4}\lim_{T\to\infty}\lim_{V\to\infty}\frac{|F(\mathbf{k},\omega)|^2}{16TV} \tag{10.107}$$

This permits us to find the value of the average power considering both time and spatial variations together. In terms of the transformed correlation function, Eq. (10.106) becomes

$$\begin{aligned}\langle f^2(\mathbf{r}, t)\rangle = \frac{1}{(2\pi)^8}\int_{-\infty}^{+\infty}\int_{-\infty}^{+\infty}\int_{-\infty}^{+\infty}\int_{-\infty}^{+\infty} \langle F(\mathbf{k},\omega)F^*(\mathbf{k}',\omega')\rangle e^{j(\omega-\omega')t} \\ \times e^{j(\mathbf{k}-\mathbf{k}')\cdot\mathbf{r}}\,d\omega\,d\omega'\,d\mathbf{k}\,d\mathbf{k}'\end{aligned} \tag{10.108}$$

We again add the additional variables to (10.107) by use of the delta functions. The result is

$$\langle f^2(\mathbf{r}, t)\rangle = \int_{-\infty}^{+\infty}\int_{-\infty}^{+\infty}\int_{-\infty}^{+\infty}\int_{-\infty}^{+\infty} \mathscr{S}(\mathbf{k},\omega)\,\delta(\omega-\omega')\,\delta(\mathbf{k}-\mathbf{k}')\,d\omega\,d\omega'\,d\mathbf{k}\,d\mathbf{k}' \tag{10.109}$$

We are able to set the two integrands in (10.109) and (10.108) equal to each other. The result is a solution for the transform of the temporal and spatial correlation function of

$$\langle F(\mathbf{k},\omega)F^*(\mathbf{k}',\omega')\rangle = (2\pi)^8\mathscr{S}(\mathbf{k},\omega)\,\delta(\omega-\omega')\,\delta(\mathbf{k}-\mathbf{k}') \tag{10.110}$$

This development means that, once that we know the correlation function, we may find the spectral power density, and vice versa.

Having obtained the theoretical development of correlation functions, a word is necessary as to their physical meaning and value. We first note that we may compute either the transformed *auto*correlation function or the *cross*-correlation function, depending upon whether we utilize the same or different functions to obtain the transforms $F(\omega)$ and $F^*(\omega')$. However, only when the autocorrelation function is used is the physical concept of spectral power density valid.

We may compute the time autocorrelation function $C(\tau)$ of a function $f(\mathbf{r}, t)$ by use of the following integral.

$$C(\mathbf{r}, \tau) = \lim_{T\to\infty} \frac{1}{2T} \int_{-T}^{+T} f(\mathbf{r}, t) f(\mathbf{r}, t+\tau)\, dt \tag{10.111}$$

The spatial autocorrelation function is

$$C(\mathbf{r}_0, t) = \lim_{V\to\infty} \frac{1}{8V} \int_{-\Delta x}^{+\Delta x} \int_{-\Delta y}^{+\Delta y} \int_{-\Delta z}^{+\Delta z} f(\mathbf{r}, t) f(\mathbf{r} + \mathbf{r}_0, t)\, dx\, dy\, dz \tag{10.112}$$

where V is the product $\Delta x\, \Delta y\, \Delta z$. The variable of integration goes over a box whose volume is $(2\,\Delta x)(2\,\Delta y)(2\,\Delta z)$, hence the factor $8V$. We may conveniently combine the two integrals if desired. Note that these correlation functions are *not* Fourier transforms, but are actual functions of variables having the dimensions of time and position.

The cross-correlation function may be defined similarly. We will write the time cross-correlation function for two functions f_1 and f_2 as

$$C_{12}(\mathbf{r}, \tau) = \lim_{T\to\infty} \frac{1}{2T} \int_{-T}^{+T} f_1(\mathbf{r}, t) f_2(\mathbf{r}, t+\tau)\, dt \tag{10.113}$$

The spatial cross-correlation function is

$$C_{1,2}(\mathbf{r}_0, t) = \lim_{V\to\infty} \frac{1}{8V} \int_{-\Delta x}^{+\Delta x} \int_{-\Delta y}^{+\Delta y} \int_{-\Delta z}^{+\Delta z} f_1(\mathbf{r}, t) f_2(\mathbf{r} + \mathbf{r}_0, t)\, dx\, dy\, dz \tag{10.114}$$

The combined time and spatial cross-correlation function is

$$C_{1,2}(\mathbf{r}_0, \tau) = \lim_{\substack{T\to\infty \\ V\to\infty}} \frac{1}{8V} \int_{-\Delta x}^{+\Delta x} \int_{-\Delta y}^{+\Delta y} \int_{-\Delta z}^{+\Delta z} \times \frac{1}{2T} \int_{-T}^{+T} f_1(\mathbf{r}, t) f_2(\mathbf{r} + \mathbf{r}_0, t+\tau)\, dx\, dy\, dz\, dt \tag{10.115}$$

Equation (10.115) is the most general formulation.

We now imply that if a function has an autocorrelation function that is not zero, then there exists some periodicity in the function; that is, it is not completely random. If a cross correlation exists between two functions, it implies that their fluctuations and/or variations have something in common.

As an example, let us compute the time autocorrelation function of a function. We will use the function $f(t) = \sin \alpha t$ for this example. The autocorrelation function integrand is

$$C(\tau) = \lim_{T\to\infty} \frac{1}{2T}\int_{-T}^{+T} \sin(\alpha t)\sin\alpha(t+\tau)\,dt = \cos(\alpha\tau) \qquad (10.116)$$

Equation (10.116) states that this function, which is periodic, has an autocorrelation function that depends upon τ. This implies that if τ is picked to be 0 or 2π, we have a correlation of 1, but if τ is other than this, the correlation may vary between $+1$ and -1, *passing through zero*. In other words, the time interval between "measurements" is important.

Experimentally, we may measure the autocorrelation function in time by first measuring the desired function $f(t)$ at time t and then at time $t + \tau$ over a long period of time (T), where τ is fixed. The product is taken and the result is then integrated, either electronically or graphically over the time interval T. The integrated product, divided by T, may then be placed on a graph as the value of $C(\tau)$ for the specified value of τ. Then τ is changed and the process repeated. Usually, $C(\tau)$ for each specified value of τ is determined from an "ensemble" average, that is, the average of the value of $C(\tau)$ obtained over many trials. From this result, the spectral power density may be obtained by use of Eq. (10.111).

A similar measurement process can be done in space as well as time. In addition, it now becomes clear that the cross-correlation function may be similarly developed from experimental measurements as well.

The previous work has now resulted in an expression for the spatial and temporal correlation of variables with themselves, or with other variables (e.g., $\langle E(\omega)H^*(\omega')\rangle$, etc). The usefulness of this approach can be demonstrated if it is remembered that the fluctuations of these variables in a plasma are associated with dissipation. For example, the mean value of open-circuit voltage fluctuations observed across a resistor is

$$\langle \mathscr{V}^2\rangle = \frac{2}{\pi}\,kTR\,\Delta\omega \qquad (10.117)$$

where k is Boltzmann's constant, T is the temperature, R the resistance, and $\Delta\omega$ the band of frequencies over which the noise voltage is measured. Note that the magnitude of the fluctuations is proportional to the resistance and the temperature. That is, the more dissipation, the greater the fluctuations.

A *fluctuation-dissipation theorem* can be determined for a plasma. It can be stated as

$$\langle f^2(t)\rangle = -\frac{kT}{\pi}\int_{-\infty}^{+\infty} \frac{\operatorname{Im}\chi(\omega)\,d\omega}{\omega} \qquad (10.118)$$

where T is the temperature and Im $\chi(\omega)$ is the imaginary part of the susceptibility of the medium. The foregoing statement is valid for low energies (nonquantum-mechanical conditions). If quantum mechanics applies, then Eq. (10.118) becomes

$$\langle f^2(t)\rangle = -\frac{\hbar}{2\pi}\int_{-\infty}^{+\infty} \operatorname{Im}\chi(\omega)\coth\left(\frac{\hbar\omega}{kT}\right) d\omega \tag{10.119}$$

This fluctuation-dissipation theorem may then be related to the spectral power density previously derived. That is

$$\langle F(\omega)F^*(\omega')\rangle = -\frac{4\pi}{\omega} kT \operatorname{Im}\chi(\omega)\,\delta(\omega-\omega') \tag{10.120}$$

or

$$\langle F(\mathbf{k},\omega)F^*(\mathbf{k}',\omega')\rangle = -(2\pi)^8 \frac{kT}{\pi\omega} \operatorname{Im}\chi(\mathbf{k},\omega)\,\delta(\omega-\omega')\,\delta(\mathbf{k}-\mathbf{k}') \tag{10.121}$$

depending upon where spatial as well as temporal variations are considered.

If the susceptibility is known, then the averaged fluctuations may be found, or vice versa. Note that a *temperature* must exist for dissipation.

Knowledge of the correlations and fluctuations in plasma phenomena, which determine the frequency spectra, help to determine much of the nature and behavior of the particles making up the plasma.

Not only electromagnetic emission, but fluctuations in density, diamagnetism, and so forth, may be used to produce correlation functions.

Suggested Reading

Bekefi, G., "Radiation Processes in Plasmas," Wiley, New York, 1966. This is a definitive work covering much of the material discussed in this chapter.

Greim, H. R., "Plasma Spectroscopy," McGraw-Hill, New York, 1964. A good coverage of both spectral line emission and the radiation processes is presented here.

Heitler, W., "The Quantum Theory of Radiation," 3rd ed., Oxford Univ. Press, London and New York, 1954. This is a treatment of both quantum-mechanical and classical radiation.

Jackson, J. D., "Classical Electrodynamics," Wiley, New York, 1962. This work gives equally good coverage of radiation phenomena.

Kadomstev, B. B., "Plasma Turbulence," Academic Press, New York, 1965.

Panofsky, W., and Phillips, M., "Classical Electricity and Magnetism," 2nd ed., Addison-Wesley, Reading, Massachusetts, 1965. This is a basic treatment of radiation phenomena.

Papoulis, A., "The Fourier Integral and Its Applications," McGraw-Hill, New York, 1962. This book covers correlation processes and Fourier integrations.

Papoulis, A., "Probability, Random Variables and Stochastic Processes," McGraw-Hill, New York, 1965.

Shkarofsky, I. P., Bachinsky, T. W., and Watson, M. P., "The Particle Kinetics of Plasmas," Addison-Wesley, Reading, Massachusetts, 1967. This is a complete reference work on many aspects of kinetic theory and radiation.
Sommerfeld, A., "Atombau und Spektrallinen," Friedr. Viewig & Son, Braunschweig, Germany, 1931.

Problems

1 Determine the energy radiated for one revolution of an electron orbiting about a magnetic field line if its velocity = $0.5c$. Note that some nonrelativistic approximations have been made in the development about the orbiting particle. Be sure that these are corrected (i.e., the mass and hence the cyclotron frequency changes, etc.). In what direction is most of the radiated energy concentrated?

2 If an electron were moving about a proton (hydrogen atom) with an orbit radius of a, what would happen if only classical radiation fields were considered? How long would it take before a "collision" with the nucleus would occur? Why does this not happen?

3 Is it possible that a charged particle orbiting in a dc magnetic field might lose all of its energy by radiation and stop? Show the reasoning semiquantitatively. Hint: Consider the Schroedinger wave equation for an orbiting particle in a dc magnetic field. Write the equation and explain your reasoning based on the solution to the equation.

4 Find the autocorrelation function (a function of τ, not ω) of a plasma for electric fields, if the plasma is characterized by a temperature T and a dielectric constant of

$$\frac{\varepsilon}{\varepsilon_0} = \left[1 - \frac{\omega_{\mathrm{Pe}}^2}{\omega(\omega + j\nu_m)}\right]$$

Hint: The susceptibility of a medium χ is related to the permittivity as shown in the relation

$$\mathbf{D} = \varepsilon_0(1 + \chi)\mathbf{E} = \varepsilon\mathbf{E}$$

5 Find the total energy per unit volume for a blackbody that obeys the Planck radiation law (Eq. (10.9) with $n_r^2 = 1$). The total energy should go as the fourth power of the temperature. This result is the Stefan–Boltzmann law.

6 Consider a charge that moves under the influence of an ac electric field $E_0 \cos \omega t$, pointing in the z direction. Determine the spectrum, angular distribution of radiation, and total energy lost per unit time for this problem.

7 We must consider the conditions under which a plasma may be considered a blackbody.

(a) For microwave emission, considering typical values, is a glow discharge (size about 10 cm), the sun, or the ionosphere a blackbody?

(b) Do part (a) for optical wavelengths. Refer to Fig. 1.1. Discuss.

8 Plot the spectra of the first five cyclotron harmonics of a nonrelativistic charged particle orbiting in a uniform magnetic field at 90° with respect to the magnetic field.

9 What are the units of η_ω, $\eta_{\omega s}$, and $\eta_{\omega A}$?

10 Obtain Eqs. (10.21) and (10.22).

11 Find the quadrupole and hexupole terms in the expression for the radiation flux (10.26).

12 Assume that $\dot{u}$ is a constant for Δt units of time and zero otherwise. Plot the spectrum of bremsstrahlung from an electron using (10.55) and (10.56). Is there a difference in the shape of the spectrum as the angle of observation is varied?

13 Show that if two electrons collide, the dipole contribution to the radiation from both particles cancels, at least to first order.

14 Show that (10.61) is an energy function.

15 Integrate Eq. (10.67).

16 Determine the shape of the cyclotron spectrum. Is it the same in all directions, relative to the magnetic field?

17 Obtain Eqs. (10.54), (10.55) and (10.56).

Appendix

A. FREQUENTLY USED PHYSICAL CONSTANTS

c	Speed of light in vacuum	2.99793×10^{8} meter/second
ε_0	Permittivity of vacuum	8.854×10^{-12} farad/meter
μ_0	Permeability of vacuum	1.257×10^{-6} henry/meter
h	Planck constant	6.625×10^{-34} joule-second
k	Boltzmann constant	1.3804×10^{-23} joule/°K
q_p	Charge of proton	1.6021×10^{-19} coulomb
eV	Electron volt	1.6021×10^{-19} joule
q_p/k		11,609°K/volt

Particle	Mass (kg)	Rest energy (MeV)
Electron	9.108×10^{-31}	0.511
Neutron	1.674×10^{-27}	939.512
^{1}H atom	1.673×10^{-27}	938.730
^{2}H atom (deuterium)	3.343×10^{-27}	1876.017
^{3}H atom (tritium)	5.006×10^{-27}	2809.272
^{4}He atom	6.643×10^{-27}	3728.189

B. USEFUL VECTOR IDENTITIES

$$\mathbf{A} \times \mathbf{B} = -(\mathbf{B} \times \mathbf{A})$$
$$\mathbf{A} \times (\mathbf{B} \times \mathbf{C}) = (\mathbf{A} \cdot \mathbf{C})\mathbf{B} - (\mathbf{A} \cdot \mathbf{B})\mathbf{C}$$
$$(\mathbf{A} \times \mathbf{B}) \times \mathbf{C} = (\mathbf{A} \cdot \mathbf{C})\mathbf{B} - (\mathbf{B} \cdot \mathbf{C})\mathbf{A}$$
$$(\mathbf{A} \times \mathbf{B}) \cdot (\mathbf{C} \times \mathbf{D}) = (\mathbf{A} \cdot \mathbf{C})(\mathbf{B} \cdot \mathbf{D}) - (\mathbf{A} \cdot \mathbf{D})(\mathbf{B} \cdot \mathbf{C})$$
$$(\mathbf{A} \cdot \nabla)\mathbf{B} = \mathbf{A} \cdot (\nabla \mathbf{B})$$
$$\nabla(\phi + \psi) = \nabla\phi + \nabla\psi$$
$$\nabla(\phi\psi) = \phi\nabla\psi + \psi\nabla\phi$$
$$\nabla \cdot (\phi\mathbf{A}) = \phi\nabla \cdot \mathbf{A} + \mathbf{A} \cdot \nabla\phi$$
$$\nabla \times (\phi\mathbf{A}) = \phi\nabla \times \mathbf{A} + \nabla\phi \times \mathbf{A}$$
$$\nabla \cdot (\mathbf{A} \times \mathbf{B}) = \mathbf{B} \cdot \nabla \times \mathbf{A} - \mathbf{A} \cdot \nabla \times \mathbf{B}$$
$$\nabla \times (\mathbf{A} \times \mathbf{B}) = (\mathbf{B} \cdot \nabla)\mathbf{A} - (\mathbf{A} \cdot \nabla)\mathbf{B} + \mathbf{A}(\nabla \cdot \mathbf{B}) - \mathbf{B}(\nabla \cdot \mathbf{A})$$
$$\nabla(\mathbf{A} \cdot \mathbf{B}) = (\mathbf{A} \cdot \nabla)\mathbf{B} + (\mathbf{B} \cdot \nabla)\mathbf{A} + \mathbf{A} \times (\nabla \times \mathbf{B}) + \mathbf{B} \times (\nabla \times \mathbf{A})$$
$$\nabla \times \nabla\phi = 0$$
$$\nabla \cdot (\nabla \times \mathbf{A}) = 0$$
$$\nabla \times (\nabla \times \mathbf{A}) = \nabla(\nabla \cdot \mathbf{A}) - \nabla^2\mathbf{A}$$

C. UNITS AND DIMENSIONS

The following table† lists a series of conversion factors. They are to be used whenever it is needed to convert an expression written in cgs units to mks units, which are the ones used in this book. The factors are simply written as replacements for the corresponding quantities in the other system of units. For example, in converting from electric field **E** expressed in cgs units, write $(4\pi\varepsilon_0)^{1/2}\mathbf{E}$, where **E** is now written in mks units.

	cgs	mks
Velocity of light	c	$\dfrac{1}{(\mu_0 \varepsilon_0)^{1/2}}$
Electric field, potential, or voltage	$\mathbf{E}, \phi, \mathscr{V}$	$(4\pi\varepsilon_0)^{1/2}(\mathbf{E}, \phi, \mathscr{V})$
Electric displacement	$\mathbf{D}$	$\left(\dfrac{4\pi}{\varepsilon_0}\right)^{1/2}\mathbf{D}$
Charge, charge density, current density, polarization	$\mathscr{q}, \rho_E, \mathbf{J}_E, \mathbf{P}$	$\dfrac{1}{4\pi\varepsilon_0}(\mathscr{q}, \rho_E, \mathbf{J}_E, \mathbf{P})$
Magnetic field	$\mathbf{B}$	$\left(\dfrac{4\pi}{\mu_0}\right)^{1/2}\mathbf{B}$
Magnetic displacement	$\mathbf{H}$	$(4\pi\mu_0)^{1/2}\mathbf{H}$
Magnetization	$\mathbf{M}$	$\left(\dfrac{\mu_0}{4\pi}\right)^{1/2}\mathbf{M}$
Conductivity	σ	$\dfrac{\sigma}{4\pi\varepsilon_0}$
Dielectric constant	ε	$\dfrac{\varepsilon}{\varepsilon_0}$
Relative permeability	μ	$\dfrac{\mu}{\mu_0}$
Impedance	Z	$4\pi\varepsilon_0 Z$
Inductance	L	$4\pi\varepsilon_0 L$
Capacitance	C	$\dfrac{1}{4\pi\varepsilon_0} C$

† From J. D. Jackson, "Classical Electrodynamics," © John Wiley & Sons, Inc., New York, 1962.

Index

A

Absolute instability, 232–234
Absorption, 287–288
AC electric field, motion of particles in, 52
Acceleration of plasma, 144–146
Acoustic waves, 196–199
Action, 106–108
  conditions for conservation of, 108
Adiabatic compression, 108–109
Adiabatic invariants, 101–107, *see also* Constants of motion
  action, 106–108
  magnetic moment, 103–105
Allis, W. P., 65, 163, 199, 244
Alfvén, H., 163
Alfvén waves, 146–149, 165
  phase velocity of, 149
Ambipolar diffusion, 31, *see also* Diffusion, Mobility
  coefficient of, 32
    in magnetic field, 33
Analytic continuation, 221–226, 235
  of dispersion relation, 223–226
  theory, 221–222
Angular momentum
  conservation in collision, 19, 25
  of orbiting charged particle, 130
Angular relaxation time, 154
Anisotropic media, 40, 60, *see also* Permittivity, Plasmas
Anisotropic velocity distribution, 96, 131
Apse line, 20
Artsimovich, L. A., 36
Aurora, 11
Autocorrelation function, 314

B

Bachynsky, T. W., 285, 319
Baldwin, D. E., 244
Balescu–Lenard Equation, 283
"Ball" lightning, 164
Barn, 16
Barnett, C. F., 36
BBGKY hierarchy, 265–284
Bekefi, G., 318
Bennett, W. H., 163
Bers, A., 163, 199, 244
Bers and Briggs criteria for instabilities, 232–234
Bessel functions
  in harmonics of cyclotron radiation, 309, 311
  in position space damping, 241
β-parameter, 141
Bishop, A. S., 12
Blackbody radiation function, 289–290
  for photons, 289
  for plasma, 290
Bogoliubov, Born, Green, Kirkwood, Yvon hierarchy, *see* BBGKY hierarchy
Bogoliubov, N. N., 284
Bohm, D., 35, 36
Bohm diffusion, 35, 243, *see also* Diffusion
Boltzmann equation, 97–98, 115, 127–129, 256

collisionless, *see* Vlasov equation
development from Liouville's theorem, 97–98
with collisions, 115, 127–129
constants of motion in, 109–111
diffusion and mobility from, 119–124
for electrostatic waves, 161–163
moments of, 124–129
Bose–Einstein statistics, 88–93, 289–290, *see also* Distribution function
Boyd, T. J., 12
Branch lines, 232, 235–236
Breakdown, *see* Ionization
Bremsstrahlung, 23, 291, 300–306
electron ion, 300
integration over impact parameters, 305
over velocity distribution, 305
nonrelativistic, 303–304
quantum-mechanical effects, 305
relativistic, 305–306
spectrum of, 302
Briggs, R. J., 244
Briggs and Bers criteria for instabilities, 232–234
Brown, S. C., 36
Buchsbaum, S., 163, 199, 244

C

Canonical transformation, 112–113
constants of integration, 112
generating function for, 112
Cauchy integral theorem, 217-220
principal value, 218, 226
Center of mass velocity, conservation of, 18
CGS–MKS conversion table, 323
Chandrasekhar, S., 65
Chapman, S., 36
Charge, conservation of, 82
Charge exchange, 3
Charge separation
in toroidal fields, 143–144
in uniform magnetic field, 44
Charged particle, motion of, *see* Motion
Charged particle drifts, *see* Drifts
χ scattering angle, *see* Scattering angle
Churchill, R., 244
Clark, M., 12, 36, 130, 163
Cluster expansion, 274–284
CMA diagram, 179–183, 200–201
Cobine, J. D., 12
Coefficients, *see also* Diffusion
of absorption, 287–288
of spontaneous emission, 287–288
of stimulated emission, 287–288
Cold plasma
dielectric constant, 63–67
dispersion relation for monochromatic waves, 170–179
waves in, 167–201
Collective phenomena, *see* Fluctuations, Fluid mechanical approximation, Instabilities, Kinetic theory, Plasma statistical mechanics, Waves
Collision cross section, *see* Cross section
Collision frequency, 16
Collision integrals, 116, 119–129, 249–263, *see also* Fokker–Planck methods
Collision plane, 19, 20
Collisional ionization, 3
Collisional models, validity of, 262–265
Collisional processes, 8, *see also* Collisions
attachment, 8
charge exchange, 8
deexcitation, 8
excitation, 8
ionization, 3, 8
momentum transfer, 8
recombination, 3, 8
sputtering, 8
Collisions, 9, 13–38
Coulomb, 22–26, 251
cross section for, 14–16, 21
elastic, 9, 17, 249
Fokker–Planck model for, 115–119
frequency of, 16
inelastic, 9
mean free path for, 14
in moments of Boltzmann equation, 125–129
Rutherford model, 18
statistics of, 120–121

Conductivity of plasma, 60–65, 173–176, *see also* Mobility, Permittivity
Confinement of plasma, 5, 139–141, 143–144, 242–244
Conservation
of angular momentum, *see* Angular momentum
of charge analogy in Liouville's theorem, 82
of energy, *see also* Energy conservation
in microcanonical ensemble, 83
of momentum, *see* Momentum
Conservative laws for systems, 17, 75–77
Constants
of motion, 101–103, 109–111, *see also* Adiabatic invariants
action, 103, 106–108
angular momentum of charged particle, 130
in Boltzmann equation, 109–111
energy, 109–110
magnetic moment, 103–105
values of physical, 321
Constitutive equations, 6, 168
Contact ionization, 8
Containment of plasma, *see* Confinement
Conte, S. D., 214, 244
Continuity equation, 137
Continuum model, *see* Fluid approximation
Contour integrations, 217
principal value integrals, 217–219
Controlled thermonuclear fusion, 4
Convective instability, 232–234
Conversion factors, 322
Convolution integrals, 281
Correlation functions, 126
for fluctuations in plasmas, 311–318
pair, 274–285
triplet, 274
Coulomb collision cutoff parameter Λ, 154, 252–253
Coulomb collisions, 22–26, 152–153, *see also* Collisions, Cross section for collisions, Rutherford scattering
Coulomb scattering, 26, *see also* Collisions, Cross section for collisions, Rutherford scattering
Cowling, T. G., 36, 130, 163
Criteria for waves and instabilities, 230–234
Cross-correlation function, 315–318, *see also* Correlation functions
Cross section for collisions, 14–16, 251
Coulomb, 26, 153
differential, 21, 153
momentum transfer, 23, 153
Rutherford, 26
truncation of, 153
Crossed fields, motion in, 41–44
Current density, 136, *see also* Electric current density
Cusped fields, *see* Magnetic mirror, Ioffe containment system
Cutoffs, 180–201
in propagation through plasma, 157
Cyclic coordinates, 111–112
Cyclotron emission, 291, *see also* Cyclotron radiation
Cyclotron frequency, 40
Cyclotron radiation, 291, 306–311
broadening of, 310–311
harmonics of, 309–311
spectrum of, 307–311
Cyclotron radius, 40, 53, 306
Cyclotron resonance, 58, 67, 68

D

Damping, *see also* Landau damping
by collisions, 61–65, 194
in position space, 236–242
DC conductivity of plasma, 62
DC electric field, motion of particles in, 57
Debye length, 150-152, 264
particles inside sphere of radius λ_D, 152
shielding distance, 264
truncation for impact parameters, 152
Deexcitation, 8
Deflection angle, 20–26, 153–154
averages over, 249–253
Degree of freedom, 70
Delcroix, J. L., 65, 244
Delta function
in correlations, 313–315
in two stream instability, 245

Denisse, J. F., 244
Detailed balance, 289–290
Diamagnetism, 48, 50, 140–143
Dirac delta function, 245, 313–315
Dielectric constant, *see* Permittivity
Dielectric tensor, *see* Permittivity
Differential collision cross section, 21, 153, 251, *see also* Cross section
Diffusion, 27–38, 50, *see also* Collisions, Drifts, Mobility
  ambipolar, 31–33
  Bohm, 35, 243
  from Boltzmann equation, 119–124
  coefficient of, 29
  flux, 121
  free, 27–29, 37–38, 119–124
  in hydromagnetics, 146–148
  length, 35
  in magnetic field, 29–34
  particle current, 122
  short circuit effect, 33–34
  time, 35
  time dependence of, 34–35, 37–38
  in velocity space, 118, 249, 255–257
Dipole approximation, 294–299
Discharge, 8
Dispersion function, 215–216
Dispersion relation, 159, *see also* Eigenvalues
  for cold plasma in magnetic field, 175–178
  for electrostatic waves, 161–163, 213–218, 223–230
  for plane monochromatic waves, 159
  plots of, 179–201
  in position space, 238
  for transverse waves, 245
  for two stream instability, 245
Distribution function, 69, 78–99
  anisotropic, 96
  blackbody, 289
  Bose–Einstein, 92
  energy, 95
  equilibrium, 95
  Fermi–Dirac, 92
  Maxwell–Boltzmann, 92
  nonequilibrium, 96, 248, 257
  single-particle, 98
  speed, 95
  for system, 97
  velocity, 94
Distribution laws, 85–99
Drifts of charged particles, 41–44
  in AC electric field, 57–60
  in crossed electric and magnetic fields, 41–44
  of guiding center, 52–56
  in magnetic field gradients, 44, 46, 66–67
  in toroidal magnetic fields, 144
Dupree, T. H., 284
Dupree, method of, 277–284
Dyadics, 116–117, 126, 255–257, 285
Dynamical friction, 117–118, 249–254

E

Earth's magnetic field, 68
Eigenvalues, *see also* Dispersion relation
  of ordinary differential equations, 170
  of partial differential equations, 170–173
  representation of, 179–201
  of *RLC* circuit, 160, 169–170, 206, 208
  of wave equation for cold plasmas, 173
Einstein relation, 290
Elastic collisions, 8–9, 17–26, 249–263, *see also* Collisions
  conservation of energy, momentum and angular momentum in, 17
  in nonequilibrium statistical mechanics, 249-263
Electric charge density, 136
Electric current density, 136
Electron cyclotron resonance, 58, 67–68
Electron plasma frequency, *see* Plasma frequency
Electrostatic analogy, 254–262
Electrostatic approximation, 211–212
Electrostatic waves, 161–162, 213–218, 223–230
Emission coefficients, 287–288
Emission processes, *see* Radiation
Energy conservation, *see also* Constants of motion, Adiabatic invariants
  in elastic collision, 17–18
Energy equation, 129
Energy spectrum, 298–300

Ensemble, 70, 79, 83
averages, 79, 124–129, 313
microcanonical, 83
theory applied to plasma, 96–99
Equations of motion, *see* Motion
Equilibrium, 92, 96, 242–244
hydrodynamic, 138–139, 242–244
microinstabilities in, 242–244
thermal, 138
Equilibrium surfaces, 138–139
Error function, 263, *see also* Dispersion function
Eulerian coordinates, 146
Evanescent waves, criteria for, 232–234
Excitation of atoms, 8
Extraordinary wave, 183

F

Fälthammer, G. G., 163
Fermi–Dirac statistics, 88–93, 100
Finite difference procedure, 13, 66–67
Finite energy signals, 297–299
Finite power signals, 312–318
Fluctuation–dissipation theorem, 317–318
Fluctuations in plasmas, 311–318
power density of, 312
Fluid mechanical methods, 8, 133–149
from moments of Boltzmann equation, 124, 129
Flux conservation, 46–48, 67
Fokker–Planck equation, 118, 249, 264
Fokker–Planck methods, 115–119
electrostatic analogy, 256–263
velocity averaging, 251–254
Fourier transform, 203–204, 297–306
of Boltzmann equation, 161–162, 210–211
of energy flux radiated by charged particle, 297–306
of fluctuations, 307–311
for spectrum of bremsstrahlung, 301–311
of cyclotron radiation, 207–211
Free diffusion, 27, *see also* Diffusion
Free-diffusion coefficient, 29, 34–45
Frequency of interactions, 8, *see also* Collision frequency, Collisions
Fried, B. D., 214, 244
Frozen field lines, 147–149, *see also* Alfvén waves
Fusion, controlled thermonuclear, 4, 5

G

γ-space, 86
Gartenhaus, S., 199
Gas discharges, 8
Gaunt factor, 305–306
Gauster, W. B., 36
Generalized coordinates, 72–74
Generating function, in Hamilton–Jacobi theory, 112–113
Glasstone, S., 12
Glow discharge, 8
Goldstein, S., 98, 130
Gradients, 44, 46, *see also* Drifts, Magnetic fields, Shear
of magnetic field, 44, 46, 66, 67
in position space, 102
in velocity space, 102
Gravitational drift, 43–44
Gravitational fields, motion in, 43–45
Greim, H. R., 318
Group velocity, 185–186
Growing waves, criteria for, 232–234
Guernsey's kinetic equation, 284
Guiding center, 53, *see also* Drifts
Gyro frequency, 40

H

Hall effect, 61
terms in plasma conductivity, 61–65
Hamiltonian, 74, 99
charged particle, 99
system, 74
time average of, 107
transformation Hamilton–Jacobi theory, 111–112
Hamilton–Jacobi theory, 111–115, *see also* Canonical transformation, Cyclic coordinates
Hamilton's equations, 74, 81
Hamilton's principal function, 113
Harmonics of cyclotron frequency, 309–311
Haskell, R. E., 12

Heald, M. A., 163
Heat
of fusion, 1
of vaporization, 1
Heating, of plasma, 10, *see also* Cyclotron resonance
Heitler, W., 318
Holonomic system, 70
nonholonomic system, 70
Holt, E. H., 12
Hot plasma, waves in, 202–247
Huddlestone, R. H., 163
Hybrid resonances, 191
Hydrodynamic instabilities, 242–244
Hydromagnetic acceleration, 144–146, 165
Hydromagnetic channel, 145
Hydromagnetic diffusion, 146
Hydromagnetic equations, 133
Hydromagnetic equilibrium, 138
Hydromagnetics, 134, 138–165, 242–244
acceleration in, 144–146, 165
equilibrium in, 138–139

I

Ignat, D. W., 244
Ignorable coordinates, 111–112
Impact parameter, 20
in bremsstrahlung, 301–306
truncation at Debye length, 152–154
Index of refraction, 289–290, *see also* Permittivity
Inelastic collisions, 8
Inhomogeneous magnetic fields, 44–51, 166–67
Instabilities, 9, 34
absolute, 232
convective, 232
criteria for, 230–234
hydrodynamic, 242–244
micro-, 242–244
resistive, 242
two stream, 245
Interactions between particles, *see also* Collisions
momentum transfer, 8
recombination, 8
Interferometer, 158
Inverse Fourier transform, 203–204
Inverse Laplace transform, 203–208
Ioffe containment, 51
Ion acoustic waves, 195–199
Ion motion, *see* Motion of particles
Ion plasma frequency, 156
Ion sound waves, 195–199
Ionization, 1, 3, 8
collisional, 3
contact, 9
photo-, 9
thermal, 3
Ionosphere, 11

J

Jackson, J. D., 318, 323

K

k versus ω plots of dispersion relation, 187–201
Kadomtsev, B. B., 318
Kennard, E. H., 36
Kinetic stress tensor, 136
Kinetic theory, 249–285, *see also* Plasma statistical mechanics
Kittel, C., 99

L

Lagrange multipliers, 91
Lagrange's equations, 72, 74
Lagrangian, 71–72
coordinates, 146
of element, 71
of system, 72, 74
Landau, L., 244
Landau damping, 213–217, 227–230
of longitudinal electrostatic waves, 227–230
Langmuir probe, 166
Laplace transform, 203–208
of Boltzmann equation, 213–216
modified, 205
of wave equation, 209–210
Leighton, R. B., 99
Left circularly polarized waves, 183, 187–190

Legendre polynomials, 261
Lenard–Balescu equation, 283
Leonard, S. L., 163
Lienard–Wiechart potentials, 291
Linhart, J. G., 65
Liouville's theorem, 80–82, 101, *see also* Plasma statistical mechanics
integration of, 266–270
in nonequilibrium statistical mechanics, 265–274
Long range forces, *see* Coulomb force
Long-wavelength approximation, 211–212
Longitudinal adiabatic invariant, *see* Action
Longitudinal electrostatic waves in plasma, 213–217
Landau damping of, 227–230
Longmire, C. L., 65
Lorentz gauge, 7, 211–212, 246
Loveberg, R. H., 12
Loss-cone distribution, 131, 143
Loss cones, 49
Low frequency approximation, *see also* Long-wavelength approximation
in hydromagnetics, 147–148
Lower hybrid resonance, 191

M

MacDaniel, E. W., 36
Magnetic bottle, 48, 108
Magnetic field, *see also* Gradients, Magnetic mirror
shear in, 144
spatially varying, 105
time varying, 104
Magnetic field lines, 48
Magnetic flux, 47
Magnetic mirror, 45, 47, 143
loss cones, 49
mirror machine, 49, 50
mirror ratio, 49, 108
reflection of particles in, 47
throat, 48
Magnetic moment, 46, 49–50, 68
action in cusped geometry, 103
conservation of, 103–105
Magnetic pressure, 140
Magnetoacoustic waves, 195–199
Magnetofluid dynamics, *see* Hydromagnetics
Magnetohydrodynamics, *see* Hydromagnetics
Markoff process, 116
formulation for f, 277
Mass current, 136
Mass density, 136
Massey, H. S. W., 36
Maxwell–Boltzmann statistics, 88–96, 99, 100
Maxwellian velocity distribution, 99–100, 216, 242
Maxwell's equations, 6, 211
Mayer cluster expansion, 274–283
Mean free path, 15
MHD accelerator, 145
approximation, *see* Fluid mechanical methods
equations, 133–138
generator, 145
waves, 146–149, 165
Microcanonical ensemble, 83
Microinstabilities, 242–244, *see also* Resistive instabilities
Microwave cavity resonances, 200–201
Microwave interferometer, 158
Mirror machine, 49, 50
Mirror ratio, 49, 108–109
MKS–CGS conversion table, 323
Mobility, 27, 58, *see also* Conductivity, Permittivity, Diffusion
in AC electric fields, 59–61
in Boltzmann equation, 119–129
flux, 121
in magnetic fields, 59–61
tensor, 60–65
Moments of Boltzmann equation, 124–129
Momentum transfer, *see also* Collisions
collision frequency, 23
cross section, 23, 153
Momentum–conservation equation, 6, 18, 39, 52, 72, 77, 102, 129, 137
Monochromatic waves, 156–160, 167–201
Montgomery, D. C., 244, 284
Motion, of charged particles, 6, 39, *see also* Drifts, Diffusion, Gradients, Mobility, Radiation

basic equations of, 6, 39
effects of ac electric fields on, 52–67
guiding center, 52–54
reflection of particles in magnetic mirror, 47
Mott, N. F., 36
μ-space, 85–88, 114–115

N

Newton's second law, 6, 72, *see also* Momentum equation
Nonequilibrium conditions, 96, 97
Nonequilibrium statistical mechanics, 248–285
Nonholonomic system, 70
Normal modes, 160–163, *see also* Eigenvalues
Northrup, T. G., 65

O

Ohm's law in a plasma, 6, 145–146
ω versus k plots of dispersion relations, 187–201
Orbits of charged particles, *see* Motion of particles
Ordinary wave, 183

P

Pair correlation function, 274–285
Panofsky, W., 318
Papoulis, A., 318
Paramagnetism, 142
Parseval's theorem, 298–299
Particle current, 27, 136
in Boltzmann equation, 121–124
Permittivity of plasma, 60–65, 173–176, *see also* Mobility, Conductivity
Phase space, 70
for elements, 85
for system, 85
Phase velocity, 185–186
Phillips, M., 318
Photoionization, 10
Physical constants, 321, 323
Pinch effect, 163, 165
Plane waves, 156–160, 167–201
Plasma, 1
acceleration, 144–146
collective phenomena, 69–100
conductivity, 60–65, 173–176, *see also* Mobility, Permittivity
confinement, 5, 138, 144
continuity equation for, 137
diamagnetism, 140–143
frequency, 65, 155–163
of electrons, 65, 155–163, 173, 215
of ions, 65, 155–167, 173
heating, 9, 10
instabilities, 9
mobility, *see* Mobility
Ohm's law for, 146
paramagnetism, 142
permittivity, *see* Permittivity
production, 9, 11
AC discharge, 10
arc, 9
contract ionization, 9
electron cyclotron resonance, 10
glow discharge, 9
injection, 11
photoionization, 10
shock wave, 9
thermal, 9
turbulent heating, 10
simulation, 7, 13–14
solid state, 11
Plasma dispersion function, 215–216, *see also* Dispersion relation
Plasma radiation, 286–310
Plasma rocket engine, 11
Plasma statistical mechanics, 69–100
fundamental postulate of, 84
Poisson bracket, 75, 266–267
Polarization of waves, 183, 187–190
Poles in contour integration, 217–227, 230–247
Position space damping, 236–242
Position space diffusion, *see* Diffusion
Power spectrum, *see also* Energy spectrum
for spatial fluctuations, 314
for temporal fluctuations, 312
Poynting vector for radiation fields, 295
Present, R. D., 99
Pressure tensor, 136

Principal resonances, 180–201, *see also* Resonances
Principal value, 217, 219, 223–226
of dispersion relation, 223–226
Probability distribution, 78, *see also* Distribution function
Propagation of waves, *see* Wave propagation
Propagation vector, 158–162

Q

Quantum effects, 304–306, 312, 318
Quasi-static approximation, 211–213

R

Radiation from charged particle, 291, *see also* Bremsstrahlung, Cyclotron radiation
effects of acceleration on, 296–297
from fluctuations, 311–318
Lienard–Wiechart potentials, 291
outward flux of, 293–297
Poynting vector, 293–297
radiation field of, 292–293
Radiation ionization, 3
Radius of gyration, 40, 53, 306
Random velocity, 134
Ratio of specific heats, 196
Ray, J. A., 36
Recombination, 3, 8
Reduced mass, 24, 251
Reflection by magnetic mirror, 47
Relaxation time, 154
Representative point, 70
Residues, 219–221
Resistive instabilities, 242
Resonances, 180–201
principal, 180
upper and lower hybrid, 191
Resonant frequency, 160–163, *see also* Eigenvalues
Retarded time, 293
RF discharge, 10
Right circularly polarized waves, 183 187–190
Rose, D. J., 12, 36, 130, 163
Rosenbluth potentials, 256, 257
Rutherford scattering, 18, 23–26, *see also* Collisions
cross section for, 26

S

Sanderson, J., 12
Scattering
of particles in collision, 19
of radiation, *see* Fluctuations
Scattering angle, 20–26, 153–154
averages over, 249–253
Schmidt, G., 65, 130, 163, 199, 244, 285
Schneider, H. M., 244
Screening, *see* Shielding, Debye length
Sears, F. W., 99
Shear in magnetic fields, 144
Sheaths, 151–152
Shkarofsky, I. P., 285, 319
σ (collision cross section), *see* Cross section
Simulation, of plasmas, 7, 13–14
Slow wave approximation, 211–212
Solid state plasma, 11
Sommerfeld, A., 319
Spectral density function, 314
Spectrum, *see also* Energy spectrum, Power spectrum
for bremsstrahlung, 302
for cyclotron radiation, 307–311
of fluctuations in plasmas, 312–318
of radiation from charged particle, 297
Spherical harmonics, 260
Spitzer, L., Jr., 12, 36, 65, 130
Spontaneous emission, 287–288
Sputtering, 8
States of matter, 1
Stationary coordinates, 69
Statistical mechanics, *see also* Kinetic theory, Plasma statistical mechanics
fundamental postulate of, 84
Statistical methods, 7
Stefan–Boltzmann law, 319
Stimulated emission, 287–288
Stix, T. H., 163, 199, 244
Superparticle, 7, 13, 14
Synchrotron emission, 291
System, 70
angular momentum of, 76, 77
conservation laws for, 75

elements in, 70, 71
Hamiltonian of, 74
holonomic and nonholonomic, 70
Lagrangian of, 72
linear momentum of, 76, 77
macroscopic state of, 88–90
microscopic state of, 89, 90
phase space for, 85, 86

T

Tananbaum, B. S., 12, 163
Target particle, 14, 15, 251–263
Temperature, 92–100, 141
in ion-acoustic waves, 195–199
Test particle, 251–263
Thermal equilibrium, 138
Thermal ionization, 3
Thermonuclear processes, *see* Controlled fusion
Thompson, W. B., 130, 163
Tidman, D. A., 284
Time average, 78, 106
Time dependent diffusion equation, 147
Time stationary process, 313
Tolman, R. C., 99
Toroidal multipoles, 51
system, 143
Transform theory, 203–208, *see also* Contour integration
applications to wave equation, 209
Transformed time correlation function, 314
Transport phenomena, *see* Conductivity, Diffusion, Instabilities, Permittivity, Waves
Transverse gauge, 7, 211–212, 276
Transverse waves, dispersion relation for, 245
Triplet correlation function, 274
Truncation of BBGKY equation, 272–274, *see also* Cluster expansion
Two-body collisions, 14
Two-stream instability, 245

U

Uman, M., 12, 130
Unit tensor, 63
Upper hybrid resonance, 191

V

Vandenplas, P. E., 199, 244
Vector identities, 322
Velocity distribution, 92–96, *see also* Distribution functions
Velocity space diffusion, 118, 249
averages for, 255–257
Vlasov equation, 98, 101, 161, 203, 248, *see also* Boltzmann equation

W

Warm plasma, waves in, 202–247
Watson, M. P., 285, 319
Wave equation
for Alfvén waves, 148
for cold plasma, 167–178
for plane monochromatic waves, 157–162
in warm plasma, 202–247
Wave normal surface, 179–180, 200
Wave number, 158–163
Wave propagation
cutoff of, 157, 158
experiment, 158
introduction, 157–162
normal modes, 160
principal resonance, 180
resonance, 180
Waves
acoustic, 196
Alfvén, 146–149
in cold plasmas, 167–201
in magnetic field, 173–193
electrostatic, 191
finite temperature effects, 193–199
in hot plasmas, 202–247
hydromagnetic, 146–149
ion-acoustic, 197
left circularly polarized, 183, 187–190
ordinary and extraordinary, 183
plane monochromatic, 157
right circularly polarized, 183, 187–190
spatial damping (growth), 195
temporal growth, 195
Wharton, C. B., 163
Whistler mode, 188–191